Control and
Dynamic Systems

Neural Network Systems Techniques and Applications

Edited by **Cornelius T. Leondes**

VOLUME 1. Algorithms and Architectures

VOLUME 2. Optimization Techniques

VOLUME 3. Implementation Techniques

VOLUME 4. Industrial and Manufacturing Systems

VOLUME 5. Image Processing and Pattern Recognition

VOLUME 6. Fuzzy Logic and Expert Systems Applications

VOLUME 7. Control and Dynamic Systems

Control and Dynamic Systems

Edited by

Cornelius T. Leondes
Professor Emeritus
University of California
Los Angeles, California

VOLUME **7** OF

Neural Network Systems
Techniques and Applications

ACADEMIC PRESS
San Diego   London   Boston   New York   Sydney   Tokyo   Toronto

Academic Press
a division of Harcourt Brace & Company
525 B Street, Suite 1900, San Diego, California 92101-4495, USA
http://www.apnet.com

Academic Press Limited
24-28 Oval Road, London NW1 7DX, UK
http://www.hbuk.co.uk/ap/

Library of Congress Card Catalog Number: 97-80441

International Standard Book Number: 0-12-443867-9

PRINTED IN THE UNITED STATES OF AMERICA
97 98 99 00 01 02 ML 9 8 7 6 5 4 3 2 1

Contents

Contributors   xiii
Preface   xv

Orthogonal Functions for Systems Identification and Control

Chaoying Zhu, Deepak Shukla, and Frank W. Paul

I. Introduction   1

II. Neural Networks with Orthogonal Activation Functions   2
    A. Background   2
    B. Neural System Identification and Control   4
    C. Desired Network Properties for System Identification and Control   5
    D. Orthogonal Neural Network Architecture   7
    E. Gradient Descent Learning Algorithm   12
    F. Properties of Orthogonal Activation Function-Based Neural Networks   14
    G. Preliminary Performance Evaluation of Orthogonal Activation Function-Based Neural Networks   21

III. Frequency Domain Applications Using Fourier Series Neural Networks   25
    A. Neural Network Spectrum Analyzer   25
    B. Describing Function Identification   33
    C. Fourier Series Neural Network-Based Adaptive Control Systems   35

IV. **Time Domain Applications for System Identification**
and Control   **47**
A. Neural Network Nonlinear Identifier   47
B. Inverse Model Controller   55
C. Direct Adaptive Controllers   58
V. **Summary**   **71**
References   **72**

Multilayer Recurrent Neural Networks for Synthesizing and Tuning Linear Control Systems via Pole Assignment

Jun Wang

I. **Introduction**   **76**
II. **Background Information**   **77**
III. **Problem Formulation**   **79**
IV. **Neural Networks for Controller Synthesis**   **85**
V. **Neural Networks for Observer Synthesis**   **93**
VI. **Illustrative Examples**   **98**
VII. **Concluding Remarks**   **123**
References   **125**

Direct and Indirect Techniques to Control Unknown Nonlinear Dynamical Systems Using Dynamical Neural Networks

George A. Rovithakis and Manolis A. Christodoulou

I. **Introduction**   **127**
A. Notation   129
II. **Problem Statement and the Dynamic Neural**
Network Model   **130**
III. **Indirect Control**   **132**
A. Identification   132
B. Control   134
IV. **Direct Control**   **139**
A. Modeling Error Effects   140
B. Model Order Problems   148

V. Conclusions   154
References   154

A Receding Horizon Optimal Tracking Neurocontroller for Nonlinear Dynamic Systems

Young-Moon Park, Myeon-Song Choi, and Kwang Y. Lee

I. Introduction   158
II. Receding Horizon Optimal Tracking Control
Problem Formulation   159
   A. Receding Horizon Optimal Tracking Control Problem
      of a Nonlinear System   159
   B. Architecture for an Optimal
      Tracking Neurocontroller   162
III. Design of Neurocontrollers   163
   A. Structure of Multilayer Feedforward
      Neural Networks   163
   B. Identification Neural Network   164
   C. Feedforward Neurocontroller   168
   D. Feedback Neurocontroller   170
   E. Generalized Backpropagation-through-
      Time Algorithm   171
IV. Case Studies   176
   A. Inverted Pendulum Control   176
   B. Power System Control   180
V. Conclusions   187
References   188

On-Line Approximators for Nonlinear System Identification: A Unified Approach

Marios M. Polycarpou

I. Introduction   191
II. Network Approximators   193
   A. Universal Approximators   194
   B. Universal Approximation of Dynamical Systems   197
   C. Problem Formulation   199

III. Learning Algorithm   200
  A. Weight Adaptation   202
  B. Linearly Parametrized Approximators   206
  C. Multivariable Systems   208
IV. Continuous-Time Identification   210
  A. Radial-Basis-Function Network Models   213
  B. Multilayer Network Models   223
V. Conclusions   228
References   229

The Determination of Multivariable Nonlinear Models for Dynamic Systems

S. A. Billings and S. Chen

I. Introduction   231
II. The Nonlinear System Representation   233
III. The Conventional NARMAX Methodology   235
  A. Structure Determination and
    Parameter Estimation   236
  B. Model Validation   245
IV. Neural Network Models   246
  A. Multilayer Perceptrons   247
  B. Radial Basis Function Networks   248
  C. Fuzzy Basis Function Networks   251
  D. Recurrent Neural Networks   252
V. Nonlinear-in-the-Parameters Approach   254
  A. Parallel Prediction Error Algorithm   255
  B. Pruning Oversized Network Models   257
VI. Linear-in-the-Parameters Approach   259
  A. Regularized Orthogonal Least-Squares Learning   262
  B. Enhanced Clustering and
    Least-Squares Learning   267
  C. Adaptive On-Line Learning   271
VII. Identifiability and Local Model Fitting   271
VIII. Conclusions   273
References   275

High-Order Neural Network Systems in the Identification of Dynamical Systems

Elias B. Kosmatopoulos and Manolis A. Christodoulou

I. **Introduction**   **279**

II. **RHONNs and g-RHONNs**   **281**

III. **Approximation and Stability Properties of RHONNs and g-RHONNs**   **284**
   A. Stability and Robustness Properties of g-RHONNs   286

IV. **Convergent Learning Laws**   **289**
   A. Robust Adaptive Learning Laws   290
   B. Learning Laws That Guarantee Exponential Error Convergence   292

V. **The Boltzmann g-RHONN**   **294**

VI. **Other Applications**   **298**
   A. Estimation of Robot Contact Surfaces   298
   B. RHONNs for Spatiotemporal Pattern Recognition and Identification of Stochastic Dynamical Systems   300
   C. Universal Stabilization Using High-Order Neural Networks   301

VII. **Conclusions**   **304**
   References   **304**

Neurocontrols for Systems with Unknown Dynamics

William A. Porter, Wie Liu, and Luis Trevino

I. **Introduction**   **307**

II. **The Test Cases**   **309**

III. **The Design Procedure**   **313**
   A. Using Higher Order Moments   314
   B. Embedding the Controller Design in $H(v)$ 315
   C. HOMNA Training Algorithms   316
   D. Tensor Space Matchups with the HOMNA Calculations   318

IV. **More Details on the Controller Design**   **318**

V. More on Performance   320
   A. Disturbance Rejection   321
   B. Propulsion System Application   325
VI. Closure   331
   References   331

On-Line Learning Neural Networks for Aircraft Autopilot and Command Augmentation Systems

Marcello Napolitano and Michael Kincheloe

I. Introduction   333
II. The Neural Network Algorithms   336
   A. Extended Back-Propagation Training Algorithm   339
III. Aircraft Model   341
IV. Neural Network Autopilots   342
   A. Phase I: Flight Envelope Performance   343
   B. Phase II: Neural Autopilot Controllers under Linear and
      Nonlinear Conditions   348
   C. Conclusions   350
V. Neural Network Command Augmentation Systems   353
   A. Phase I: Statistics of Learning and Adaptation   357
   B. Phase II: Multiple Model Following Capabilities   368
   C. Conclusions   373
**VI. Conclusions and Recommendations for
Additional Research   379**
   A. Conclusions   379
   References   380

Nonlinear System Modeling

Shaohua Tan, Johan Suykens, Yi Yu, and Joos Vandewalle

I. Introduction   383
II. RBF Neural Network-Based Nonlinear Modeling   385
   A. The Neural Modeling Problem   385
   B. Two Basic Issues in Neural Modeling   386
   C. RBF Neural Network Structure   386

      D. Suitability of the RBF Neural Network for
         Nonlinear Modeling   388
      E. Excitation-Dependent Modeling   390
      F. RBF Neural Network Modeling with a Fixed
         Model Structure   391

  III. On-Line RBF Structural Adaptive Modeling   394
      A. The Need for On-Line Structural Adaptation   394
      B. On-Line Structure Adaptation Technique   395
      C. Remarks   398

  IV. Multiscale RBF Modeling Technique   399
      A. Basic Motivation   399
      B. Structure of a Multiscale RBF Neural Network   400
      C. Coarse-to-Fine Residue-Based Modeling Idea   401
      D. Construction of Multiscale RBF Model   402
      E. Model Validation   405
      F. Extension to Recurrent Neural Modeling   406

  V. Neural State–Space–Based Modeling Techniques   406
      A. Basic Motivation   406
      B. Neural State–Space Models   407

  VI. Dynamic Back-Propagation   409

  **VII. Properties and Relevant Issues in State–Space
      Neural Modeling   412**
      A. Uncertain Linear System Representations   412
      B. Linear Models as the Starting Point   416
      C. Imposing Stability   417

  VIII. Illustrative Examples   419
      A. RBF Neural Network Models   419
      B. Neural State–Space Models   424
      References   431

Index   435

Contributors

Numbers in parentheses indicate the pages on which the authors' contributions begin.

S. A. Billings (231), Department of Automatic Control and Systems Engineering, University of Sheffield, Sheffield S1 3JD, England

S. Chen (231), Department of Electrical and Electronic Engineering, University of Portsmouth, Portsmouth PO1 3DJ, England

Myeon-Song Choi (157), Department of Electrical Engineering, Myong Ji University, Yongin 449-728, Korea

Manolis A. Christodoulou (127, 279), Department of Electronic and Computer Engineering, Technical University of Crete, 73100 Chania, Crete, Greece

Michael Kincheloe (333), Lockheed Martin Electronics and Missiles, Orlando, Florida 32800

Elias B. Kosmatopoulos (279), Department of Electrical and Computer Engineering, University of Victoria, Victoria, British Columbia, V8W 3P6 Canada

Kwang Y. Lee (157), Department of Electrical Engineering, Pennsylvania State University, University Park, Pennsylvania 16802

Wie Liu (307), Department of Electrical and Computer Engineering, University of Alabama, Huntsville, Alabama 35899

Marcello Napolitano (333), Department of Mechanical and Aerospace Engineering, West Virginia University, Morgantown, West Virginia 26506-6106

Young-Moon Park (157), Department of Electrical Engineering, Seoul National University, Seoul 151-742, Korea

Frank W. Paul (1), Center for Advanced Manufacturing, Clemson University, Clemson, South Carolina 29634-0921

Marios M. Polycarpou (191), Department of Electrical and Computer Engineering, University of Cincinnati, Cincinnati, Ohio 45221-0030

William A. Porter (307), Department of Electrical and Computer Engineering, University of Alabama, Huntsville, Alabama 35899

George A. Rovithakis (127), Department of Electronic and Computer Engineering, Technical University of Crete, 73100 Chania, Crete, Greece

Deepak Shukla (1), Department of Mechanical Engineering, Clemson University, Clemson, South Carolina 29634-0921

Johan Suykens (383), Department of Electrical Engineering, Katholieke Universiteit Leuven, Heverlee, Belgium

Shaohua Tan (383), Department of Electrical Engineering, National University of Singapore, Singapore 119260

Luis Trevino (307), EP-13, Bldg. 4666, Marshall Space Flight Center, MSFC, Alabama 35812

Joos Vandewalle (383), Department of Electrical Engineering, Katholieke Universiteit Leuven, Heverlee, Belgium

Jun Wang (75), Department of Mechanical and Automation Engineering, The Chinese University of Hong Kong, Shatin, New Territories, Hong Kong

Yi Yu (383), Institute of Systems Science, National University of Singapore, Singapore 119597

Chaoying Zhu (1), Bell Communication Research, Piscataway, New Jersey 08854

Preface

Inspired by the structure of the human brain, artificial neural networks have been widely applied to fields such as pattern recognition, optimization, coding, control, etc., because of their ability to solve cumbersome or intractable problems by learning directly from data. An artificial neural network usually consists of a large number of simple processing units, i.e., neurons, via mutual interconnection. It learns to solve problems by adequately adjusting the strength of the interconnections according to input data. Moreover, the neural network adapts easily to new environments by learning, and can deal with information that is noisy, inconsistent, vague, or probabilistic. These features have motivated extensive research and developments in artificial neural networks. This volume is probably the first rather comprehensive treatment devoted to the broad areas of algorithms and architectures for the realization of neural network systems. Techniques and diverse methods in numerous areas of this broad subject are presented. In addition, various major neural network structures for achieving effective systems are presented and illustrated by examples in all cases. Numerous other techniques and subjects related to this broadly significant area are treated.

The remarkable breadth and depth of the advances in neural network systems with their many substantive applications, both realized and yet to be realized, make it quite evident that adequate treatment of this broad area requires a number of distinctly titled but well-integrated volumes. This is the seventh of seven volumes on the subject of neural network systems and it is entitled *Control and Dynamic Systems*. The entire set of seven volumes contains

Volume 1: *Algorithms and Architectures*
Volume 2: *Optimization Techniques*
Volume 3: *Implementation Techniques*
Volume 4: *Industrial and Manufacturing Systems*
Volume 5: *Image Processing and Pattern Recognition*
Volume 6: *Fuzzy Logic and Expert Systems Applications*
Volume 7: *Control and Dynamic Systems*

The first contribution to this volume is "Orthogonal Functions for Systems Identification and Control," by Chaoying Zhu, Deepak Shukla, and Frank W. Paul. This contribution presents techniques for orthogonal activation function-based neural network (OAFNN) architectures with the objective of their application in identification and adaptive control of unknown dynamic systems. The activation function for the proposed OAFNNs can be any orthogonal function that permits flexibility. Such functions have attractive properties for system identification and control. These properties include the absence of local minima and fast parameter convergence. The single epoch learning capabilities of the harmonic as well as Legendre polynomial-based OAFNNs are presented. The Fourier series neural network is used for frequency domain applications, and time domain applications of the OAFNNs for system identification and control are also presented. These applications include neural network spectrum analyzers, transfer function identification, describing function identification, neural self-tuning regulators, and a neural model reference adaptive system. In the time domain, the schemes for OAFNN-based discrete nonlinear system identification and direct adaptive control schemes are presented. The overall stability is shown by using a Lyapunovlike analysis. Experimental results are presented that demonstrate the feasibility of the developed neural controllers. These experimental results clearly demonstrate that the neural model reference adaptive controller (frequency domain) and direct adaptive neural controllers (time domain) meet real time control objectives and track a desired trajectory with a low tracking error. The superior model learning capability of the OAFNNs is demonstrated through simulation as well as experimental results.

The next contribution is "Multilayer Recurrent Neural Networks for Synthesizing and Tuning Linear Control Systems via Pole Assignment," by Jun Wang. This contribution presents two classes of two-layer and two classes of four-layer recurrent neural networks for synthesizing and tuning linear feedback control systems in real time. The synthesis techniques for the design of these neural network systems are presented in detail, as are techniques for the synthesis of state observers and state estimators by means of two-layer recurrent neural networks and four-layer recurrent neural networks. The substantive effectiveness of the techniques presented in this contribution are clearly demonstrated by numerous illustrative examples.

The next contribution is "Direct and Indirect Techniques to Control Unknown Nonlinear Dynamical Systems Using Dynamical Neural Networks," by George A. Rovithakis and Manolis A. Christodoulou. This contribution presents two techniques for adaptive control of unknown nonlinear dynamical systems through the use of neural network systems. In the indirect control case, that is, where there is no system modeling error and only parametric uncertainties are present, methods are presented that guarantee the convergence of the system control error to zero with boundedness of all other system variables in the closed loop. In the direct control case, that is, when both modeling errors and unmodeled dynamics are present, techniques are presented for system neural network architectures to guarantee a uniform ultimate boundedness of the system control error and a uniform ultimate boundedness of all other state variables in the closed loop system. Illustrative examples are presented throughout this contribution that demonstrate the effectiveness of the techniques presented.

The next contribution is "Receding Horizon Optimal Tracking Neurocontroller for Nonlinear Dynamic Systems," by Young-Moon Park, Myeon-Song Choi, and Kwang Y. Lee. A neural network training algorithm called the generalized backpropagation through time (GBTT) algorithm is presented to deal with a cost function defined in a finite horizon. Multilayer neural networks are used to design an optimal tracking neurocontroller (OTNC) for discrete-time nonlinear dynamic systems with quadratic cost function. The OTNC is composed of two controllers. Feedforward neurocontroller (FFNC) and feedback neurocontroller (FBNC). The FFNC controls the steady-state output of the plant, whereas the FBNC controls the transient-state output. The FFNC is designed using a highly effective inverse mapping concept by using a neuro-identifier. The GBTT algorithm is utilized to minimize the general quadratic cost function for the FBNC training. The techniques presented are useful as an off-line control method where the plant is first identified and then a controller is designed for it. Two case studies for a typical plant and a power system with nonlinear dynamics clearly manifest the high degree of effectiveness of these techniques.

The next contribution is "On-Line Approximators for Nonlinear System Identification: A Unified Approach," by Marios M. Polycarpou. Due to their approximation capabilities and inherent adaptivity features, neural networks can be employed to model and control complex nonlinear dynamical systems. This contribution presents a unified approximation theory perspective to the design and analysis of nonlinear system identification schemes using neural network and other on-line approximation models. Depending on the location of the adjustable parameters, networks are classified into *linearly and nonlinearly* parametrized networks. Based on this classification, a unified procedure for modeling discrete-time and continuous-time dynamical systems using on-line approximators is presented and analyzed using Lyapunov stability theory. A projection algorithm guarantees the stability of the overall system even in the presence of approximation errors. This contribution presents illustrative examples which clearly manifest the substantive effectiveness of the presented techniques.

The next contribution is "Determination of Multivariable Nonlinear Models for Dynamic Systems," by S. A. Billings and S. Chen. A general principle of system modeling is that the model should be no more complex than is required to capture the underlying system dynamics. This concept, known as the parsimonious principle, is particularly relevant in nonlinear model building because the size of a nonlinear model can become, to a great extent, overly large. An overly complicated model may simply conform to the noise in the training data, resulting in overfitting. An overfitted model does not capture the underlying system structure well and can, in fact, perform badly on new data. In neural network system technology, the model is said to have a poor generalization property. This contribution is a rather comprehensive treatment of methods and techniques for the determination of multivariable nonlinear models for dynamic systems using neural network systems. A rather substantive array of illustrative examples is a valuable feature of this contribution.

The next contribution is "High-Order Neural Network Systems in the Identification of Dynamical Systems," by Elias B. Kosmatopoulos and Manolis A. Christodoulou. Methods are presented for approximating and identifying a very large class of dynamical systems. Neural network system learning laws are presented that are globally convergent, stable, and robust. These neural network systems techniques are linear in the weights, and thus there is one and only one minimum in the system error functional and this is, of course, a global minimum. This is contrary to multilayer neural network systems where there are many local minima. The proposed neural network systems architectures possess other significant properties, as noted in this contribution.

The next contribution is "Neurocontrols for Systems with Unknown Dynamics," by William A. Porter, Wie Liu, and Luis Trevino. This contribution presents a methodology for the development of neural network systems controllers for systems about which no *a priori* model information is available. These neural network system design techniques presume that a finite duration input–output histogram is available. The methods presented extract from the histogram sufficient information to specify the neural network system controller, which will drive a given system along a general output reference profile (unknown during the design). It also has the capability of disturbance rejection and the capacity to stabilize unstable plants. The substantive effectiveness of the techniques presented in this contribution are demonstrated by their application to a rather complex system problem.

The next contribution is "On-Line Learning Neural Networks for Aircraft Autopilot and Command Augmentation Systems," by Marcello Napolitano and Michael Kincheloe. Flight control system design for high performance aircraft is an area where there exists an increasing need for better control system performance. The design of suitable flight control systems for these vehicles is challenging due to the coupled, nonlinear, and time-varying dynamics that lead directly to uncertainties in modeling. Classical and modern control law design methods rely on linearized models to compute the controller gains and interpolation algorithms to schedule the gains in an attempt to meet performance specifications throughout the flight envelope. A better approach can be implemented through a flight control system capable of "learning" throughout the flight envelope, where the controller updates its free parameters based on a function approximation used to map the current flight condition to an appropriate controller response. At first glance, this appears similar to gain scheduling, but there is a critical distinction: the learning controller adapts on-line based on the actual aircraft system rather than a model. A completely new approach to the design and implementation of control laws for modern high performance aircraft can be found in artificial neural network technology. Among the many features presented for this problem are the robustness of neural network systems aircraft flight control systems and their robustness to system nonlinearities. The significant effectiveness of the techniques presented in this contribution are clearly manifest by their application to a number of illustrative examples.

The final contribution to this volume is "Nonlinear System Modeling," by Shaohua Tan, Johan Suykens, Yi Yu, and Joos Vandewalle. Modeling nonlinear dynamical systems using neural networks is now well established as a distinct and important system identification paradigm with diverse and expanding applications. The choice of model structure is a key issue in neural modeling, and there exist many such choices with their respective pros and cons. This contribution selects two specific and very significant general types of neural network model structures: one is the radial basis function (RBF) neural network with a single hidden layer, and the other is the multilayer backpropagation neural network. This focus enables the detailed formulation of the modeling problem and the modeling techniques, and also allows precise treatment of the critical issues at both the problem formulation stage and the stage of modeling algorithm development. Accordingly, this contribution is an in-depth treatment of the utilization of RBF neural networks for nonlinear modeling, and also the utilization of the state space formulation with the multilayer backpropagation neural network model structure for the nonlinear modeling problem. Two detailed illustrative examples of each of these two broad classes of neural network systems are presented.

This volume on control and dynamic systems clearly reveals the effectiveness and essential significance of the techniques available and, with further development, the essential role they will play in the future. The authors are all to be highly commended for their splendid contributions to this volume which will provide a significant and unique reference for students, research workers, practitioners, computer scientists, and others on the international scene for years to come.

Cornelius T. Leondes

Orthogonal Functions for Systems Identification and Control

Chaoying Zhu
Bell Communication
Research
Piscataway,
New Jersey 08854

Deepak Shukla
High Technology
Corporation
Hampton, Virginia 23666

Frank W. Paul
Center for Advanced
Manufacturing
Clemson University
Clemson,
South Carolina 29634-0921

I. INTRODUCTION

This chapter addresses development of a class of neural networks, that have properties especially attractive for the task of identification and control of dynamic systems. These networks employ orthogonal functions as neuron activation functions.

The chapter is divided into three sections. In Section II, the architecture and properties of the class of orthogonal activation function-based neural networks (OAFNN) are developed and discussed. A detailed development of the Fourier series neural network (FSNN) architecture, which is a member of the class of OAFNNs, is presented. Superior learning capabilities of the OAFNNs are demonstrated using preliminary simulation results. Section III presents the frequency domain applications of the Fourier series neural network (FSNN). A neural network spectrum analyzer and several frequency domain schemes for system identification and adaptive control are presented. Section IV presents time domain applications of OAFNNs for system identification and control. Schemes for identification of discrete nonlinear systems are developed and then evaluated using simulations. Direct adaptive control schemes for continuous nonlinear systems are also devel-

1

oped and discussed in this section. These neural adaptive controllers permit real time, on-line implementation while the overall stability is guaranteed using Lyapunov analysis. Experimental results are presented for frequency domain-based neural model reference-adaptive controllers as well as for time domain-based direct adaptive neural controllers. A summary of the work is presented in Section V.

II. NEURAL NETWORKS WITH ORTHOGONAL ACTIVATION FUNCTIONS

A. BACKGROUND

An artificial neural network is an information processing and modeling system which mimics the learning ability of biological systems in understanding an unknown process or its behavior. The knowledge about the unknown process, stored as neural network weights, can then be utilized to analyze and control the process. In general, a neural network is composed of a large number of interconnected neurons that are stimulated by input signals (activation) to produce an output signal. The output signal is weighted before it is received as an input by the connected neurons and is reflected by the values of these weights. Neural network learning is achieved by modifying the weights in a dynamic training process subject to a learning rule for approaching a desired specification.

The behavior of a neural network is determined by three basic elements of the neural network: the neurons (activation functions), the connecting pattern of the neurons, and the learning rule. The basic requirement for selecting an activation function is that the neural network be able to mathematically map a wide range of linear and nonlinear input–output relations for the targeted processes. The most commonly used activation functions for existing neural network designs include sigmoidal functions, Boolean functions, radial basis functions, or combinations thereof [1]. The neuron interconnecting pattern is determined by the link topology and signal paths of the network.

Two elementary network patterns are feedforward and recurrent. In a feedforward network all the signals flow from the input layer to the output layer without feedback paths, whereas in a recurrent network there are many signal loops that provide feedback, which is employed to improve learning stability and correctness. On the other hand, these loops may introduce computational complexity and local minima problems. Some neural network designs use partial feedback in a subnetwork to gain the advantages of a recurrent network locally, without sacrificing performance of the whole network [2]. Neural network learning is a self-organizational process in the sense that the network can form its own representation based on the information received through training. Such a learning characteristic permits neural networks to identify an unknown process with little *a priori* knowledge about the process. For example, in supervised learning [1], neural network weights are tuned to make the outputs as close to those of the identified process when responding to the same inputs.

When all weights converge to an equilibrium state at which the neural model approximates the actual process's behavior, the neural network is said to be globally stable. If the weights converge to incorrect values, the neural network model is considered to converge to a local minimum and may not represent the actual system. A neural network is said to be unstable if the weights diverge during the training process. A properly designed learning rule should guarantee the global convergence of the network learning process.

Although the development of neural networks is motivated by different discipline applications, the following four metrics are often adopted to evaluate neural network performance: learning convergence, learning speed, model generalization, and interpretation.

First, the learning convergence is most important because the neural network model is not accurate if the learning process is not guaranteed to globally converge. A well known problem regarding the convergence of neural network designs is the local minima phenomenon. Mathematically, neural network learning is based on minimizing or maximizing a cost function in terms of a desired specification. This is accomplished by modifying weights so that the neural network best emulates the target process. If the cost function has a multiminimum surface, as shown in Fig. 1, the process of finding the minimum may find a local minimum (m_1 in the figure) closer to the initial state, thereby yielding a poor model. The local minima issue exists with many neural network designs and has attracted numerous research efforts [3–5].

Second, neural network learning is a reinforcement process where the training data set excites the neural network repetitively in each training cycle. It is desirable that the neural network converges with a minimum of training samples, a smaller number of training cycles, and with shorter training time, that is, higher learning speed. A high learning speed is essential for real time applications, where it is required that the model modifies itself at a fast rate.

Third, the generalization of a neural network measures the robustness of a trained neural network. For example, a neural network trained to represent an unknown process is robust if it can correctly predict the response of the process to the inputs not used for training [1]. The neural network intelligence is built on the generalization that the network is able to handle unfamiliar independent information after being trained.

The fourth performance measurement focuses on interpretation of weights of a trained neural network. Most current networks have complex structures which lead to complexities in translating the neural model to an understandable expres-

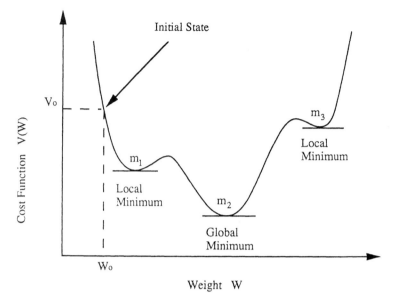

Figure 1 Simplified cost function.

sion for a given application. Control engineers need to translate neural networks to "control engineering languages" such as differential equations, state variables, transfer functions, and frequency response. Addressing this problem can facilitate control system design and controller tuning via the neural network model.

B. Neural System Identification and Control

There are numerous real world dynamic systems about which little or no knowledge is available. Knowledge of a system's dynamics is a key issue in prediction of the system response as well as in adaptive control. The principles of identification and control of linear systems are well developed, and standard procedures are available to identify and control any linear system. However, there are limited procedures available for identification and control of unknown, nonlinear systems. Such knowledge of the system dynamics is necessary for nonlinear controller design. Two widely used procedures for nonlinear system identification are the describing function (DF) analysis method in the frequency domain and parameter estimation methods in the time domain. The DF analysis is based on the assumption that the signal passes through low pass filter(s) before entering the nonlinearity [6], which is not always true. Such an analysis may be inadequate for accurate prediction of system response and therefore for adaptively controlling the system. However, the DF analysis provides qualitative understanding of nonlinear system dynamics. Time domain parameter estimation methods require extensive preliminary analysis to determine an estimated regression matrix based on knowledge of system dynamics. This often limits their application for unknown systems.

Neural networks provide a means for identification of nonlinear systems because they are tools which can approximate nonlinear functions to a desired level of accuracy, given an adequate network structure and sufficient training. Furthermore, the neural-network based system identification does not require an extensive preliminary analysis to determine a regression matrix, unlike the parameter estimation method.

C. DESIRED NETWORK PROPERTIES FOR SYSTEM IDENTIFICATION AND CONTROL

Neural networks provide a flexible tool for identification and control of dynamic systems. However, they have not received widespread acceptance in the dynamic system community due to a number of limiting features. One of these limitations is the general complexity of a neural network which makes it difficult to acquire an insight into the learned model. It is difficult for a control designer to accept a model about which little physical information is available [7]. Another major factor that limits the use of neural networks for adaptive control is the lack of guaranteed stability for the neural controller. The guaranteed overall stability is especially important if the network parameters are tuned on-line, in real time, where they are a part of the control system. The neural networks may require long training times (requiring a few hundred to a few thousand training cycles) and may not always converge to optimum network weights due to a poorly understood learning mechanism as well as possible local minima in the performance surface. This may lead to high training cost (time and computing resources) without a guarantee for optimum learning.

Neural networks can be employed effectively for system identification and control if the foregoing shortcomings are addressed. The desired properties of a network which make it attractive for application in the area of identification and control of dynamic systems are enumerated and discussed in the following list:

1. *Parallel architecture.* Neural networks derive their computational advantage from their parallel structure which permits their implementation on parallel processors. This is especially advantageous for real time applications where the computations have to be carried out on-line and in real time.

2. *Function approximation capability.* The universal function approximation capability is a fundamental requirement for a neural network. This capability enables approximation of the functional model of the dynamics of any system with a desired accuracy. Thus the network must be able to model any nonlinear function with a desired accuracy and a sufficient number of neurons.

3. *No local minima.* The presence of local minima in the neural network error surface may result in premature termination of learning before the network weights reach a global minimum. This causes expensive training efforts and limits the ability to determine whether the encountered minimum is local or global. There are methods (e.g., momentum method [8]) available to address local minima, but the results are not guaranteed. Under these conditions, it is desirable to have a network with no local minima in its performance surface to ensure that the first minimum which the network encounters is a global minimum. This also eliminates the computational cost for the algorithms (e.g., momentum method) needed to overcome local minima.

4. *Guaranteed stable learning.* A few heuristics exist for choosing the stable learning rate for popular, sigmoid activation function-based multilayer perceptrons (MLP). This results in a potential instability of the network weight adaptation mechanism. *A priori* knowledge of stable learning rates can guarantee stable learning and therefore is a desirable property for neural networks.

5. *Fast parameter convergence.* The sigmoid neural networks generally learn for a hundred to a few thousand training cycles before the weights converge to a small training error. This results in very high training costs because of long computational time, and is especially undesirable if the network is being trained on-line in real time for a system which is time varying. Ideally the network can be trained, with reasonably high and acceptable accuracy, in a small number (less than four or five) of training cycle.

6. *Provable stability for on-line implementation.* The parameters of a neural network need to be tuned on-line for systems with time varying characteristics or for systems in which the training inputs do not necessarily excite the total input–output space of the network. The architecture of the network should be simple enough to establish proof of the stability of the overall system for on-line implementation. Neural networks with complex and poorly understood internal dynamics and without proof of their stability may not prove acceptable for on-line implementation.

7. *Easy insight into the learned model.* Neural networks are complex structures and often do not permit any insight into the "learned model." Such a learned model cannot be used confidently to control a dynamic system. An important objective is to have a simple and organized network structure which easily can provide insight into the learned model.

D. Orthogonal Neural Network Architecture

The orthogonal activation function based neural network (OAFNN) [9, 10] is a three-layer neural structure with an input layer, an output layer, and a hidden layer as shown in Fig. 2 for a multi-input–single-output (MISO) system. A stack or parallel configuration (as shown in Fig. 3) of MISO networks can be employed to construct multi-input–multi-otput (MIMO) systems. The hidden layer consists of neurons with orthogonal (preferably orthonormal) activation functions. The activation functions for these neurons belong to the same class of orthogonal functions, and no two neurons have the same order activation function. This ensures that no cross-correlation occurs among the neurons in the hidden layer. The input and output layers consist of linear neurons. The weights between input and hidden layers are fixed and depend on the orthogonal activation function. These weights are equal to frequency numbers for harmonic activation functions (as explained for FSNN in Section II.D.2) and are unity for the other activation functions. The nodes on the right of the orthogonal neurons implement the product (π) operation. Each π node has m input signals from m different input blocks. These π nodes are considered a part of the hidden layer because there is no weighing operation between the orthogonal neurons and the π nodes. The output of the network $[\hat{y}(\bar{x}, \widehat{w})]$ is given by the weighted linear combination of activation functions

$$\hat{y}(\bar{x}, \widehat{w}) = \sum_{n_1=0}^{(N_1-1)} \sum_{n_m=0}^{(N_m-1)} \widehat{w}_{n_1 \cdots n_m} \phi_{n_r \cdots n_m}(\bar{x}) = \overline{\Phi}^T(\bar{x})\widehat{w}, \tag{1}$$

where $\bar{x} = [x_1 x_2 \cdots x_m]^T$ is the m-dimensional input vector, N_i is the number of neurons associated with the ith input, and $\widehat{w}$s are weights between hidden and output layers. The functions $\phi_{n_1 \cdots n_m}(\bar{x})$ are orthogonal functions in m-dimensional space, given by

$$\phi_{n_1 \cdots n_m}(\bar{x}) = \prod_{i-1}^{m} \phi_{n_i}(x_i), \tag{2}$$

where the ϕ_is are one-dimensional orthogonal functions implemented by each hidden layer neuron. $\overline{\Phi}(\bar{x})$ and $\widehat{w}$ are the transformed input vector and weight vector, respectively, in n-dimensional network weight space, where n is given by

$$n = \prod_{i=1}^{m} N_i. \tag{3}$$

Chaoying Zhu et al.

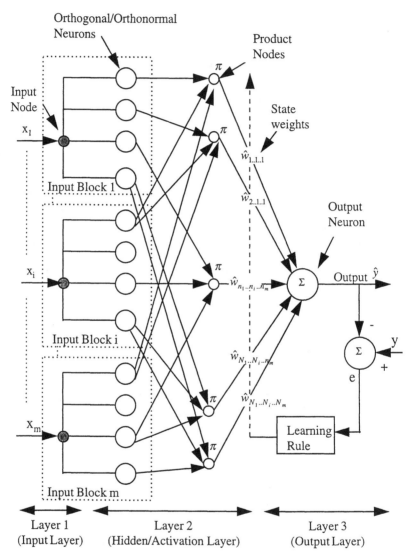

Orthogonal/Orthonormal Neurons

Product Nodes

Input Node

x_1

Input Block 1

π

π

State weights

$\hat{w}_{1..1..1}$

$\hat{w}_{2..1..1}$

Output Neuron

x_i

Input Block i

π

$\hat{w}_{n_1..n_i..n_m}$

Output $\hat{y}$

Σ

$-$

y

$+$

e

$\hat{w}_{N_1..N_i..n_m}$

$\hat{w}_{N_1..N_i..N_m}$

x_m

π

π

Input Block m

Learning Rule

Layer 1 (Input Layer)

Layer 2 (Hidden/Activation Layer)

Layer 3 (Output Layer)

Figure 2 Proposed orthogonal activation function-based neural network.

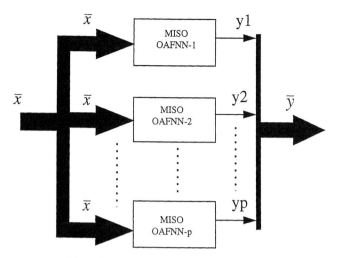

Figure 3   A MIMO architecture of an OAFNN.

The properties of OAFNNs are developed in the following section using the specific case of harmonic activation functions. The Legendre activation function also is considered later in this work.

1. Orthonormal Activation Functions

The orthonormal activation functions demonstrate a fast parameter convergence. The set of one-dimensional orthonormal functions $\phi_i^{(N)}(x)$ are defined as

$$\int_{x_1}^{x_2} \phi_i^{(N)}(x)\phi_j^{(N)}(x)\,dx = \delta_{ij}, \tag{4}$$

where δ_{ij} is the Kronecker delta function and $[x_1,\ x_2]$ is the domain of interest. Several examples of orthonormal functions are the normalized Fourier (harmonic) functions, Legendre polynomials, Chebychev polynomials, and Laguerre polynomials [11]. Only Fourier and Legendre activation functions are discussed in this work. These activation functions, with modifications for system identification, are shown in Table I.

2. Fourier Series Neural Network Architecture

The Fourier series neural network (FSNN) is a member of the class of OAFNNs where the harmonic functions are taken as activation functions and the requirement of normalization of the activation functions has been relaxed. Thus

Table I

Harmonic and Legendre Orthonormal Activation Functions

Type	Definition	Remarks
Harmonic	$a_0(x) = \frac{1}{\sqrt{x_2-x_1}}$, $a_n(x) = \sqrt{\frac{2}{x_2-x_1}}\sin(nu)$ or $a_n(x) = \sqrt{\frac{2}{x_2-x_1}}\cos(nu)$, $u = \frac{2\pi(x-x_1)}{x_2-x_1}$, where n is a positive integer	x_1 and x_2 are estimated lower and upper bounds of input x; u is linearly scaled input such that $u \in [0, 2\pi]$ corresponds to $x \in [x_1, x_2]$
Legendre	$a_n(x) = \sqrt{\frac{2n+1}{x_2-x_1}}L_n(u)$, $u = \frac{2x-(x_1+x_2)}{x_2-x_1}$, where n is a nonnegative integer and $L_n(u)$ is a Legendre polynomial of nth order.	x_1 and x_2 are estimated lower and upper bounds of input x; u is linearly scaled input such that $u \in [-1, 1]$ corresponds to $x \in [x_1, x_2]$

the FSNN design [10] is based on the property of the Fourier series and specifically aims to provide intelligent system identification and control tools in the frequency domain with extensions to the time domain.

A function always can be approximated mathematically by a Fourier series as long as the function is piecewise continuous and satisfies the Dirichlet condition [12], implying that a neural network using a harmonic function has the potential for function approximation. The FSNN's neurons are connected in such a way as to simulate a multiple Fourier series topology. Figure 4 shows the architecture of a MISO FSNN which maps a function $y = f(x_1, x_2, \ldots, x_m)$. It is a feedforward network with the input layer containing m nodes receiving inputs $x_1, x_2, \ldots, x_m$. In the activation layer there are N_i harmonic neurons associated with the ith input x_i for $i = 1, 2, \ldots, m$, resulting in total $\sum_{i=1}^{m} N_i$ neurons. Each neuron activates the input with a harmonic activation function $h(\cdot)$. The frequency of the harmonic activation function is defined by the frequency weight between the input node and the harmonic neuron. These frequency weights are selected to equal an integer n_i times a base frequency $\omega_{01} = 2\pi/T_i$ when the corresponding input can be scaled to a variable range of $[0, T_i]$ or has a period T_i, for all $i = 1, 2, \ldots, m$. The nodes on the right of the harmonic neurons implement the product (π) operation with each π node having m input signals from the m different input blocks. A total of $\prod_{i=1}^{m} N_i$ such π nodes are needed to complete the connections for all the harmonic neurons. The π nodes are considered a part of the middle layer because there is no weighing operation between the harmonic neurons and these π nodes. The output layer has only one neuron for a MISO FSNN configuration. This neuron sums the weighted outputs from the middle layer plus a bias $\widehat{w}_{0\cdots0}$, yielding the output $\hat{y}$ of the neural network. A MIMO FSNN, which is able to map func-

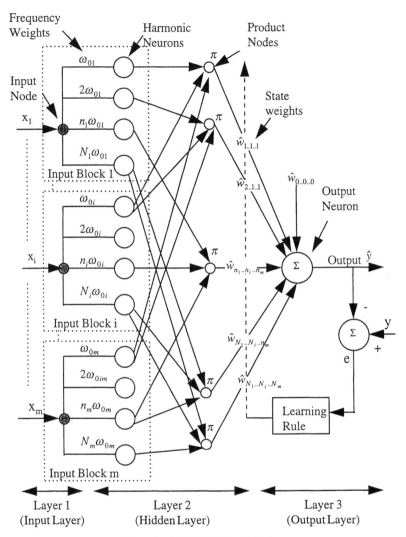

Figure 4 A MISO FSNN architecture.

tion $\bar{y} = f\bar{x}$, with $\bar{y} = [y_1, y_2, \ldots, y_k]^T$ and $\bar{x} = [x_1, x_2, \ldots, x_m]^T$, can be constructed using a parallel stack of k MISO FSNNs.

The weights $\widehat{w}_{n_1 \cdots n_m}$ between the output and middle layer are the state variables of the FSNN, defined as the state weights. These state weights are dynamically modified with a learning rule, providing FSNN the capability to learn and model an unknown process. All the frequency weights are constant during any

FSNN training cycle while the state weights are adjusted. The relationship between the input $\bar{x} = [x_1, x_2, \ldots, x_m]^T$ and the output $\hat{y}$ of the MISO FSNN model is formalized by

$$\hat{y}(\bar{x}, \widehat{w}) = \sum_{n_1=0}^{N_1} \cdots \sum_{n_m=0}^{N_m} \widehat{w}_{n_1 \cdots n_m} h_{n_1}(x_1) \cdots h_{n_m}(x_m), \tag{5}$$

where $h(\cdot)$ is a complex harmonic function with the $\widehat{w}$s as the state weights. The FSNN model defined by Eq. (5) approximates a multiple Fourier series representation if weights $\widehat{w}_{n_1 \cdots n_m}$ are trained to approach the coefficients of the Fourier series. The FSNN model with real sine and cosine activation functions can be expressed as

$$\hat{y}(\bar{x}, \widehat{w}) = \sum_{n_1=0}^{N_1} \cdots \sum_{n_m=0}^{N_m} \left[\widehat{w}_{n_1 \cdots n_m}^c \cos(\overline{w}_{n_1 \cdots n_m} \cdot \bar{x}) + \widehat{w}_{n_1 \cdots n_m}^s \sin(\overline{w}_{n_1 \cdots n_m} \cdot \bar{x}) \right], \tag{6}$$

where $\overline{w}_{n_1 \cdots n_m} = [n_1 \omega_{01}, \ldots, n_m \omega_{01}]^T$ and the multi dot between $\overline{w}_{n_1 \cdots n_m}$ and $\bar{x}$ denotes the dot product operation of the two vectors.

E. GRADIENT DESCENT LEARNING ALGORITHM

The learning in neural networks is performed by adapting the network weights ($\widehat{w}$s) such that the expected value of the mean squared error between network output and training output is minimized. The gradient descent-based learning algorithms are popular training algorithms for neural networks in supervised learning. In addition, in direct adaptive controllers, the learning rules are determined from Lyapunov-like stability analysis and were observed [13–16] to be similar to those given by gradient descent techniques. This is the basis for using gradient descent-based learning rules to train proposed networks.

The cost function for evaluation of learning performance is given by

$$J = E[(y - \hat{y})^2] = E\left[(y - \overline{\Phi}^T \widehat{w})^2\right] = E(y^2) + \widehat{w}^T \overline{R} \widehat{w} - 2\overline{P}^T \widehat{w}, \tag{7}$$

where y is the training output, $\widehat{w}$ is the estimate of network weight vector, $\overline{R} = E(\overline{\Phi} \overline{\Phi}^T)$ is the autocorrelation matrix of activation functions, and $\overline{P} = E(y\overline{\Phi})$ denotes the cross-correlation vector of training outputs and activation functions. For orthonormal activation functions, the autocorrelation matrix is an nth-order diagonal matrix with equal elements on the diagonal and therefore the error contour projection surface (J) is a hyper-spheroid in n-dimensional network weight space. The perfect symmetry of the error contour projection surface contributes to the fast convergence discussed in this research.

1. Geometric Significance of Performance Surface

The error contour projection surface (J) can be rewritten in terms of the error between optimum and actual network weights as

$$J = J_{\min} + (\widehat{\overline{w}} - \overline{w})^T \overline{R}(\widehat{\overline{w}} - \overline{w}), \tag{8}$$

where $\overline{w}$ is the optimum network weight vector. This performance surface, in general, is a hyperellipsoid in n-dimensional network weight space [17] and is illustrated in Fig. 5a for a two-dimensional weight space. The principal axes of this hyperellipsoid are in the same proportion as the eigenvalues of the autocorrelation matrix. For orthogonal activation functions the autocorrelation matrix is a diagonal positive definite matrix and therefore these principal axes are aligned with the network weight axes. For orthonormal activation functions, the auto correlation matrix is a diagonal matrix (of order n) with all equal eigenvalues. These equal eigenvalues result in *hyperspheroid* error contour projection surface (J) in n-dimensional network weight space, as illustrated in Fig. 5b. The perfect sym-

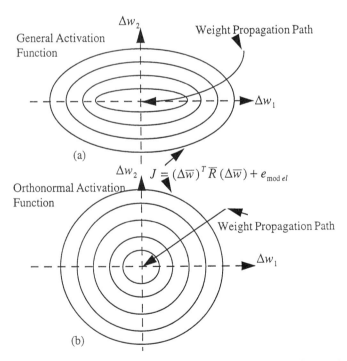

Figure 5 Performance surface: (a) nonorthonormal activation functions; b) orthonormal activation functions.

metry of the error contour projection surface contributes to fast parameter convergence when a gradient descent-based learning algorithm is employed.

In gradient descent methods, the weights are updated normal to the error contour projection surface, which is the direction of the negative error gradient vector. A typical propagation path of the weight vector from a random initial state to the optimal weight state is shown in Fig. 5. For orthonormal activation functions, the path between a random initial weight state and the final optimal weight state is a straight line. A straight line is the shortest path between any two points, resulting in fast parameter convergence. A straight line weight update path (for true gradient descent) enhances the possibility for *single training cycle learning*, because the gradient direction of the optimal weight state is known from any random initial weight state.

2. True versus Instantaneous Gradient Descent

There are many learning algorithms used to minimize the training cost function, but the two simplest and most utilized learning algorithms are the true gradient descent learning algorithm and the instantaneous gradient descent learning algorithm. In the first learning algorithm, the gradient descent is calculated for the expected value of mean squared error (MSE) and weights are updated along the negative direction of this gradient. This rule employs all the training data at each update and therefore it is only suitable for off-line or time delayed system identification methods. In instantaneous gradient descent training algorithm applications, the weights are updated at each training step toward the negative gradient of the instantaneous performance surface which makes it suitable for on-line and real time applications. This method may result in gradient noise which slows the weight convergence. However, the instantaneous gradient descent learning algorithm approximates the true gradient descent rule if the weight update step is sufficiently small. Thus the characteristics of the true gradient descent learning algorithm can be applied to instantaneous gradient descent methods.

F. PROPERTIES OF ORTHOGONAL ACTIVATION FUNCTION-BASED NEURAL NETWORKS

1. Function Approximation Capability

The orthogonal functions are established mathematically complete sets that are capable of approximating any nonlinear, bounded, piecewise continuous function $[y(\bar{x})]$ with simple discontinuities to an arbitrary accuracy, that is,

$$\lim_{n \to \infty} \left[y(\bar{x}) - \sum_{i=1}^{n} w_i \phi_i(\bar{x}) \right]^2 \to 0, \tag{9}$$

where w_is are optimum weights for the network [11, 12]. Thus given a sufficient number of neurons in the hidden layer, any arbitrary function (at least piecewise continuous) can be approximated by the network to any desired level of accuracy. The approximation error is inversely proportional to the number of neurons. This validates our choice of network architecture given in Fig. 2.

2. Computational Parallelism

The OAFNN possesses a computational parallelism defined by its topological structure. Within an OAFNN learning step, the training data are sampled, the computational process proceeds, and the OAFNN state weight modification is calculated. The computations performed in the OAFNN neurons are independent and each neuron functions without interaction with other neurons. Therefore, every input signal takes an equal amount of computational time to flow from the input node to the output node without waiting time, thus providing state weight tuning synchronization. This computational parallelism allows the OAFNN to have very fast learning with parallel processing computer (transputer) hardware. Increasing the number of the inputs and outputs of an MIMO system essentially does not increase the computational time if enough processors are added. This OAFNN approach potentially permits the learning process to occur within a sampling time interval if enough high speed processors are used. A comparison between the FFT and FSNN in this regard is discussed in Section III.A.

3. Learning Stability for the Gradient Descent Rule

The gradient descent approach, also known as the delta rule, adjusts network weights in such a way that the square of the neural network learning error (quadratic learning error) always changes in a negative gradient direction. A general gradient descent rule for network weight modification is given by

$$\Delta \widehat{w} = -\delta \frac{\partial J}{\partial \widehat{w}}, \tag{10}$$

where $\Delta \widehat{w}$ is the weight variation, δ is a constant called the learning rate, $\widehat{w}$ represents the state weight before the modification, and J is the cost function. For the OAFNN, the cost function can be defined by

$$J \equiv \tfrac{1}{2} e^2 = \tfrac{1}{2} (y - \hat{y})^2, \tag{11}$$

where learning error e is the difference between the actual output y and the FSNN output $\hat{y}$.

The Lyapunov approach is used to investigate the learning stability of the FSNN when applying the delta rule to govern the learning. The cost function J is chosen as the Lyapunov energy function for the investigation. Assume that the system to be identified can be represented by a Fourier series with real coefficients $A_{n_1 \cdots n_m}$ and $B_{n_1 \cdots n_m}$ as

$$y(\bar{x}) = \sum_{n_1=0}^{\infty} \cdots \sum_{n_m=0}^{\infty} \left[A_{n_1 \cdots n_m} \cos(\overline{\omega}_{n_1 \cdots n_m} \cdot \bar{x}) + B_{n_1 \cdots n_m} \sin(\overline{\omega}_{n_1 \cdots n_m} \cdot \bar{x}) \right]. \quad (12)$$

This equation has a form similar to the FSNN model defined in Eq. (6) with sine and cosine harmonic neurons. The FSNN learning error e in a series form can then be written as

$$\begin{aligned}
e = &\sum_{n_1=0}^{N_1} \cdots \sum_{n_m=0}^{N_m} \left[(A_{n_1 \cdots n_m} - \widehat{w}_{n_1 \cdots n_m}^c) \cos(\overline{\omega}_{n_1 \cdots n_m} \cdot \bar{x}) \right. \\
&\left. + (B_{n_1 \cdots n_m} - \widehat{w}_{n_1 \cdot n_m}^s) \sin(\overline{\omega}_{n_1 \cdots n_m} \cdot \bar{x}) \right] \\
&+ \sum_{n_1=N_1+1}^{\infty} \cdots \sum_{n_m=N_m+1}^{\infty} \left[(A_{n_1 \cdots n_m} \cos(\overline{\omega}_{n_1 \cdots n_m} \cdot \bar{x}) \right. \\
&\left. + B_{n_1 \cdots n_m} \sin(\overline{\omega}_{n_1 \cdots n_m} \cdot \bar{x}) \right].
\end{aligned} \quad (13)$$

The second multiple summation in Eq. (13) contains higher order terms which may be neglected if N_is ($i = 1, \ldots, m$) are large. The inaccuracy due to neglecting this summation is inversely proportional to N_is in terms of the uniform convergence property of the Fourier series [12]. After neglecting the higher-order terms, the learning error e is approximated by

$$\begin{aligned}
e \cong &\sum_{n_1=0}^{N_1} \cdots \sum_{n_m=0}^{N_m} \left[(A_{n_1 \cdots n_m} - \widehat{w}_{n_1 \cdots n_m}^c) \cos(\overline{\omega}_{n_1 \cdots n_m} \cdot \bar{x}) \right. \\
&\left. + (B_{n_1 \cdots n_m} - \widehat{w}_{n_1 \cdots n_m}^s) \sin(\overline{\omega}_{n_1 \cdots n_m} \cdot \bar{x}) \right].
\end{aligned} \quad (14)$$

The delta rule for adjusting state weights $\widehat{w}_{n_1 \cdots n_m}^c$ and $\widehat{w}_{n_1 \cdots n_m}^s$ in the FSNN model can be deduced by substituting Eq. (14) into Eq. (11) and then taking the partial derivative of Eq. (11) with respect to a state weight. This gives the network cosine and sine weight modification as

$$\Delta \widehat{w}_{n_1 \cdots n_m}^c = \delta e \cos(\overline{\omega}_{n_1 \cdots n_m} \cdot \bar{x}), \qquad \Delta \widehat{w}_{n_1 \cdots n_m}^s = \delta e \sin(\overline{\omega}_{n_1 \cdots n_m} \cdot \bar{x}). \quad (15)$$

The variation of the learning error $\Delta e = e^a - e$ (where e^a is the new error after the modification) is obtained by substituting Eq. (15) into Eq. (14), giving

$$\Delta e = -\delta e (N_1 + 1)(N_2 + 1) \cdots (N_m + 1). \quad (16)$$

The variation of the Lyapunov energy function is then given by

$$\Delta J = e \Delta e = -\delta (N_1 + 1)(N_2 + 1) \cdots (N_m + 1) e^2. \tag{17}$$

Equation (17) indicates that ΔJ is negative if the learning rate δ is selected to be positive. This means that the quadratic learning error changes in the negative gradient direction when the FSNN learning advances, using the delta learning rule with a learning rate δ that has a lower bound of zero.

Learning rate δ also determines the step size of the state weight modification $\Delta \widehat{w}$. A large δ may cause excessive modification of the weights, resulting in the possible learning divergence, although the learning error moves in the negative gradient direction. An upper bound of δ must also be chosen properly to ensure that the quadratic learning error decreases asymptotically. Equation (16) can be reorganized to find this bound as

$$\left| \frac{e^a}{e} \right| = \left| 1 - \delta (N_1 + 1)(N_2 + 1) \cdots (N_m + 1) \right|, \tag{18}$$

where e^a is the new error after the modification. From this equation, it can be shown that the δ value which causes the absolute value of the learning error e to decline is bounded by

$$0 < \delta < \frac{2}{(N_1 + 1)(N_2 + 1) \cdots (N_m + 1)}. \tag{19}$$

Thus, the asymptotic stability or convergence of the FSNN learning process is guaranteed by Eq. (19) when the delta rule (15) is used for the FSNN training.

In general, for an OAFNN, the weight update law for the instantaneous gradient descent algorithm is given by

$$\widehat{w}(t) = \widehat{w}(t-1) + \delta e \overline{\Phi}(t), \tag{20}$$

where $e = y - \hat{y}$ is the error between actual and estimated output, $\overline{\Phi}(t)$ is the transformed input vector consisting of orthogonal functions, and δ is the learning rate. The range for stable learning rate δ is derived [10, 18] as

$$0 < \delta < \frac{2}{\overline{\Phi}^T \overline{\Phi}}. \tag{21}$$

For the true gradient descent learning rule, the gradient of the error surface is given by

$$\overline{\Delta} = \frac{\partial J}{\partial \widehat{w}} = 2 \overline{R} \widehat{w} - 2 \overline{P} = -2E(e\overline{\Phi}). \tag{22}$$

The optimum weight vector $\overline{w}$, which makes the gradient zero, is $\overline{w} = \overline{R}^{-1}\overline{P}$ which is the same result as determined when using a linear regression analysis method. However, researchers have pointed out the advantages of using iterative learning of neural networks over conventional linear regression methods [18, 19]. The weight adaptation principle is given as

$$\widehat{\overline{w}}(k) = \widehat{\overline{w}}(k-1) - \frac{\delta}{2}\overline{\Delta} = \widehat{\overline{w}}(k-1) + \delta E(e\overline{\Phi}(t)), \tag{23}$$

where δ is the learning rate. It can be shown that learning is stable for values of δ lying in the range given by [18]

$$0 < \delta < \frac{2}{\lambda_{max}}, \tag{24}$$

where λ_{max} is the maximum eigenvalue of the autocorrelation matrix, $\overline{R}$.

4. Absence of Local Minima

The error surface (cost function J) for an OAFNN is a hyperparabola in n-dimensional space. Such an error surface has no local minima irrespective of the choice of activation functions. If the activation functions are linearly independent and training is sufficiently exciting, there exists a unique global minimum [7, 18]; otherwise there are an infinite number of possible global minima. The choice of proposed orthogonal activation functions ensures the linear independence of activation functions for the majority of the input space. Thus the proposed choice of orthogonal activation functions increases the probability that the network reaches the unique global minimum.

5. Parameter Convergence

In the absence of local minima and with stable learning rate, the parameter weights converge to the optimum weight space given by global minima of the error surface. In the case of nonunique global minima, global minimum space (also termed null space) is given by the space spanned by the eigenvectors corresponding to the zero eigenvalues of the autocorrelation matrix. For an arbitrary set of activation functions this global minimum may not be unique, but for orthogonal activation functions there exists a unique global minimum because the autocorrelation matrix ($\overline{R}$) is positive definite (no zero eigenvalue; therefore empty null space). Thus the *parameters (weights) in a network, using orthogonal activation functions with sufficient training always converge to the unique global minimum defined by the optimum weight vector* provided no modeling and measurement errors exist. In the presence of these errors, the weights converge to a minimal capture zone [18] in the neighborhood of the global minimum. The size of this zone is small for orthonormal activation functions due to symmetric (hyperspheroid) error contours. The property of parameter convergence for OAFNNs is further illustrated in the following text using a proof of parameter convergence for FSNN.

The surface contour of the Lyapunov energy function of the FSNN is considered. The analysis is performed by taking the partial derivative of J with respect to the state weight W and setting the derivative equal to zero to find the extremum of the Lyapunov function. This results in

$$e\cos(\overline{\omega}_{n_1\cdots n_m} \cdot \bar{x}) = 0, \qquad e\sin(\overline{\omega}_{n_1\cdots n_m} \cdot \bar{x}) = 0. \qquad (25)$$

Whereas, in general, $\cos(\cdot)$ and $\sin(\cdot)$ are not zero, the sole solution of Eq. (25) is $e = 0$, or $\widehat{w}^c_{n_1\cdots n_m} = A_{n_1\cdots n_m}$ and $\widehat{w}^s_{n_1\cdots n_m} = B_{n_1\cdots n_m}$ in terms of Eq. (14). The corresponding extremum of the Lyapunov function is $J = 0$. This extremum is a minimum because the Lyapunov function is nonnegative. Consequently, the FSNN learning approaches the global or only minimum at the learning error $e = 0$ with the resulting FSNN model approximating the Fourier series coefficients as the learning converges. Such a convergence of the FSNN learning process is guaranteed by the delta rule with a learning rate bounded by Eq. (19), because the Lyapunov function of the FSNN is parabolic, which is free from local minima.

6. Rate of Convergence

The OAFNNs have a much higher learning speed than the multilayered feedforward networks. The details of the comparison results are presented in later sections along with an evaluation of network performance. The high leaning speed of OAFNNs can be attributed to the absence of cross-coupling between orthogonal neurons. The rate of convergence is expected to improve if orthonormal activation functions are used in OAFNNs as discussed subsequently.

The rate of convergence of a linearly parameterized network is inversely proportional to the condition number $[C(\overline{R})]$ of the autocorrelation matrix for activation functions. The condition number is defined as the ratio of the maximum eigenvalue to the minimum positive eigenvalue of $\overline{R}$ [18] as given by

$$C(\overline{R}) = \frac{\lambda_{\max}\{\overline{R}\}}{\lambda_{\min,+}\{\overline{R}\}}. \qquad (26)$$

It can be shown that the minimum condition number possible is equal to unity which occurs when the minimum positive eigenvalue ($\lambda_{\min,+}\{\overline{R}\}$) equals the maximum eigenvalue ($\lambda_{\max}\{\overline{R}\}$). The autocorrelation matrix is a positive semidefinite matrix, in general, which has zero or positive eigenvalues. The zero eigenvalue corresponds to the nonunique global minimum. However, the autocorrelation matrix for orthonormal activation functions is a diagonal, positive definite matrix with all equal eigenvalues. Thus the condition number is a minimum (equals unity) and *results in the fastest convergence for the orthonormal activation functions.*

7. Model Generalization

Generalization is the capability which allows a trained neural network model to respond appropriately when presented with unfamiliar inputs in the sense that they are not used for the neural network training. The OAFNN model is generic if enough training data are sampled in the input variable range corresponding to the highest order orthogonal activation function using the constraint analogous to the Shannon sampling theorem [20]. For example, for a FSNN, the highest frequency of the frequency weights for the input variable x_i is $N_i w_{0i}/(2p) = N_i/T_i$. If training data are uniformly sampled within a range T_i, the total number of the training samples $N_s^{(i)}$ within the range T_i can be determined by

$$N_s^i = S_i N_i, \qquad i = 1, 2, \ldots, m, \qquad (27)$$

where S_i is the number of required sampling points in the range T_i corresponding to the highest frequency. A minimum value of $S_i = 2$ according to the Shannon sampling theorem is required, but an S_i value of 5 or greater is recommended as a "rule of thumb" for good model construction. The generalization capability can be measured by the response of the learning error of a trained neural model to unfamiliar inputs. The generalization capability of the FSNN is evaluated quantitatively through case studies presented in [10] and in Sections II.G.1 and IV.A of this work.

8. Model Interpretation

The OAFNN model has clear physical meaning by its representation of the identified function with the stabilized OAFNN state weights that estimate the coefficients of orthogonal series expansion of the function. This technique, using FSNN, provides an alternative for the conventional fast Fourier transforms (FFT), which are used in commercial spectrum analyzers. The details of using the FSNN technique to construct a neural network spectrum analyzer (NNSA) are presented in Section III.A of this work.

9. Reasoning Structure Design

Unlike the other neural networks, the OAFNN can be designed rationally based on the specifications of the target system, such as the number of inputs and outputs, required mapping accuracy, and range of the system variables. The OAFNN has a MIMO modeling capability where the number of the input and output nodes of the OAFNN is equal to that of the system to be identified. Knowing that the truncating error of the OAFNN is inversely proportional to the number N of orthogonal neurons, then N can be determined in terms of the required accuracy. The frequency weights in a FSNN can be determined using only one parameter, which is the base frequency $\omega_{0i} = 2p/T_i$, because the frequency weight sequence is an arithmetic series. The period T_i should equal the period of the corresponding system input variable x_i if x_i is periodic, or equal the range of the variable x_i if x_i is nonperiodic.

One important variation of the FSNN design is based on the possibility of using only cosine or sine activation functions to simulate cosine or sine half range expansion of the Fourier series. The advantage of doing this lies in the reduced number of the neurons required for the computation.

G. Preliminary Performance Evaluation of Orthogonal Activation Function-Based Neural Networks

1. Boolean Function Identification Using Fourier Series Neural Networks

An exclusive-or function (abbreviated X-OR), is an elementary relationship in Boolean algebra and is often used to evaluate a neural network's capability for modeling nonlinear functions. Such a capability permits the neural network to be applied to Boolean logic pattern recognition and image processing. The work presented here evaluates the robustness of the FSNN model when it is used for Boolean logic function estimation via the identification of the X-OR function. With the signs $\cap$ and $\cup$ representing logical AND and OR operations, respectively, an X-OR relationship can be described by

$$y(x_1, x_2) = \begin{cases} -1, & \text{if } (x_1 = 0 \cap x_2 = 0) \cup (x_1 = 1 \cap x_2 = 1) \\ 1, & \text{if } (x_1 = 0 \cap x_2 = 1) \cup (x_1 = 1 \cap x_2 = 0). \end{cases} \tag{28}$$

Hence, the FSNN model should be an MISO system that has two input nodes for x_1 and x_2, and one output node for y. The FSNN harmonic neuron uses only the cosine activation function, and the number of harmonic neurons associated with each input is equal to 10. The period R of the FSNN is selected to equal 2, which covers a range of input variables $(-0.5 < x_1 < 1.5)$ and $(-0.5 < x_2 < 1.5)$. A 20×20 training data set is generated uniformly over that $R \times R$ square area using Eq. (28) and setting $y = 0$ if x_1 or x_2 is undefined in Eq. (28); that is, neither a 0 nor a 1.

Figure 6 shows the resulting FSNN map of the X-OR function after two training cycles with a learning rate of 0.01. This map is plotted with a 50×50 point mesh which contains at least 2100 unfamiliar input data points, which indicates that the FSNN model has good generalization capability. The FSNN model for this discrete X-OR function is a continuous mesh. Two peaks that have height 1 in the map correspond to $x_1 = 1 \cap x_2 = 0$ and $x_1 = 0 \cap x_2 = 1$, whereas two valleys that have depth 1 correspond to $x_1 = 0 \cap x_2 = 0$ and $x_1 = 1 \cap x_2 = 1$, respectively, which shows that the FSNN identification of the X-OR function is valid and correct.

The evaluation of the FSNN generalization shows that the FSNN model is effective, accurate, and robust when presented with unfamiliar information. The successful application of the FSNN to logical X-OR relationship identification indicates that the FSNN model is also valid for discrete logic Boolean functions, which potentially enables FSNN to be applied to the related applications.

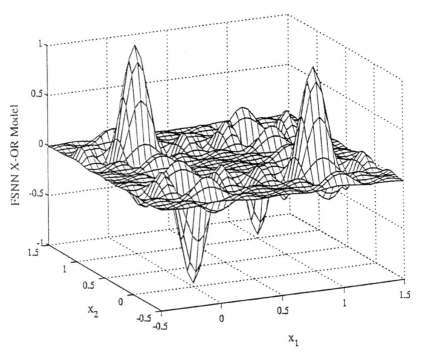

Figure 6   X-OR function estimated by the FSNN identifier.

2. Function Approximation Using Orthogonal Activation Function-Based Neural Networks

Four types of neural networks are used for simulation-based performance comparison [9]:

1. Orthonormal neural networks with harmonic neurons
2. Orthonormal neural networks with Legendre neurons
3. Functional link network using normalized polynomial (nonorthogonal) neurons
4. Three-layer perceptrons with tan sigmoidal neurons in the hidden layer

These networks are tested for approximation of a hyperbolic tangent (tan sigmoid) function given by

$$y(x) = \frac{2}{1 + e^{-2x}} - 1, \tag{29}$$

which is also the activation function for tan sigmoid neural networks. This function was selected to give sigmoid networks an added advantage.

A simple cost function is developed to compare the learning performances of the different networks. This cost function is given by the product of the number of training cycles and the sum of squared errors at that training cycle. This cost function is designed to penalize slow convergence of the training error for a network and is therefore termed training error convergence cost (TECC) function. A high TECC indicates either slow error convergence or high learning error or both, whereas a low TECC indicates fast error convergence as well as low training error.

An architecture of 1 input neuron, 11 hidden layer neurons, and 1 output neuron (1-11-1) was used for each neural network, to approximate the tan sigmoid function. The training data were generated for $x \in [-2, 2]$, whereas the estimated lower bound (x_1) and upper bound (x_2) of input were taken as -2.5 and 2.5, respectively.

The instantaneous gradient descent learning algorithm with constant learning rate of 0.091 was employed to train Legendre, harmonic, and polynomial networks, whereas the true gradient descent learning algorithm with adaptive learning rate and momentum was used for faster learning of the sigmoid network. The sigmoid network weights were initialized based on Nguyen–Widrow random weight generation algorithm for better performance. The networks were trained until either the sum squared error (SSE) reduced to 0.001 or the slope of the error dropped below 0.0001 per cycle or 500 training cycles were completed. The approximation errors for the four networks are shown in Fig. 7. None of the four networks reached the error goal (SSE = 0.001), but sigmoidal, Legendre, and harmonic networks approximated the function with reasonable accuracy (SSE < 0.005). Figure 8 shows the training error for all four networks. The Legendre and harmonic networks demonstrate fast error convergence compared to

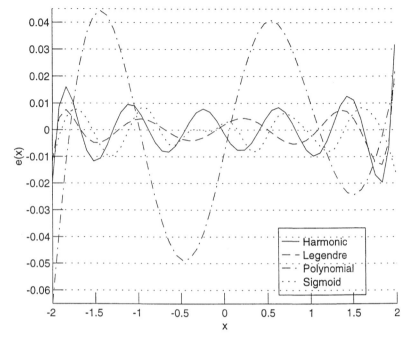

Figure 7 Function approximation error.

polynomial and sigmoid networks. The sigmoid network reached a low error of 0.0018, but took 500 cycles to reach this error level as compared to 36 cycles taken by the Legendre network to reach a comparable error level. The best and the second best learning performance were demonstrated by Legendre and harmonic networks, respectively, as shown in the Table II.

Table II

Simulation Results for Function Approximation

Type of NN	(N)	SSE (E)	TECC (N × E)
Sigmoid NN	500	*0.0018*	0.90
Polynomial NN	500	0.0433	21.67
Legendre NN	36	*0.0018*	0.07
Harmonic NN	*18*	0.0048	0.09

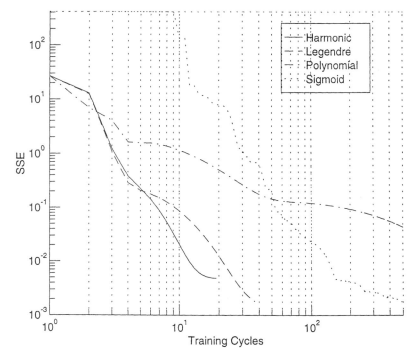

Figure 8   Training error for function approximation.

III. FREQUENCY DOMAIN APPLICATIONS USING FOURIER SERIES NEURAL NETWORKS

A. NEURAL NETWORK SPECTRUM ANALYZER

System transfer function identification is a technique which uses records of system operation to develop a Laplace representation of the system. One common approach for system transfer function identification is to use the FFT to change the time series descriptions of sampled system input and output signals into the frequency domain, thereby enabling the construction of a frequency transfer function of the system. The FSNNs possess learning ability which allows them to form the Fourier transformation of time series signal once trained. This FSNN application is called neural network spectrum analyzer [21] NNSA. Theoretically, the NNSA concept differs from the FFT concept in computational structure. The FFT has a combination of serial and parallel structure in which the transformation can be performed only after all the signal samples in a period are presented, whereas the NNSA structure is purely parallel so that the learning proceeds or the transformation is performed at every step when a new sample signal is presented. With a purely parallel structure, an NNSA can take advantage of the parallel processing hardware that computes the parallel branches of the structure simultaneously. This hardware arrangement, combined with software that is configured and timed accordingly, leads to faster identification and fault tolerance when data are missing. With these advantages, the NNSA has the potential to be used to improve the performance of spectrum analyzers if its identification accuracy is shown to be competitive with the existing FFT algorithm. The following two sections evaluate the FSNN identification accuracy when applied to frequency spectrum and transfer functions identification, respectively.

1. Frequency Spectrum Identification

If a signal $r(t)$ (where t is time) is periodic or defined within a finite time interval, it can be approximated by a Fourier series given by

$$r(t) = \frac{1}{2}a_0 + \sum_{n=1}^{N}\left[a_n\cos\left(\frac{2n\pi}{T}t\right) + b_n\sin\left(\frac{2n\pi}{T}t\right)\right], \tag{30}$$

where a and b are the coefficients of the Fourier series, T is the period or the time interval of the signal $r(t)$, and N determines the number of terms of the series. A single-input–single-output (SISO) FSNN given by

$$\hat{r}(t) = \frac{1}{2}\widehat{w}_0 + \sum_{n=1}^{N}\left[\widehat{w}_n^{\,c}\cos\left(\frac{2n\pi}{T}t\right) + \widehat{w}_n^{\,s}\sin\left(\frac{2n\pi}{T}t\right)\right], \tag{31}$$

is called NNSA because it is used to approximate the time series signal with the state weights $\widehat{w}$ approximating the frequency spectrum. To train the FSNN, samples of signal $r(t)$ and time t are introduced to the FSNN as illustrated in Fig. 9 during the training process. For the evaluation study, an FSNN that has 10 cosine and 10 sine neurons, with a base frequency of $\omega_0 = 2\pi$ rad/s, was chosen to model a mixed harmonic signal given by

$$r(t) = \sin(\omega_0 t) - 0.7\cos(5\omega_0 t) + 0.3\sin(8\omega_0 t).$$

Twenty-five points of the signal $r(t)$ were uniformly sampled in a 1-s period for training with a learning rate $\delta = 0.91$.

Table III lists the FSNN learning error and state weights after each training cycle for four cycles. As shown in the table, the mean learning error e_m decreases quickly as learning advances. It is observed from the table that after the fourth training cycle, all the state weights closely match the actual coefficients of the harmonic signal $r(t)$.

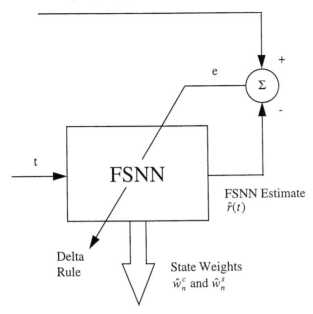

Figure 9 Frequency spectrum identification scheme using the FSNN.

2. Transfer Function Identification

The NNSA has been shown to accurately approximate the frequency spectrum of the time series signals, which indicates that the NNSA has the potential to be used as an innovative means of system transfer function identification with the advantages inherited from the FSNN. This section describes the NNSA scheme for the frequency transfer function identification of an experimental robot end effector. The performance of the NNSA is evaluated by comparing the results with those obtained by using a conventional FFT analyzer.

Figure 10 shows the transfer function identification scheme using two NNSAs—one trained with system input $x(t)$ and the other with output $y(t)$. Whereas the discrete frequency spectrum of signals $x(t)$ and $y(t)$ can be estimated by the weights $\widehat{w}_{x0}$, $\widehat{w}_{xn}^{c}$, $\widehat{w}_{xn}^{s}$, $\widehat{w}_{y0}$, $\widehat{w}_{yn}^{c}$, and $\widehat{w}_{yn}^{s}$ of the corresponding NNSA, the discrete frequency transfer function can be computed as

$$|\widehat{G}_0| = \frac{\widehat{w}_{y0}}{\widehat{w}_{x0}}, \qquad |\widehat{G}_n| = \sqrt{\frac{(\widehat{w}_{yn}^{c})^2 + (\widehat{w}_{yn}^{s})^2}{(\widehat{w}_{xn}^{c})^2 + (\widehat{w}_{xn}^{s})^2}}, \qquad n = 1, \ldots, N, \qquad (32)$$

Table III

FSNN Frequency Spectrum Identification Results for a Harmonic Signal

Training cycle	0	1	2	3	4
Learning error e_{ml}	—	11.2704	3.3719	0.9236	0.2395
Bias $\widehat{w}_0$	10.0000	4.5321	1.3911	0.3544	0.0880

Cosine weights

$\widehat{w}_1^c$	10.0000	−1.5577	−0.1381	0.0092	0.0067
$\widehat{w}_2^c$	10.0000	−0.9468	−0.1338	−0.0188	−0.0000
$\widehat{w}_3^c$	10.0000	−0.6820	−0.1416	−0.0220	−0.0012
$\widehat{w}_4^c$	10.0000	−0.4768	−0.1478	−0.0219	−0.0014
$\widehat{w}_5^c$	10.0000	−1.0032	−0.8517	−0.7207	−0.7013
$\widehat{w}_6^c$	10.0000	−0.0759	−0.1475	−0.0195	−0.0008
$\widehat{w}_7^c$	10.0000	0.1639	−0.1327	−0.0184	0.0001
$\widehat{w}_8^c$	10.0000	0.5058	−0.0905	−0.0195	0.0015
$\widehat{w}_9^c$	10.0000	0.8733	0.0412	−0.0216	0.0007
$\widehat{w}_{10}^c$	10.0000	1.4314	0.5855	0.0285	−0.0409

Sine weights

$\widehat{w}_1^s$	−10.0000	1.2382	1.5929	1.1599	1.0386
$\widehat{w}_2^s$	−10.0000	0.1788	0.2345	0.0760	0.0201
$\widehat{w}_3^s$	−10.0000	0.1003	0.1144	0.0464	0.0134
$\widehat{w}_4^s$	−10.0000	0.0036	0.0498	0.0295	0.0101
$\widehat{w}_5^s$	−10.0000	−0.0803	0.0063	0.0172	0.0082
$\widehat{w}_6^s$	−10.0000	−0.2667	−0.0198	0.0068	0.0071
$\widehat{w}_7^s$	−10.0000	−0.4738	−0.0365	−0.0025	0.0064
$\widehat{w}_8^s$	−10.0000	−0.4803	0.2661	0.2910	0.3056
$\widehat{w}_9^s$	−10.0000	−1.3264	−0.0225	0.0022	0.0047
$\widehat{w}_{10}^s$	−10.0000	−2.6152	−0.0531	0.1557	0.0305

and

$$\angle\widehat{G}_0 = 0, \qquad \angle\widehat{G} = \tan^{-1}\left(\frac{\widehat{w}_{xn}^s\,\widehat{w}_{yn}^c - \widehat{w}_{xn}^c\,\widehat{w}_{yn}^s}{\widehat{w}_{xn}^s\,\widehat{w}_{yn}^s + \widehat{w}_{xn}^c\,\widehat{w}_{yn}^c}\right), \qquad n = 1, \ldots, N, \quad (33)$$

where $\widehat{G}$ is the estimated transfer function.

To evaluate the performance of the FSNN transfer function identifier compared with an FFT spectrum analyzer, a robot end effector was used as the unknown plant for identification. The end effector uses a servo motor to adjust the position of a tool, based on the force sensor feedback, to compensate for force diversity. To design such an end effector, it is important to know the transfer function between the voltage to the servo amplifier and the end effector force. A theoretical analysis

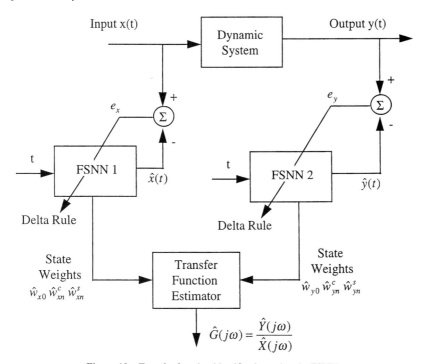

Figure 10 Transfer function identification using the FSNN.

of the end effector dynamics reveals that the end effector can be modeled by the linear transfer function

$$G_P = \frac{K_I K_B K_L / R_S}{(J + K_I K_\omega / K_a R_S)s^2 + Bs + K_F K_B K_L}, \tag{34}$$

where s is the Laplace transform, K_I is the motor torque constant, K_ω is the motor speed constant, K_a is the amplifier constant, K_B is the tool stiffness, K_L is the ball screw lead, K_F is the proportionality constant between a contact force and tool drive, J is the motor moment of inertia, B is the motor equivalent linear friction coefficient, and R_S is the grounding resistance of the motor–amplifier circuit.

Figure 11 shows a schematic diagram depicting the experimental system. The PC is equipped with an Amadeus-96 controller board through which the interfaces between the PC and the end effector are established. A digital-to-analog (D/A) converter on the controller board is used to apply input signals to the end effector, and an analog-to-digital (A/D) converter is used to receive data from the force

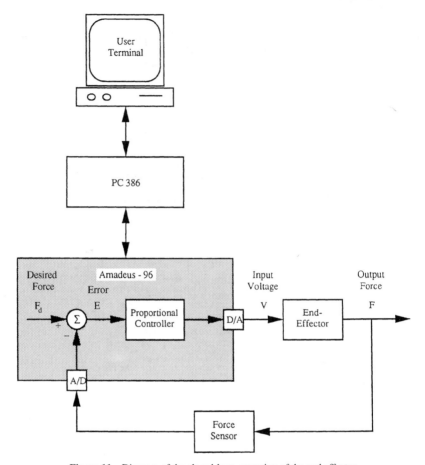

Figure 11 Diagram of the closed-loop operation of the end effector.

sensor. The error between the current force and the desired force is calculated and used in a proportional control algorithm.

The FFT spectrum analyzer used for this evaluation was an Ono Sokki CF 350 model (Ono Sokki Co. Ltd., Tokyo, Japan). A frequency transfer function identification was made with signals being sampled through a 16-bit successive approximation type A/D converter and a filter. The transfer function $G(f)$ was calculated using $G(f) = Y(f)X^*(f)/[X(f)X^*(f)]$, where $X(f)$ is the Fourier transform of the input signal, $X^*(f)$ is its complex conjugate, and $Y(f)$ is the Fourier transform of the output signal. The NNSA was implemented in an off-line manner by using stored input–output data for training. These data were acquired

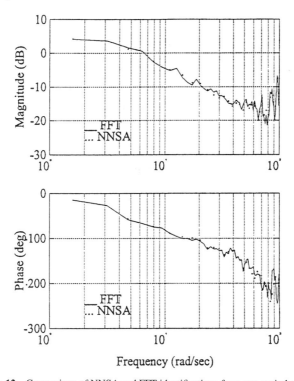

Figure 12   Comparison of NNSA and FFT identifications from one period of data.

during system operation from the FFT spectrum, which allowed for a direct comparison in the performances of the two methods. The NNSA was configured with 63 neurons to span the frequency range from 1.57 to 100 rad/s.

Each of the two identification methods was applied to estimate the open loop transfer function G_P of the end effector when operated in the closed-loop mode. The reference input to the closed loop end-effector systems had a period of 4 s. The steady-state value of the reference input force was chosen to be 5 lb. The square wave amplitudes were selected to be 0.5 lb above and below the steady-state value. The square wave was discretized such that a time sample was inputed to the closed-loop system every 0.01 s. The open-loop transfer function was determined directly by using the plant (end-effector) input and system output data during the operation. The sampling frequency used by the FFT spectrum analyzer was 256 Hz.

The first comparison between the NNSA and FFT identifications is based on only one period of data and is displayed in Fig. 12. This figure shows that the

NNSA identification made from one period of data closely matches the identification made using the FFT spectrum analyzer from the same period of data, although quantitatively the results are slightly different. The average differences between the identifications are 0.989 dB and 7.31°. These numerical differences are rationalized by acknowledging that the NNSA computations are performed with 32-bit floating point precision, whereas the FFT computations are performed with 16-bit fixed point precision. The two identifications are similarly affected by noise, as indicated by the frequent peaks at the higher frequencies.

The second comparison between the NNSA and FFT identifications is based on 16 periods of data and is displayed in Fig. 13. This figure also shows a close match between the two schemes. In this case, the average difference in magnitude is 1.06 dB and the average difference in phase is 4.53°. These results show that the frequency transfer function learned by the NNSA is very similar to the averaged frequency transfer function calculated by the FFT spectrum analyzer. The relative smoothness of the curves in Fig. 13 compared to those in Fig. 12 indicates that

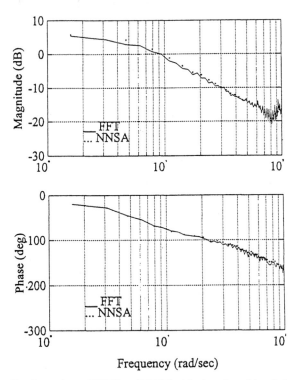

Figure 13   Comparison of NNSA and FFT identifications from 16 periods of data.

NNSA and FFT identifications made from multiple periods of data are effective to a similar degree in compensating for noise in the system.

Three factors affect the quality of an NNSA identification in terms of the experiment. First, the sampling rate only affects the NNSA identification in the same way that it affects the discrete frequency transfer function. Therefore, in the identification of a continuous-time system driven by a zero-order hold, a higher sampling rate leads to an NNSA estimation that more closely resembles the transfer function of the continuous-time system. Second, the period of the square wave input signal has a negative effect on an NNSA identification only when the settling time of the system is greater than half of the square wave period. Third, noise substantially reduces the accuracy of an NNSA identification, whereas significant improvements are obtained when training is performed with multiple periods of data. This experimental work demonstrates that after sufficient training, a frequency transfer function estimation made using the NNSA is very similar to the estimation determined by the FFT spectrum analyzer. The identified system characterization is consistent with the theoretical model derived for the end effector. These high degrees of correlation give substantial experimental reinforcement to the concept of a neural network spectrum analyzer based on the FSNN for identifying dynamic systems.

B. DESCRIBING FUNCTION IDENTIFICATION

The FSNN was applied successfully to the identification of frequency domain transfer functions in the previous section. The transfer function concept is based theoretically on the system linearity which features a sinusoidal input to the system that leads to a sinusoidal output at the same frequency. The response of a nonlinear system to a signal-frequency harmonic input contains not only the harmonic component with the same frequency, but also harmonic components at other frequencies. Consequently, it is usually difficult to apply the frequency response method to nonlinear systems. Yet, for some nonlinear systems which can be represented by a sequence of lumped nonlinear elements plus a linear transfer function, an extended version of the frequency response method, called the describing function (DF) method, can be used to approximately analyze system behavior and stability [22] and predict nonlinear phenomena such as the limit cycles [23].

The DF of a nonlinear element is defined as the complex ratio of the output fundamental component of the nonlinear element to the input harmonic function. If an input to the nonlinear element is $x(t) = b$ and the output $y(t)$ is a periodic function and can be approximated by a finite Fourier series, the amplitude and phase of the describing function $N(b_1, w)$ of the nonlinear element can be defined, respectively, as

$$|N(b_1, \omega)| = \left| \frac{\sqrt{A_1^2 + B_1^2}}{b_1} \right|, \qquad \angle N(b_1, \omega) = A \tan\left(\frac{A_1}{B_1}\right), \qquad (35)$$

where A_1 and B_1 are the coefficients of the fundamental component of the output Fourier series. Generally, the DF depends on the input frequency ω and amplitude b_1, which distinguishes it from the linear transfer function, which is independent of the input amplitude. When the nonlinearity is single-valued, the DF is real, independent of the input frequency ($A_1 = 0$), and gives zero phase shift, whereas if it is multivalued, the phase shift is not zero.

The frequency domain nature of the DF suggests that the FSNN could be used to model the DF of an unknown nonlinearity through estimating coefficients A_1 and B_1. Figure 14 shows the DF identification scheme using the FSNN. Unlike the transfer function identification, the input to the nonlinearity for each training period is a sinusoidal signal which has only one frequency component. A family of such sinusoidal inputs with changing amplitude and frequency must be applied

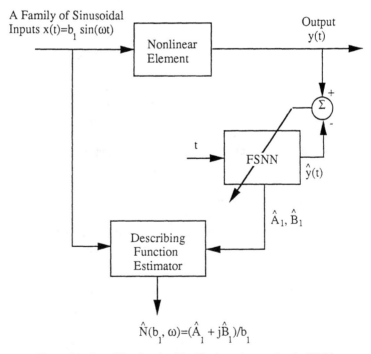

Figure 14 Describing function identification scheme using the FSNN.

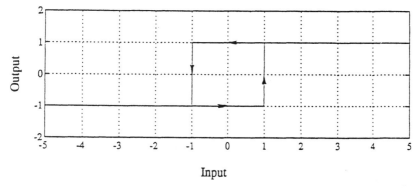

Figure 15 A backlash nonlinear element.

to the nonlinear element sequentially to model the DF which spans the desired frequency and amplitude range.

The DF identification capability of the FSNN was demonstrated for the nonlinearities of the backlash. The theoretical describing function for the backlash was given by [24]. An SISO FSNN with small size $N = 4$ was employed to carry out the estimation of the output frequency spectrum from the nonlinearity. The adoption of the small size FSNN is proper because only the weights $\widehat{w}_1^c$ and $\widehat{w}_1^s$ of the fundamental component are needed for the DF evaluation, and neglecting the higher-order terms of the FSNN model has little influence on the precision of the fundamental component estimation. The FSNN was trained for four periods when excited with each sinusoidal signal. Ten points of the output signal for each period were sampled for each training cycle.

Figure 15 shows a backlash nonlinearity which is multivalued and its DF has nonzero phase. Both the amplitude and phase of the backlash DF are independent of the input frequency. The corresponding complex DF model of the FSNN is shown in Fig. 16, which was trained using 100 training samples within each training period. All these FSNN estimations show satisfactory accuracy in comparison with its theoretical values.

C. FOURIER SERIES NEURAL NETWORK-BASED ADAPTIVE CONTROL SYSTEMS

An adaptive controller has the capability to modify its behavior in response to changes in the dynamics of the plant or process under control. The purpose of the controller modification is to achieve a desired performance criteria for the system which may not be obtained by using a conventional constant-gain controller. Two

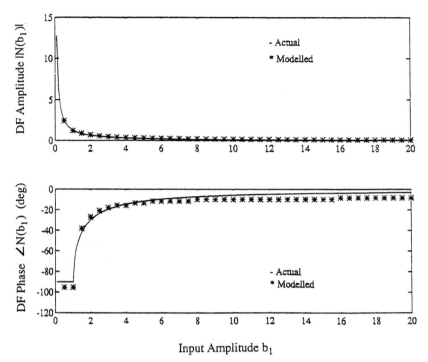

Figure 16 Describing function for backlash as estimated by the FSNN identifier.

typical adaptive control system designs are the self-tuning regulator (STR) and the model reference-adaptive system (MRAS); both are composed of a controller and an estimator [25]. The estimator is an important element which identifies the plant; it provides the guide to modify the controller parameters. A traditionally designed estimator permits only parameter estimation by assuming an *a priori* model structure. For less structured plants, the modeling process is time con- suming, and, without enough information, the presumed model structure may be deficient, which results in incorrect or erroneous estimation and control action. This work proposes two FSNN-based neural adaptive control schemes which are described in the following two sections. Unlike the traditional schemes, the self- organizational learning capability allows an FSNN estimator to perform a robust estimation for unstructured plants because the FSNN modeling requires neither *a priori* plant model structure nor parameters. The second advantage is that the FSNN model represents the system frequency response, so the controller design is carried out directly based on the system frequency response analysis. The third advantage lies in the FSNN's computational parallelism nature which potentially allows a fast computation of the on-line controller tuning algorithm with parallel processors.

1. Neural Self-Tuning Regulator

A STR, utilizing an FSNN estimator, termed a neural self-tuning regulator (NSTR), is illustrated in Fig. 17. This NSTR uses three FSNN estimators to identify the closed-loop system transfer function G and sensor transfer function H. FSNN 1, FSNN 2, and FSNN 3 identify the frequency spectrums of the system input $r(t)$, output $y(t)$, and sensor output $f(t)$, respectively. The system transfer function G can be estimated using the state weights $\widehat{w}_r$ and $\widehat{w}_y$ from FSNN 1 and FSNN 2, whereas the sensor transfer function H can be estimated using state weights $\widehat{w}_y$ and $\widehat{w}_f$ from FSNN 2 and 3, by the transfer function identification method discussed in Section III.A.2. The NSTR evaluated in this work uses a lead–lag compensator structure as the "software designer" to demonstrate the NSTR principle. Other frequency domain controller designs based on the same principle also could be adopted to implement this NSTR strategy. A basic lead–lag compensator transfer function can be defined as

$$G_c = K \frac{(S + 1/T_1)}{(S + \beta/T_1)} \frac{(S + 1/T_2)}{(S + 1/\beta T_2)}, \qquad \beta > 1, \qquad (36)$$

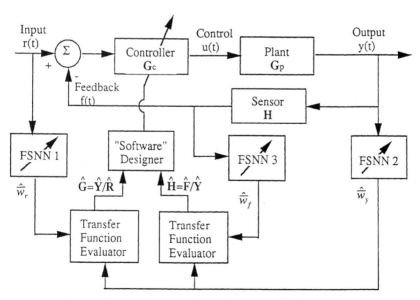

Figure 17 A neural self-tuning regulator using FSNN estimators.

where parameters K, T_1, T_2, and b are adjustable. A number of techniques have been developed to select the parameters of a lead–lag compensator off-line when the Bode diagram of the open-loop transfer function is known [26, 27]. The open-loop transfer function of the control system shown in Fig. 17 can be formulated as

$$G_P H = \frac{GH}{G_c^b(1 - GH)}, \tag{37}$$

where G_P, G_c^b, G, and H are the transfer functions of the plant, controller before tuning, the overall closed-loop system, and sensor, respectively. This relationship indicates that with a known G_c^b, the open-loop transfer function $G_P H$ can be approximated using FSNN estimates.

With the estimated open-loop transfer function, a recursive lead–lag compensator tuning algorithm, as shown in Fig. 18 and detailed in [28], is used to determine the controller parameters on-line. The parameters of the controller are selected to approach or to maintain a desired closed-loop system frequency domain performance that is measured by the error coefficients, phase margin (PM), and gain margin (GM). This tuning algorithm is evaluated through the following simulation study.

A second-order nominal plant has a transfer function $G_P = 1/s(s + 2)$. The feedback sensor transfer function $H = 1$. The lead–lag compensator is used to regulate the unknown system with an initial transfer function $G_c^b = 1$. The control system design requires a velocity error constant $K_{ed} = 15$ s^{-1}, PM$_d = 50°$, and GM$_d = 10$ dB. Two FSNNs that have 23 neurons each are used to model system input and output frequency spectrums, respectively. The system is subject to a square wave input signal that has a period of $T = 25$ and a unit amplitude. The square wave excitation is persistent because it contains all the odd frequency components. Increasing the period T of the input square wave provides a richer low frequency spectrum, whereas decreasing the period of T gives more spectral information in the high frequency domain. An appropriately selected persistent excitation with proper period T enables the FSNN frequency spectrum model to identify a specific frequency range for the system Bode diagram.

The FSNN transfer function models converged after being trained for four input periods. Table IV lists the numerical values of the open-loop transfer function $G_P H$ ($G_c^b = 1$) computed with the FSNN estimation of G. The steady-state error coefficient K_e derived from the listed data is 0.5 s^{-1}, which is considerably smaller than the desired value and results in a narrow system bandwidth and a large settling time. Therefore, a new K value of 40 (or 32 dB) is used in the tuning algorithm to raise K_e to 15 s^{-1}. All amplitude values in Table II then are increased by 32 dB when the new gain is applied. The crossover frequency ω_c in the $KG_P H$ Bode diagram after the new K is applied is 5.78 rad/s and the PM is approximately 20°. This small value of PM causes the closed-loop system to

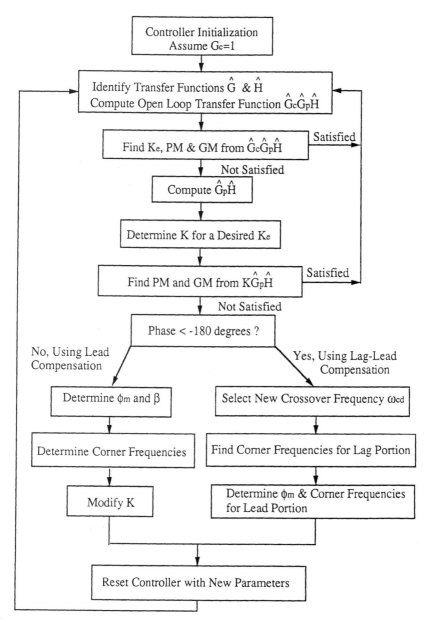

Figure 18 Recursive design flow chart for a lead–lag compensator.

Table IV

Identified Open Loop Transfer Function $\widehat{G}_p \widehat{H}$

Frequency (rad/s)	Amplitude (dB)	Phase (deg)
0.25	5.95	−97.03
0.75	−4.12	−110.26
1.26	−9.46	−121.48
1.76	−13.47	−130.41
2.26	−16.80	−137.34
2.76	−19.68	−142.71
3.27	−22.22	−146.93
3.77	−24.51	−150.30
4.27	−26.59	−153.05
4.78	−28.50	−155.35
5.28	−30.28	−157.33
5.78	−31.95	−159.06
6.28	−33.53	−160.63
6.79	−35.04	−162.09
7.29	−36.49	−163.49
7.79	−37.90	−164.87
8.29	−39.27	−166.27
8.80	−40.63	−167.72
9.30	−41.98	−169.25
9.80	−43.34	−170.88
10.30	−44.73	−172.59
10.81	−46.25	−174.33
11.31	−48.42	−177.02

have severe oscillatory response. Including lead compensation, in terms of the tuning algorithm, will eliminate the anticipated system oscillations while keeping the error coefficient $K_{ed} = 15$ s^{-1} unchanged. The compensator phase f_m is determined using the formula $f_m = \mathrm{PM}_d - \mathrm{PM}$ from the tuning algorithm with a $\mathrm{PM}_d = 55°$ (the desired phase margin) and the identified current $\mathrm{PM} = 20°$, which yields $f_m = 35°$ that is needed to approach an actual desired phase margin of 50°. The value of b computed using $b = (1 + \sin f_m)/(1 - \sin f_m)$ from the tuning algorithm is 3.7 and the new crossover frequency ω_{cd} is 7.8 rad/s. Corner frequencies $1/T_1$ and b/T_1 of the lead compensator using $1/T_1 = \omega_{cd} b^{1/2}$ and $b/T_1 = \omega_{cd}/b^{1/2}$ are given by $1/T_1 = 4.1$ rad/s and $b/T_1 = 15.0$ rad/s, respectively. Consequently, the transfer function of the lead compensator based on the recursive tuning algorithm is $148(s + 4.1)/(s + 15)$.

Figure 19 shows the closed-loop system input and output responses before and after the lead compensator parameters are reset. A faster system response is

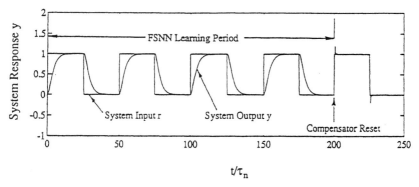

Figure 19 Closed-loop system response of the NSTR.

observed after the compensator is reset at the end of the fourth period with the parameters obtained during the last design cycle. The system after the compensation meets the design requirement with PM $= 50°$ and $K_e = 15 \text{ s}^{-1}$; the GM is in excess of the design requirement.

2. Neural Model-Reference Adaptive System

A model-reference adaptive system specifies the desired system performance using a reference model. In the MRAS, the parameters of the controller are adaptive to match the behavior of the reference model [25]. Modification of the controller parameters is guided by a function of the error measurement between the reference model and the actual system; this is often accomplished by using an estimator. The traditional design of the estimator does not account for actual system uncertainties beyond the scope of the *a priori* defined system model. The MRAS with such an estimation scheme may exhibit poor transient behavior if a mismatched system model is used in the control law computation and modification.

This section presents a MRAS strategy utilizing the FSNN as a nonparametric estimator, named the neural model-reference adaptive system (NMRAS), to gain learning intelligence, computational parallelism, and an understandable model. A related controller parameter optimization algorithm based on FSNN estimates of the system transfer functions is proposed to minimize a performance index, called the pseudo-squared transfer function error, between the closed-loop system and the reference model. The NMRAS performance in tracking the reference model output was evaluated through an experimental study.

Figure 20 shows the NMRAS strategy for a unity feedback system that uses two FSNN frequency transfer function identifiers. This strategy provides esti-

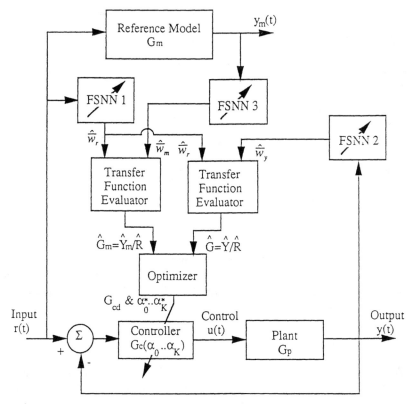

Figure 20 A neural model-reference adaptive system using FSNN estimators.

mates of the closed-loop system transfer function G and the reference model transfer function G_m if G_m is unknown. The task of the optimizer is to determine a set of optimal controller parameters based on the FSNN estimates for the adaptive controller. The plant transfer function estimate $\widehat{G}_p$ can be computed using $\widehat{G}$ as

$$\widehat{G}_p = \frac{\widehat{G}}{G_c^b(1 - \widehat{G})}, \tag{38}$$

where G_c^b denotes the transfer function of the controller before tuning. The outputs of the reference model and the plant will be identical if $G = G_m$ when subjected to the same input. The objective of the controller tuning is then to achieve the equality

$$G_{cd} = \frac{\widehat{G}_m}{\widehat{G}_p(1 - \widehat{G}_m)}, \tag{39}$$

where G_{cd} is the controller transfer function which makes the closed-loop system G equal to the model G_m. In other words, the controller G_{cd}, defined in Eq. (39), is desired and the objective of the NMRAS is to tune the controller to achieve this G_{cd}. With Eq. (38), Eq. (39) can be rewritten as

$$G_{cd} = \frac{1 - 1/\widehat{G}}{1 - 1/\widehat{G}_m} \, G_c^b. \tag{40}$$

Equation (40) defines the ideal controller based on the FSNN estimates of the closed-loop system transfer function $\widehat{G}$ and reference model transfer function $\widehat{G}_m$.

In general, preselecting a controller structure may make it impossible to achieve $G = G_m$. One way to address this issue is to use an optimal control scheme which adjusts the parameters of the controller with a fixed structure in such a way as to extremize a defined performance index. The index of performance (IP) for the NMRAS controller optimization is based on the FSNN transfer function estimates defined by

$$\text{IP} = \sum_{n=1}^{N} \left[G_{cd}^r(j\omega_n) - G_c^r(j\omega_n) \right]^2 + \left[G_{cd}^i(j\omega_n) - G_c^i(j\omega_n) \right]^2, \tag{41}$$

where G_c^r and G_c^i represent the real and imaginary portions of the controller transfer function G_c, and G_{cd}^r, and G_{cd}^i denote the real and imaginary portions of the desired controller transfer function G_{cd}, respectively. This IP measures the difference between the pseudo-squared transfer functions of the ideal and the actual controller over a frequency spectrum range of (ω_1, ω_N), and is termed the pseudo-squared transfer functions error (PSTFE).

The frequency spectrums of the system and reference model transfer functions are required to construct the PSTFE performance index in the NMRAS instead of the tracking error measurement used in a traditional MRAS. The minimization of the PSTFE performance index using the FSNN estimates is a recursive controller tuning process because the desired controller G_{cd} is determined using the previous controller parameters in G_c^d. This NMRAS design is evaluated with a classical proportional, integral, and differential (PID) controller which has a frequency domain transfer function given by

$$G_c = K_p + \frac{K_i}{j\omega} + K_d j\omega, \tag{42}$$

where the three parameters K_p, K_i, and K_d are the proportional, integral, and differential gains, respectively. The optimal gains which minimize the PSTFE index are given by [28]

$$K_p = \frac{1}{N} \sum_{n=1}^{N} G_{cd}^r(\omega_n), \tag{43}$$

$$K_i = \frac{-N \sum_{n=1}^{N} G_{cd}^i(\omega_n)\omega_n + \sum_{n=1}^{N} \omega_n^2 \sum_{n=1}^{N} (G_{cd}^i(\omega_n)/\omega_n)}{N^2 - \sum_{n=1}^{N} \omega_n^2 \sum_{n=1}^{N} (1/\omega_n^2)}, \tag{44}$$

and

$$K_d = \frac{N \sum_{n=1}^{N} (G_{cd}^i(\omega_n)/\omega_n) - \sum_{n=1}^{N} (1/\omega_n^2) \sum_{n=1}^{N} \omega_n G_{cd}^i(\omega_n)}{N^2 - \sum_{n=1}^{N} \omega_n^2 \sum_{n=1}^{N} (1/\omega_n^2)}. \tag{45}$$

Using the sufficiency condition given by [29] it can be shown that the controller parameters selected using Eqs. (43)–(45) locate the minimum value for the PSTFE. Restrictions can be applied to the range of controller gains based on considerations related to control effort, cost, and stability. The low bound for the PID controller gains is zero because negative gains of the PID controller destabilize the closed-loop system.

The feasibility of the NMRAS control strategy is investigated in the real time position control of a dc servo motor that has unknown dynamics. An advanced experimental system with a parallel computational capability was developed for this investigation. The experimental system is composed of four functional units: a dc motor (as the unknown plant), a TMS320C30 real time control system [30], a PC 486 host computer, and an i860-based parallel processing platform [31] for NMRAS. Figure 21 shows the system layout and corresponding task assignments for each functional unit. A metal bar with an unknown moment of inertia is mounted on the shaft of the dc motor. This results in a totally unknown plant because the motor dynamics without load is also *a priori* unknown. The initial PID controller gains are assumed to be $K_p^b = 20.0$ and zero K_i^b and K_d^b, where the superscript b denotes the gains before tuning. The input signal $Q_i(t)$ to the closed-loop system is a square wave signal that has a period $T = 5$ s and an amplitude of p rad. The reference model, whose output $Q_m(t)$ in response to the square wave input defines the desired motor motion trajectory, has a first-order transfer function given by $G_m = 1/(\tau s + 1)$ with time constant $t = 0.1$ s. The period of the C30 interrupt timer is set to 1 ms which determines the C30 input–output data sampling period and the control cycle period. The C30 samples the system state variables, computes the control law, and sends the resulting control signal u to drive the motor within each control cycle of 1 ms.

The FSNN transfer function model has 8 odd cosine and sine harmonic neurons ($n = 1, 3, \ldots, 15$) and is trained with 126 data samples of $Q_i(t_k)$ and $Q_0(t_k)$,

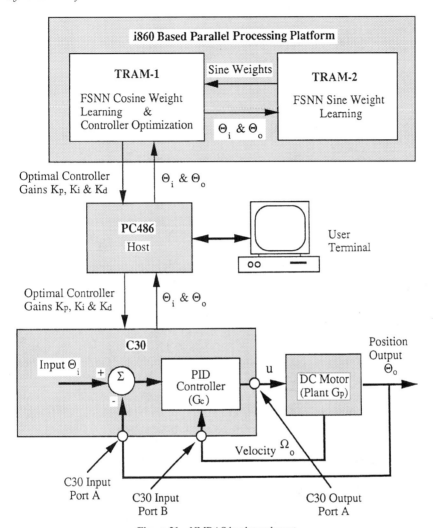

Figure 21 NMRAS hardware layout.

for $k = 0, 1, 2, \ldots, 125$ in each period $T = 5$ s. These training data are uniformly distributed in the 5-s interval with $t_{k+1} - t_k = 40$ ms. The FSNN model is trained four times using the training data set sampled during each period T, and the resulting FSNN state weights are then used to construct a desired controller G_{cd}. The software optimizer computed the optimal controller gains based on the desired controller G_{cd}. These controller gains are plotted in Fig. 22. The

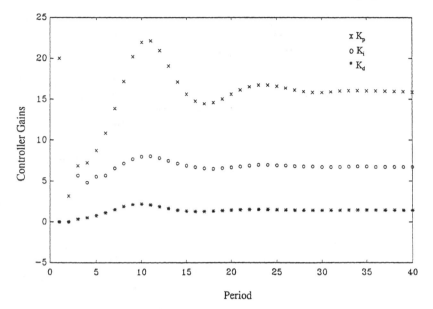

Figure 22 NMRAS controller gain values.

controller was reset with these new gains at the end of every 5-s input signal period T.

Figure 23 shows the experimental motor angular motion trajectory $Q_0(t)$
(NMRAS output) compared with the desired trajectory $Q_m(t)$ (reference model
output) during the first six periods. The figure shows the response curve in the first
period of 5 s and that the initial system before tuning has a significant dynamic
overshoot and a large tracking error during the transitional period. Therefore the
NMRAS slowed down the system response as shown in the second period by
using a set of new controller gains $K_p = 3.16$, $K_i = 0$, and $K_d = 0.05$. An
apparent tracking error still existed in the second period because the FSNN transfer function model was still not accurate enough to match the commanded steady
state. The tracking performance was improved greatly as the FSNN learning and
NMRAS controller gain optimization processes advanced into the third period
as shown in the figure; this implies that the FSNN modeling accuracy increased.
The controller gains provided by the software optimizer for the third period were
$K_p = 6.86$, $K_i = 5.68$, and $K_d = 0.37$. The motor motion trajectory $Q_0(t)$
during the following periods in the figure more closely match the reference model
output $Q_m(t)$ as the FSNN modeling converges.

The experimental results show that the NMRAS is capable of identifying such
unknown systems and determining a set of optimal controller gains independent of

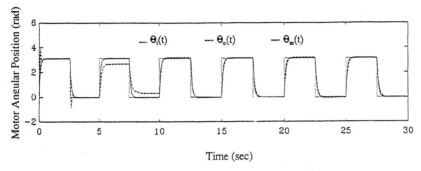

Figure 23 Motor motion trajectory in the first six periods.

their initial values. The resulting NMRAS with these optimal controller gains exhibits the generalization capability of tracking unfamiliar and noisy input signals. The experimental results also show that the NMRAS achieves control robustness when presented with system time varying uncertainties. All these laboratory experiments demonstrate that the FSNN intelligent identification and control technology is applicable to a wide range of unknown and time variant systems.

IV. TIME DOMAIN APPLICATIONS FOR SYSTEM IDENTIFICATION AND CONTROL

A. NEURAL NETWORK NONLINEAR IDENTIFIER

Major progress has been made in the identification of time invariant linear systems that have unknown parameters via model structures selected on well established linear system theory. In contrast to linear system identification, nonlinear system identification is more difficult because many properties of linear systems are no longer valid. Most existing nonlinear identifiers are based on parameter estimation, where the unknown system model structure must be estimated. The determination of the model structure is often based on system evaluation via experimental analysis. Neural network techniques potentially provide an effective means to approach the maximal robustness when modeling nonlinear systems because neural network modeling is adaptive to parametric and structural variations of the system. This section presents the identification scheme using the OAFNN for nonlinear systems. The performance of the OAFNN identifier is evaluated through its application to the identification of system difference equations in the time domain and describing functions in the frequency domain.

1. Difference Equation Identification Using Fourier Series Neural Networks

The method of describing dynamic systems by difference equations is currently well established in system theory and applies to a large class of engineering systems and processes. A valid model of a difference equation provides an insight into the system dymanics because the difference equation is a discrete-time representation of the system differential equations. In a time domain adaptive controller, an identified model of the system difference equations is able to predict system behavior, which guides the on-line controller adaptive tuning. Thus, the identification of a system difference equation using neural networks is an important issue for developing a time domain neural adaptive controller.

A general difference equation model of an SISO system can be expressed as

$$y(k+1) = f\big[y(k), y(k-1), \ldots, y(k-p+1), x(k), x(k-1), \ldots, x(k-q+1)\big], \tag{46}$$

where $y(k)$ is the system output at time k, $x(k)$ is the system input at time k, and so on. The function $f(\cdot)$ defines the difference equation for a linear or a nonlinear system. Hence, the task of system identification is to construct MISO model that approximates the function $f(\cdot)$. Another useful difference equation model, which is a subset of system model (46) and is particularly suited for control applications, is given by

$$\begin{aligned} y(k+1) &= g\big[y(k), y(k-1), \ldots, y(k-p+1)\big] \\ &\quad + h\big[x(k), x(k-1), \ldots, x(k-q+1)\big], \end{aligned} \tag{47}$$

where $g(\cdot)$ is related to the output y and $h(\cdot)$ is related to the input x. The output $y(k+1)$ depends on the linear combination of the two functions $g(\cdot)$ and $h(\cdot)$. The task of the identification in this case becomes to independently model functions $g(\cdot)$ and $h(\cdot)$. In both of the preceding system difference equation models, the number of the FSNN input nodes is determined based on numbers p and q.

The identification schemes presented here are for these two types of nonlinear system models. The principle demonstrated can be extended easily to identification of systems defined by other types of difference equations. The FSNN used to model system (46) has a MISO structure as shown in Fig. 4 with $(p + q)$ input nodes receiving the training samples $y(k), y(k-1), \ldots, y(k-p+1), x(k), x(k-1), \ldots, x(k-q+1)$ and the output is $y(k+1)$. Such an FSNN model, with N cosine neurons connected to each input node, is given by

$$\hat{y}(k+1) = \sum_{n_1=0}^{N} \cdots \sum_{n_{p+q}=0}^{N} \widehat{w}_{n_1 \cdots n_{p+q}} \cos \frac{n_1 \pi y(k)}{R_y} \cdots \cos \frac{n_p \pi y(k-p+1)}{R_y}$$

$$\times \cos \frac{n_{p+1} \pi x(k)}{R_x} \cdots \cos \frac{n_{p+q} \pi x(k-q+1)}{R_x}, \tag{48}$$

where $\widehat{w}_{n_1\cdots n_{p+q}}$s are the state weights and R_y and R_x define the range of the bounded system output and input, respectively. The learning rule for this cosine FSNN is given by

$$\Delta \widehat{w}_{n_1\cdots n_{p+q}} = \delta e \cos \frac{n_1 \pi y(k)}{R_y} \cdots \cos \frac{n_p \pi y(k-p+1)}{R_y}$$
$$\times \cos \frac{n_{p+1} \pi x(k)}{R_x} \cdots \cos \frac{n_{p+q} \pi x(k-q+1)}{R_x}, \qquad (49)$$

where e_1 is the learning error of the FSNN model defined by $e = y(k+1) - \hat{y}(k+1)$.

To evaluate the performance of the FSNN identifier, a simulation study was conducted by applying the FSNN identifier to model a nonlinear system defined by the nonlinear difference equation

$$y(k+1) = \frac{0.5y(k)}{1 + y(k) + 0.8y^2(k)} + u^3(k). \qquad (50)$$

The input to the system is a harmonic signal defined by $x(k) = \sin(2pk/25) - \cos(2pk/10)$, which varies within the bounds $[-2, 2]$. Figure 24 shows the system input $x(t)$ and the response $y(t)$ which is bounded by $[-8, 8]$.

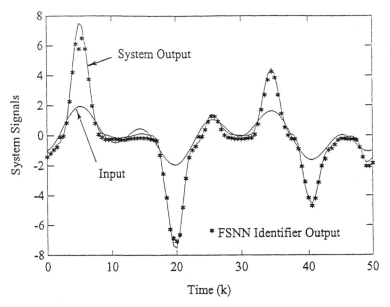

Figure 24 Outputs of the system and the FSNN identifier after training.

The numbers p and q of the FSNN identifier for this system both equal 1, which results in the FSNN having two input nodes and one output node. There were 10 cosine neurons associated with each input node, that is, $N = 10$ in FSNN model (46), which resulted in a total of 20 cosine neurons. The learning rate δ selected for this identification was 0.01. The FSNN input training data were uniformly sampled over a training cycle of 50 s with a sampling frequency of 2 rad/s. All 20 state weights of the FSNN were modified once for each mapping pair of input–output training samples.

Figure 24 shows the response of the FSNN identifier after it was trained for 40 cycles with a total of 4000 pairs of training samples. This is much faster than the identification using a multilayered backpropagation neural network described in [2], where 100,000 training samples were used. The resulting mean error over the 100 sampling points in the 50-s period was 0.14, which is 1.8% of the peak magnitude of the output signal $y(t)$. The error does not completely vanish because the FSNN model with a finite number of neurons only approximates the Fourier series representation. The accuracy of the FSNN identifier can be improved by increasing the number of neurons. The robustness of the identified neural model was examined by feeding in an unfamiliar input signal $x(k) = \sin^2(2pk/50) + \sin(2pk/10) - 1$ that was not used for training. Figure 25 shows the responses of

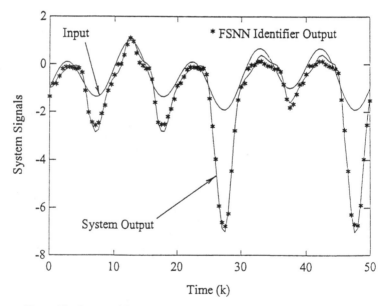

Figure 25 Outputs of the system and the FSNN identifier with unfamiliar input.

the system and the FSNN identifier. A close match between the two responses is observed, which demonstrates the robustness of the trained FSNN model.

This FSNN identification scheme also was evaluated when applied to the system model given by the difference equation (47). The same system, harmonic input signal, and sampling frequency as for the previous case were used for this evaluation. Whereas the functions $g(\cdot)$ and $h(\cdot)$ are separable in (47), the identification was performed by using two independent FSNN identifiers: one for $g(\cdot)$ and the other for $h(\cdot)$ estimation. In this case, each identifier was a SISO FSNN, which resulted in the identifier diagram as shown in Fig. 26. Each FSNN of the identifier had 10 cosine neurons and the learning rate was 0.03.

The identification process converged after four training cycles with a total of 400 training samples, which was 10 times faster than the previous process. The trained FSNN model had an output that closely matched the system output shown in Fig. 27. The mean error over the 100 sampling points in the 50-s period was 0.19, which was 2.4% of the peak magnitude of output signal y. The robustness of the identified neural model also was examined by feeding an unfamiliar in-

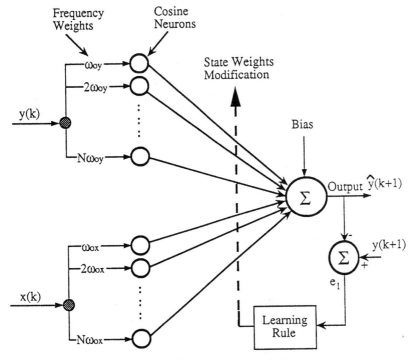

Figure 26   FSNN identifier for model (47).

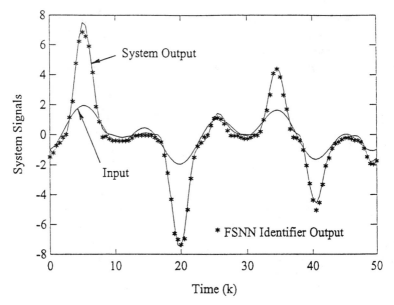

Figure 27   Outputs of the system and the FSNN identifier after training.

put signal as used for the previous evaluation. The FSNN model exhibited a satisfactory robustness; the response nicely approximated the system response as
shown in Fig. 28. Use of the difference equation model (47), if applicable, for
FSNN nonlinear system identification is preferred due to the higher convergence
speed.

2. Performance Evaluation of Orthogonal Activation Function-Based Neural Networks as Nonlinear Identifier

The nonlinear, discrete, dynamic system described by the difference equation

$$y(k+1) = \frac{y(k)}{1 + y(k) + 0.8y(k)^2} + 0.5u(k)^3 \tag{51}$$

was simulated with a random input in the range of -2 to 2 for time steps $k = 1$–
1000 and $y(0) = 0$. The input [$u(k)$ and $y(k)$] and output data [$y(k+1)$] were
used to train the networks. The estimated bounds of $u(k)$ and $y(k)$ were taken as
$[-3, 3]$ and $[-6, 6]$, respectively. The same four types of networks, namely, harmonic OAFNN, Legendre OAFNN, polynomial, and sigmoid, which were employed in Section II for performance comparison, are used for system identification. Each network was trained with 2 input neurons, 30 hidden layer neurons (15

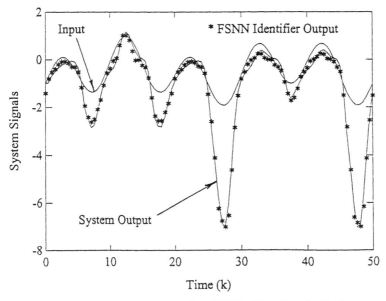

Figure 28 Outputs of the system and the FSNN identifier with unfamiliar input.

for each input), and 1 output neuron. An adaptive learning rate scheme was used for faster learning. In this scheme the learning rate was increased by a factor of 1.05 if the learning error decreased and the learning rate was decreased by a factor of 0.7 if the error increased. The networks were trained until either the SSE was less than 0.1, the rate of change of error was below 0.01 per cycle, or 500 training cycles (epochs) were completed.

Figure 29 shows training errors for the four networks. Once again no network reached the error goal, but the harmonic OAFNN converged to a stabilized error of SSE = 1.56 in 32 cycles, which resulted in the lowest TECC, as shown in Table V. The Legendre OAFNN converged after 46 cycles to a stabilized error (SSE) of 3.32. The sigmoidal network converged to a low error of 1.58, but took 500 cycles to reach this error level. The polynomial network converged fastest (in 27 cycles), but the learning error was high (SSE = 76.94). The TECC for Legendre and harmonic OAFNNs was the lowest for the entire training period after the first three training cycles as shown in Fig. 30.

The error in the responses predicted by the trained networks to an unfamiliar input given by $u(k) = \sin^2(2\pi k/50) - 0.5$ and new initial condition $y(0) = 1$ is shown in Fig. 31. The low prediction errors for sigmoidal, Legendre, and harmonic networks demonstrate acceptable learning generalization for these networks. Figure 32 shows responses to this unfamiliar input after a *single* training

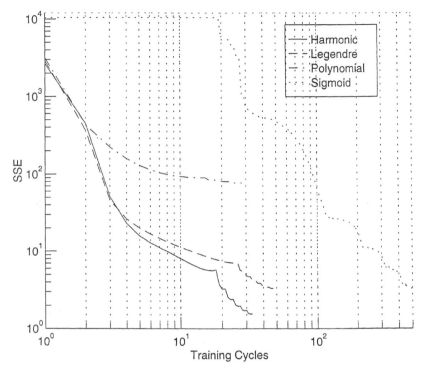

Figure 29   Training error for system identification.

cycle. The Legendre and harmonic OAFNNs demonstrate a close approximation of the system behavior after a single cycle training, whereas the sigmoid network and the polynomial network (the non-OAFNNs) demonstrate poor or no learning after just one training cycle. Close approximation with minimal training is an

Table V

Simulation Results for System Identification

Type of NN	(N)	SSE (E)	TECC ($N \times E$)
Sigmoid NN	500	3.181	1590
Polynomial NN	27	76.939	2077
Legendre NN	46	3.316	153
Harmonic NN	32	1.5635	50

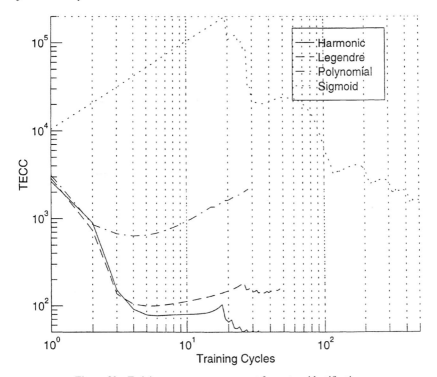

Figure 30   Training error convergence cost for system identification.

important ability for neural networks used in real time, on-line adaptive control systems.

B. INVERSE MODEL CONTROLLER

A discrete, nonlinear, nth-order dynamic system can be represented by a non-linear difference equation

$$x_k = f(\bar{x}_{k-1}, \bar{u}_{k-1}), \tag{52}$$

where $\bar{x}_{k-1} = [x_{k-n}, x_{k-n+1}, \ldots, x_{k-1}]^T$ and $\bar{u}_{k-1} = [u_{k-m}, u_{k-m+1}, \ldots, u_{k-1}]^T$. An inverse model neural network controller (IMC) was implemented as an approximation of the inverse nonlinear function $\bar{u}_{k-1} = \bar{g}(x_k, \bar{x}_{k-1})$, where the neural network is employed to approximate the inverse function, $\bar{g}(x_k, \bar{x}_{k-1})$. The neural network is trained off-line with system states $(x_k, x_{k-1}, \ldots, x_{k-n})$ as

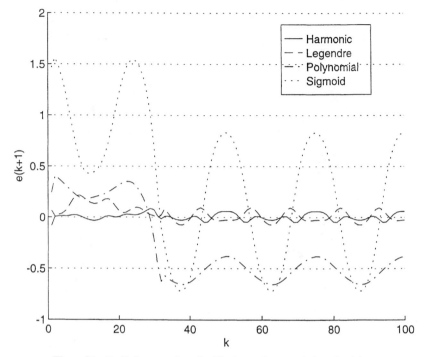

Figure 31   Prediction error for unfamiliar input after completion of training.

network input and system inputs $(u_{k-1}, u_{k-2}, \ldots, u_{k-m})$ as network output. The training continues until either the learning error is below a specified goal or the learning curve flattens with no further decrease in learning error. Once the network is trained, it is used to calculate the input signal to the system so that the system follows a desired state space trajectory. Fast learning of the network can reduce training time extensively and therefore reduce training cost.

The nonlinear, discrete, dynamic system used for simulations is a first-order system given by

$$x(k) = \frac{x(k-1)}{1 + x(k-1) + 0.8x(k-1)^2} + 0.5u(k-1)^3. \tag{53}$$

This is the same system that was used for system identification in the previous section. Once again the four networks (harmonic OAFNN, Legendre OAFNN, polynomial, and sigmoid) were employed for performance comparison. The input to the four networks is $x(k-1)$ and $x(k)$ and the output is $u(k-1)$. A random input of 1000 sample sequences is used as the system input to generate the training

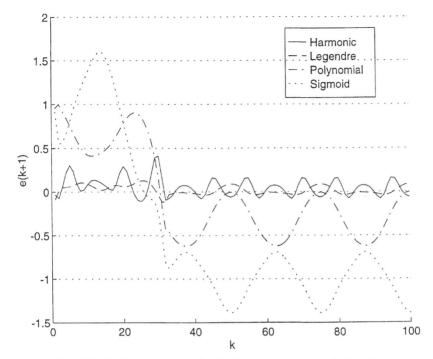

Figure 32 Prediction error for unfamiliar input after a single epoch of training.

data set. A total of 58 neurons (29 for each input) are used for each network. The desired trajectory is a sinusoidal trajectory.

The training error for the OAFNNs (Legendre and harmonic networks) and the polynomial network decays at a high rate during the initial phase of learning and then the error slowly stabilizes, as shown in Fig. 33. The trajectory tracking performance of the Legendre, harmonic, and polynomial network controllers after three training cycles is shown in Fig. 34. The OAFNN controllers follow the trajectory with reasonable accuracy whereas the polynomial network controller shows significant tracking error. The sigmoid neural network showed minimal learning after three cycles and therefore is not presented in the figure.

The performance at the end of the learning phase is shown in Fig. 35. None of the four controllers reached the desired trajectory tracking accuracy level, including sigmoid network. However, sigmoid networks improved their performance after additional training. Both OAFNNs demonstrated the least error in the least training time. The polynomial network learns fast, but has a high training error. The sigmoid network has a low training error, but requires a large (500) number of

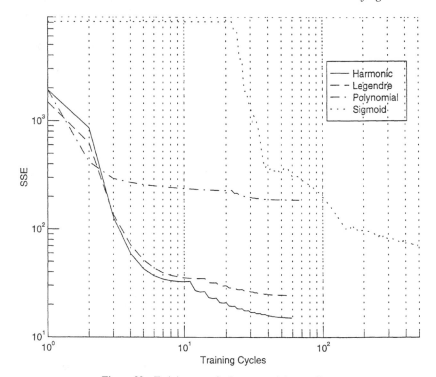

Figure 33   Training error for inverse model controllers.

training cycles to learn. The sigmoid network controller had high trajectory track-ing error despite low training error, which reflects poor generalization. The TECC is plotted in Fig. 36. The OAFNNs show a consistently low TECC throughout the training period.

C. DIRECT ADAPTIVE CONTROLLERS

The orthonormal activation function-based neural networks (OAFNNs) have properties suitable for direct adaptive control of nonlinear systems. In this sec-tion the direct adaptive control schemes are developed, using the OAFNNs with orthonormal activation functions. The goal for these controllers is to enable a nonlinear system to track a desired trajectory with minimal error.

These direct adaptive neural controllers employ OAFNNs to estimate and com-pensate for unknown system dynamics. The network compensation is either based on actual system states or the desired input trajectory. The developed controllers

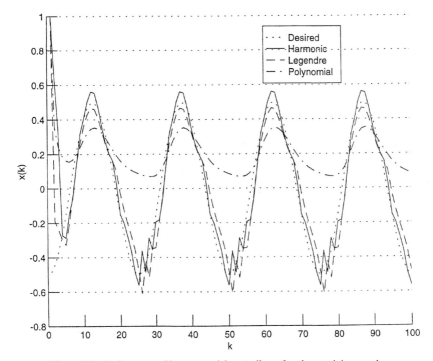

Figure 34 Performance of inverse model controllers after three training epochs.

are divided into two categories based on the manner of compensation. Those controllers in which the compensation is based on actual states are termed actual compensation adaptive law (ACAL) controllers. The other type of controllers is based on the desired trajectory, and they are termed desired compensation adaptive law (DCAL) controllers. Figures 37 and 38 show the schematics for ACAL and DCAL controllers, respectively. A detailed development of ACAL and DCAL neural controllers is presented in subsequent sections.

1. Control Objective

The objective of the developed OAFNN-based direct adaptive controllers is to enable a class of continuous nonlinear systems, represented by

$$m(x)\ddot{x} + f(x, \dot{x}) = u, \tag{54}$$

to track a desired trajectory $[x_d(t)]$. The system given in Eq. (54) follows from the general characteristics found in mechatronic and robotic systems. The tracking

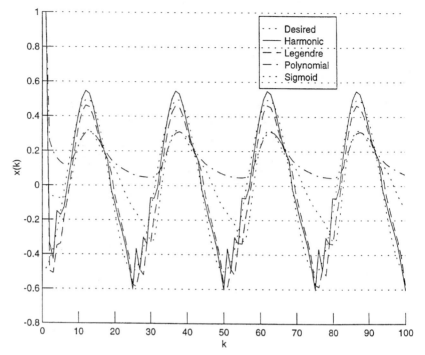

Figure 35 Performance of inverse model controllers at the end of training.

performance is quantified in terms of filtered tracking error, defined by

$$r = \dot{\tilde{x}} + \lambda \tilde{x} \tag{55}$$

where $\tilde{x} = x_d - x$ is the position tracking error and λ is a positive constant. The OAFNNs are employed in the controller for feedforward dynamic compensation of an unknown system while the weights of these networks are tuned on-line. The overall stability is guaranteed using Lyapunov-like analysis [23]. Assumptions are made about the system dynamics and the smoothness of the desired trajectory for each controller, and are discussed along with each controller development.

2. Orthogonal Activation Function-Based Neural Network as a Feedforward Compensator

The OAFNNs with normalized harmonic activation functions are employed in each direct adaptive controller to estimate and compensate for system dynamics. The dynamic functions for the class of continuous nonlinear systems given in

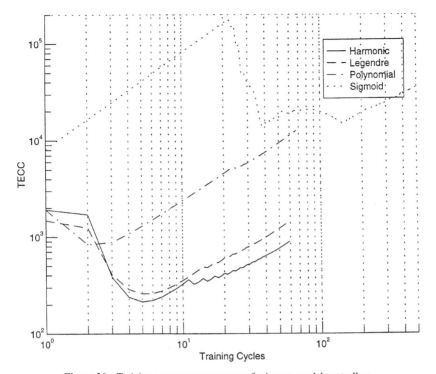

Figure 36 Training error convergence cost for inverse model controllers.

Eq. (54) are $m(\cdot)$ and $f(\cdot)$. Implementation of the DCAL controller also requires an estimate of the function $h(\cdot)$, where $h(\cdot)$ represents the space gradient of $m(\cdot)$ with respect to position, defined as

$$h(x) \equiv \frac{\partial m}{\partial x}. \tag{56}$$

a. Actual Compensation Adaptive Law Controller

The ACAL controller requires network estimates of $m(\cdot)$, $f(\cdot)$, and $h(\cdot)$, and these are denoted by $\widehat{m}$, $\hat{f}$, and $\hat{h}$, respectively. These network outputs are defined in terms of their activation functions and weights as

$$\widehat{m} = \overline{\Phi}(x)^T \hat{a}, \tag{57}$$

$$\hat{f} = \overline{\Psi}(x, \dot{x})^T \hat{b}, \tag{58}$$

$$\hat{h} = \overline{\Theta}(x)^T \hat{c}, \tag{59}$$

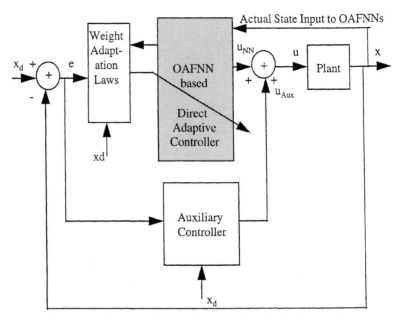

Figure 37   Schematic of the actual compensation adaptive law controller.

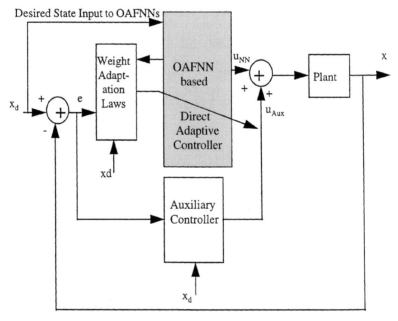

Figure 38   Schematic of the desired compensation adaptive law controller.

where $\overline{\Phi} = [\phi_1\phi_2\cdots\phi_p]^T$, $\overline{\Psi} = [\psi_1\psi_2\cdots\psi_q]^T$ and $\overline{\Theta} = [\theta_1\theta_2\cdots\theta_s]^T$ are transformed input vectors, $\hat{a} = [\hat{a}_1\hat{a}_2\cdots\hat{a}_p]^T$, $\hat{b} = [\hat{b}_1\hat{b}_2\cdots\hat{b}_q]^T$, and $\hat{c} = [\hat{c}_1\hat{c}_2\cdots\hat{c}_s]^T$ are network weight vectors, and $\phi_i(x)$ ($i = 1, 2, \ldots, p$), $\psi_j(x, \dot{x})$ ($j = 1, 2, \ldots, q$) and $\theta_k(x)$ ($k = 1, 2, \ldots, s$) are activation functions for hidden layer neurons for the respective networks.

b. Desired Compensation Adaptive Law Controller

In the DCAL control scheme, the OAFNNs estimate dynamic functions in the state space spanned by the desired trajectory. These dynamic functions are $m_d(\cdot)$, $f_d(\cdot)$, and $h_d(\cdot)$ and are defined as

$$m_d = m(x_d), \tag{60}$$

$$f_d = f(x_d, \dot{x}_d), \tag{61}$$

$$h_d = \left.\frac{\partial m}{\partial x}\right|_{x=x_d}. \tag{62}$$

The estimates of the OAFNNs for $m_d(\cdot)$, $fd(\cdot)$, and $h_d(\cdot)$ are termed $\hat{m}_d$, $\hat{f}_d$, and $\hat{h}_d$, respectively. These network outputs can be represented in terms of network activation functions and weights as

$$\hat{m}_d = \overline{\Phi}_d(x_d)^T\hat{a}, \tag{63}$$

$$\hat{f}_d = \overline{\Psi}_d(x_d, \dot{x}_d)^T\hat{b}, \tag{64}$$

$$\hat{h}_d = \overline{\Theta}_d(x_d)^T\hat{c}, \tag{65}$$

where $\overline{\Phi}_d = [\phi_{d_1}\phi_{d_2}\cdots\phi_{d_p}]^T$, $\overline{\Psi}_d = [\psi_{d_1}\psi_{d_2}\cdots\psi_{d_q}]^T$, and $\overline{\Theta}_d = [\theta_{d_1}\theta_{d_2}\cdots\theta_{d_s}]^T$ are transformed input vectors, $\hat{a} = [\hat{a}_1\hat{a}_2\cdots\hat{a}_p]^T$, $\hat{b} = [\hat{b}_1\hat{b}_2\cdots\hat{b}_q]^T$, and $\hat{c} = [\hat{c}_1\hat{c}_2\cdots\hat{c}_s]^T$, are weight vectors, and $\phi_{d_i}(x_d)$ ($i = 1, 2, \ldots, p$), $\phi_{d_j}(x_d, \dot{x}_d)$ ($j = 1, 2, \ldots, q$), and $\theta_{d_k}(x_d)$ ($k = 1, 2, \ldots, s$) are the desired compensation-based activation functions for hidden layer neurons.

3. Choice of Number of Networks

Theoretically it is possible to use just one network instead of two or three networks in the control structure. However, the linearly parameterized networks suffer from the curse of dimensionality [13], which requires the network size to increase in geometric proportion to an increase of the network input dimension. Having a single OAFNN to compensate for overall system dynamics necessitates additional inputs to the network input space and hence a substantially larger network to accomplish the control task. Whereas the total computational load increases in arithmetic proportion to increased number of networks, the motivation to minimize the number of inputs for each network leads to a choice of multiple networks in the controller designs in this research.

4. Model Mismatch Error

Theoretically an infinite number of neurons are needed for a neural network to model all well behaved nonlinear functions. Thus, in general, a model mismatch exists between the exact function and the network optimum (best) estimate. The neural network-based controllers need to be designed with robustness to perform satisfactorily despite this model mismatch error. The model mismatch for ACAL controllers is given by

$$\varepsilon_m = m(x) - \overline{\Phi}(x)^T \bar{a}, \tag{66}$$

$$\varepsilon_f = f(x, \dot{x}) - \overline{\Psi}^T(x, \dot{x})\bar{b}, \tag{67}$$

$$\varepsilon_h = h(x) - \overline{\Theta}(x)^T \bar{c}; \tag{68}$$

for DCAL controllers as

$$\varepsilon_m(x_d) = m_d(x_d) - \overline{\Phi}_d(x_d)^T \bar{a}, \tag{69}$$

$$\varepsilon_f(x_d, \dot{x}_d) = f_d(x_d, \dot{x}_d) - \overline{\Psi}_d^T(x_d, \dot{x}_d)\bar{b}, \tag{70}$$

$$\varepsilon_h(x_d) = h_d(x_d) - \overline{\Theta}_d(x_d)^T \bar{c}. \tag{71}$$

The $\bar{a}$, $\bar{b}$, and $\bar{c}$ vectors are the optimum weights for the respective OAFNNs.

5. Actual Compensation Adaptive Law Controller

This controller has a simple structure [15] and is designed to demonstrate the advantages of using OAFNNs in direct adaptive control schemes. The controller employs two OAFNNs to estimate dynamics functions $m(\cdot)$ and $f(\cdot)$. Simplicity of the controller structure is one of the objectives of this controller. The properties of orthonormal activation functions are exploited wherever possible in designing this as well as subsequent controllers.

a. Assumptions

The system dynamics functions $m(x)$, $f(x, \dot{x})$, and $\dot{m}(x, \dot{x})$ are bounded for bounded states $(x, \dot{x})$. For bounded states, $|\dot{m}(\cdot)| < M$, where M is a bounding constant. Function $m(x)$ is of fixed sign (taken as positive without loss of generality) and therefore its lower bound is a positive constant m_0. The desired trajectory (x_d) and its first two derivatives $(\dot{x}_d, \ddot{x}_d)$ are bounded.

b. Control Law

The controller employs two OAFNNs to estimate the system dynamics separately for functions $m(\cdot)$ and $f(\cdot)$. The control law is given as

$$u = kr + \widehat{m}\gamma + \hat{f}(x, \dot{x}), \tag{72}$$

where k is a positive constant gain, $\widehat{m}$ and $\hat{f}$ are estimates by the two networks for functions $m(\cdot)$ and $f(\cdot)$, respectively, and γ is a modified acceleration term defined as

$$\gamma = \ddot{x}_d + \lambda\dot{\tilde{x}}. \tag{73}$$

c. Weight Adaptation Laws

The adaptation laws for network weight vectors are

$$\dot{\hat{a}} = \alpha\bar{\phi}\gamma r, \tag{74}$$

$$\dot{\hat{b}} = \beta\bar{\psi}r, \tag{75}$$

where α and β are positive constants referred to as learning rates in the terminology of neural networks.

d. Lyapunov Stability Analysis

Using a Lyapunov function such as

$$V = \frac{1}{2}mr^2 + \frac{1}{2\alpha}\tilde{a}^T\tilde{a} + \frac{1}{2\beta}\tilde{b}^T\tilde{b}, \tag{76}$$

the derivative of this Lyapunov function can be shown to be upper bounded as

$$\dot{V} \leq -\eta_1 r^2 + \eta_2|r|, \tag{77}$$

where η_1 and η_2 are positive constants. From Eq. (77), $\dot{V} < 0$ for

$$|r| > \frac{\eta_2}{\eta_1}, \tag{78}$$

which makes the filtered error r bounded and therefore the tracking position error $\tilde{x}$ and velocity error $\dot{\tilde{x}}$ also bounded. This further strengthens the assumptions that the system states $x = x_d - \tilde{x}$ and $\dot{x} = \dot{x}_d - \dot{\tilde{x}}$ are bounded. These bounds can be made as small as desired by adjusting the controller gain k. The detailed development as well as the proof of bounded network weights were given by Shukla [16].

6. Desired Compensation Adaptive Law Controller

The DCAL controller employs OAFNNs to estimate and compensate for unknown system dynamics. The network compensation is based on the desired position trajectory. In DCAL controllers, the network activation functions can be computed off-line. This feature can reduce on-line computational load and may be especially advantageous when large networks are needed in a control structure. The controller requires knowledge of bounds for system dynamics functions such that $|m(x)| \leq M(x)$, $|f(x, \dot{x})| \leq F(x, \dot{x})$, and $|h(x)| \leq H(x)$, and guarantees a bounded tracking error. The boundedness of the network weights is ensured by using a projection algorithm [32]. The estimated bounds of the network weights for the projection algorithm are computed using the properties of the orthonormal activation functions.

a. Control Law

The control law is continuous and is given as

$$u = k_1 r + \widehat{m}_d \gamma + \hat{f}_d + \tfrac{1}{2} \hat{f}_d \dot{x}_d r + \rho_1 r + k_2 \rho_2^2 r + k_3 \rho_3^2 \gamma^2 r, \qquad (79)$$

where k_1, k_2, and k_3 are positive constant gains and ρ_1, ρ_2, and ρ_3 are nonlinear gains defined as

$$\rho_1 = \tfrac{1}{2} \big[H(x)|\dot{x}| + \big| \hat{h}_d \dot{x}_d \big| \big], \qquad (80)$$

$$\rho_2 = F(x, \dot{x}) + \big| \hat{f}_d \big|, \qquad (81)$$

and

$$\rho_3 = M(x) + |\widehat{m}_d|. \qquad (82)$$

b. Weight Adaptation Laws

The weight adaptation laws are based on a projection algorithm [32] and are given as

$$\dot{\hat{a}}_i = \begin{cases} 0, & \text{if } \hat{a}_i = \begin{cases} -A_i \text{ and } \phi_i \gamma r < 0, \\ A_i \text{ and } \phi_i \gamma r > 0, \end{cases} \\ \alpha \phi_i \gamma r, & \text{otherwise}, \end{cases} \qquad (83)$$

$$\dot{\hat{b}}_j = \begin{cases} 0, & \text{if } \hat{b}_j = \begin{cases} -B_j \text{ and } \psi_j r < 0, \\ B_j \text{ and } \psi_j r > 0, \end{cases} \\ \beta \psi_j r, & \text{otherwise}, \end{cases} \qquad (84)$$

and

$$
\dot{\hat{c}}_k = \begin{cases} 0, & \text{if } \hat{c}_k = \begin{cases} -C_k \text{ and } \theta_k \dot{x}_d < 0, \\ C_k \text{ and } \theta_k \dot{x}_d > 0, \end{cases} \\ \frac{1}{2} \mu \theta_k \dot{x}_d r^2, & \text{otherwise,} \end{cases} \tag{85}
$$

where A_i, B_j, and C_k are estimated bounds for the optimum magnitudes of the OAFNN weights a_i, b_j, and c_k for $i = 1, 2, \ldots, p$, $j = 1, 2, \ldots, q$, and $k = 1, 2, \ldots, s$. The computations of these weight bounds for OAFNNs are presented next.

c. Bounds for Optimum Orthogonal Activation Function-Based Neural Network Weights

Using the theory of orthonormal functions, the upper bounds for optimum OAFNN weights are given as

$$
|\hat{a}_i| \le A_i = \int_{x_{d_1}}^{x_{d_2}} |M(x)| \, |\phi_i(x)| \, dx, \qquad i = 1, 2, \ldots, p, \tag{86}
$$

$$
|\hat{b}_j| \le B_j = \int_{\dot{x}_{d_1}}^{\dot{x}_{d_2}} \int_{x_{d_1}}^{x_{d_2}} |F(x, \dot{x})| \, |\psi_j(x, \dot{x})| \, dx \, d\dot{x}, \qquad j = 1, 2, \ldots, q, \tag{87}
$$

and

$$
|\hat{c}_k| \le C_k = \int_{x_{d_1}}^{x_{d_2}} |H(x)| \, |\theta_k(x)| \, dx, \qquad k = 1, 2, \ldots, s. \tag{88}
$$

d. Lyapunov Stability Analysis

Using a Lyapunov function

$$
V = \frac{1}{2} m r^2 + \frac{1}{2\alpha} \tilde{a}^T \tilde{a} + \frac{1}{2\beta} \tilde{b}^T \tilde{b} + \frac{1}{2\mu} \tilde{c}^T \tilde{c}, \tag{89}
$$

the derivative of this Lyapunov function can be shown to be upper bounded as

$$
\dot{V} \le -k_1 r^2 + \frac{1}{4k_2} + \frac{1}{4k_3}. \tag{90}
$$

From Eqs. (89) and (90) it can be shown that filtered tracking error (r) is bounded. Integrating both sides of Eq. (90), yields that

$$
\lim_{T \to \infty} \frac{1}{T} \int_0^T r^2(\tau) \, d\tau \le \frac{1}{4k_1 k_2} + \frac{1}{4k_1 k_3}. \tag{91}
$$

Thus, in the average square integral sense, the tracking error can be arbitrarily bounded by choosing sufficiently high gains k_1, k_2, and k_3. The detailed development was given by Shukla [16].

7. Experimental Evaluation

The developed neurocontrollers were evaluated using physical experiments on a motor load system. An inertia disk mounted on a motor drive and subjected to a static–dynamic friction load was used as a test bed for the experimental evaluation. The displacement of the disk was measured using a encoder, and the velocity of the disk was determined by differentiating and filtering the displacement data. The schematic of the experimental setup is shown in Fig. 39.

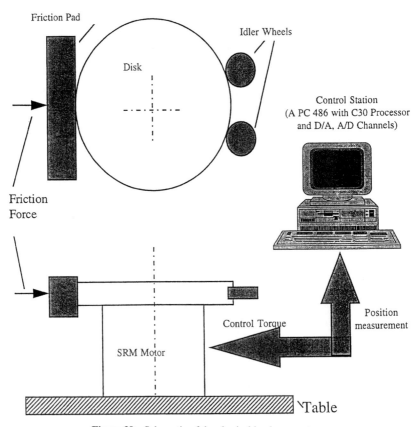

Figure 39   Schematic of the physical hardware system.

The desired tracking trajectory as shown in Fig. 40 was designed to excite the static–dynamic friction behavior of the system. The trajectory tracking errors for the ACAL and DCAL controllers are shown in the Figs. 41 and 42, respectively. Both controllers tracked the desired trajectory with a low tracking error. However, the DCAL controller gave the superior performance with a lower tracking error. The performance of the ACAL controller was degraded because of differentiation noise in the feedback of the velocity signal. The simulation results for the ACAL controller were found to be superior. Differentiation noise did not affect the DCAL controller because the desired velocity was used in the feedback.

Figure 43 shows the learned model of the friction by the OAFNN for the DCAL controller. This model is remarkably similar to textbook models of static–dynamic (stick–slip) friction in terms of its basic characteristics. The low tracking error and superior model learning capability of the OAFNN-based neural adaptive controllers provides confidence for practical, real time, on-line implementation of these controllers. A complete discussion of the experiments and other OAFNN-based on-line DCAL controllers are presented in [16].

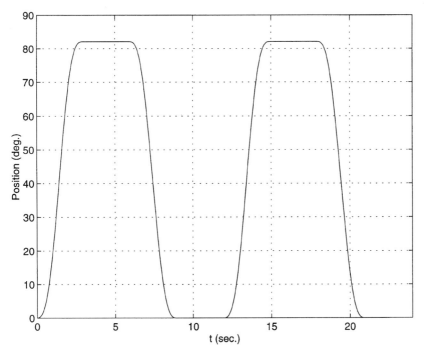

Figure 40   Start–stop trajectory.

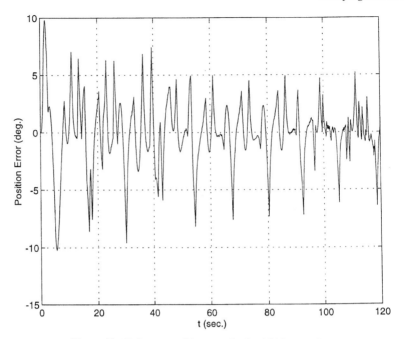

Figure 41   Trajectory tracking error for the ACAL controller.

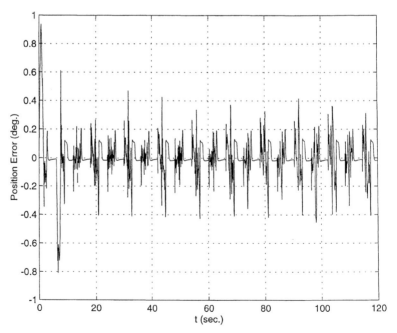

Figure 42   Trajectory tracking error for the DCAL controller.

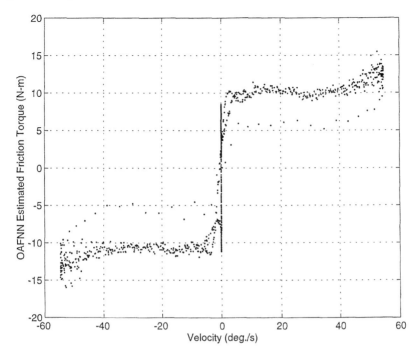

Figure 43   OAFNN estimates of friction characteristics for the DCAL controller.

V. SUMMARY

An orthogonal activation function-based neural network (OAFNN) architecture was developed with the objective of its application in identification and adaptive control of unknown dynamic systems. The activation function for the proposed OAFNNs can be any orthogonal function that permits flexibility. These functions have attractive properties for system identification and control that include absence of local minima and fast parameter convergence. The single epoch learning capability of the harmonic as well as the Legendre polynomial-based OAFNNs was demonstrated.

Schemes for frequency domain as well as time domain applications of the OAFNNs for system identification and control were developed. The Fourier series neural network (FSNN) was used for frequency domain applications that include neural network spectrum analyzer, transfer function identification, describing function identification, neural self-tuning regulator, and a neural model-reference adaptive system. In the time domain, schemes for OAFNN-based discrete nonlinear system identification and direct adaptive control were developed. The overall stability was guaranteed by using a Lyapunov-like analysis. Experiments were conducted to evaluate the practical feasibility of the developed neural controllers. These experimental results showed that the neural model-reference adaptive controller (frequency domain) and direct adaptive neural controllers (time domain) met the real time control objectives and were able to track the desired trajectory with a low tracking error. The superior model learning capability of the OAFNNs was demonstrated through simulation as well as experimental results.

REFERENCES

[1] A. J. Maren, C. Harston, and R. Pap, Eds. *Handbook of Neural Computing Applications*. Academic Press, San Diego, CA, 1990.
[2] K. S. Narendra and K. Parthasarathy. Identification and control of dynamic systems using neural networks. *IEEE Trans. Neural Networks* 1:4–27, 1990.
[3] M. A. Cohen and S. Grossberg. Absolute stability of global pattern formation and parallel memory storage by competitive neural networks. *IEEE Trans. Systems, Man, Cybernetics* SMC-13:815–826, 1983.
[4] A. Guez, V. Protopopsecu, and J. Barhen. On the stability, storage capacity, and design of nonlinear continuous neural networks. *IEEE Trans. Systems, Man, Cybernetics* 18:80–87, 1988.
[5] J. L. Johnson. Globally stable saturable learning laws. *Neural Networks* 4:47–51, 1991.
[6] A. Gelb and W. E. Vander Velde. *Multiple Input Describing Function and Nonlinear System Design*. McGraw-Hill, New York, 1968.
[7] M. Hasheminejad, J. Murata, and K. Hirasawa. System identification using neural network with parametric sigmoid function. In *Proceedings of the Artificial Neural Networks in Engineering Conference*, 1994, Vol. 4, pp. 39–44.
[8] J. M. Zurada. *Introduction to Artificial Neural Systems*. West Publishing Company, 1992.
[9] D. Shukla and F. W. Paul. Neural networks with orthonormal activation functions for non-linear dynamic system identification. In *Proceedings of the Artificial Neural Networks in Engineering Conference*, St. Louis, 1995, Vol. 5, pp. 517–524. Also invited for *Int. J. Smart Eng. Systems Design*.
[10] C. Zhu and F. W. Paul. A fourier series neural network and its application to system identification. *Trans. ASME Dynamic Systems, Measurements and Control* 117:253–261, 1995.
[11] G. Arfken. *Mathematical Methods for Physicists*. Academic Press, New York, 1970.
[12] J. S. Walker. *Fourier Analysis*. Oxford Univ. Press, New York, 1988.
[13] R. M. Sanner and J. J. Slotine. Gaussian networks for direct adaptive control. *IEEE Trans. Neural Networks* 3:837–863, 1992.
[14] C. M. Kwan, D. M. Dawson, and F. L. Lewis. Robust adaptive control of robots using neural network: Global tracking stability. In *34th Conference on Decision and Control*, 1994.
[15] D. Shukla and W. Paul. Computationally efficient control of nonlinear dynamic systems using orthonormal activation function based neural networks. In *13th World Congress of the International Federation of Automatic Control*, San Francisco, CA, 1996, Vol. K, pp. 55–60.
[16] D. Shukla. Orthonormal activation function based neural networks for adaptive control of nonlinear systems. Ph.D. Dissertation, Department of Mechanical Engineering, Clemson University, Clemson, SC, 1996.
[17] B. Widrow and S. D. Stearn. *Adaptive Signal Processing*. Prentice-Hall, Englewood Cliffs, NJ, 1985.

[18] M. Brown and C. Harris. *Neurofuzzy Adaptive Modeling and Control*. Prentice-Hall, Englewood Cliffs, NJ, 1994.

[19] N. F. Hubele and H. B. Hwarng. A neural network model and multiple linear regression: Another point of view. *Proceedings of the Artificial Neural Networks in Engineering Conference*, 1994, Vol. 4, pp. 199–203.

[20] R. G. Jacquot. *Modern Digital Control Systems*. Dekker, New York, 1981.

[21] J. F. Bouchard, C. Zhu, and F. W. Paul. A neural network spectrum analyzer. *Mechatronics* 5:603–622, 1995.

[22] D. P. Garg. Frequency-domain synthesis of controllers for absolute stability. In *Nonlinear System Analysis and Synthesis* (R. V. Ramnath, J. K. Hedrick, and H. M. Paynter, Eds.), Vol. 2. ASME, New York, 1980.

[23] J-J. E. Slotine and W. Li. *Applied Nonlinear Control*. Prentice-Hall, Englewood Cliffs, NJ, 1991.

[24] D. P. Atherton. *Nonlinear Control Engineering*. Van Nostrand-Reinhold, New York, 1975.

[25] K. J. Astrom and B. Wittenmark. *Adaptive Control*. Addison-Wesley, Reading, MA, 1989.

[26] K. Ogata. *Modern Control Engineering*. Prentice-Hall, Englewood Cliffs, NJ, 1970.

[27] C. L. Phillips and R. D. Harber. *Basic Feedback Control Systems*. Prentice-Hall, Englewood Cliffs, NJ, 1991.

[28] C. Zhu. A Fourier series neural network and its application to intelligent system identification and control. Ph.D. Dissertation, Clemson University, Clemson, SC, 1993.

[29] B. S. Gottfried and J. Weisman. *Introduction to Optimization Theory*. Prentice-Hall, Englewood Cliffs, NJ, 1973.

[30] Spectrum Signal Processing, Inc., *TMS320C30 System Board-User's Manual*, Issue 1.01, West-borough, MA, 1990.

[31] Transtech Parallel Systems Corp., *Transtech TTM110 User Guide*, Document Reference TTM110MAN1191, Ithaca, NY, 1991.

[32] S. Sastry and M. Bodson. *Adaptive Control: Stability, Convergence and Robustness*. Prentice-Hall, Englewood Cliffs, NJ, 1989.

Multilayer Recurrent Neural Networks for Synthesizing and Tuning Linear Control Systems via Pole Assignment

Jun Wang
Department of Mechanical and Automation Engineering
The Chinese University of Hong Kong
Shatin, New Territories, Hong Kong

Multilayer recurrent neural networks are presented for synthesizing linear control systems through pole assignment. Two multilayer recurrent neural networks are presented for computing the feedback gain matrix of a linear state feedback controller. The first recurrent neural network consists of two bidirectionally connected layers and each layer consists of an array of neurons. Regardless of the initial condition, the two-layer recurrent neural network is capable of determining and tuning the feedback gain matrix with specified closed-loop poles. Another recurrent neural network consists of four bidirectionally connected layers and is able to perform on-line computation of feedback gain matrices with the minimum Frobenius norm as well as desired closed-loop poles. When the states of a dynamic system are not completely known, state estimation is necessary for state feedback control. Two more multilayer recurrent neural networks are presented for on-line synthesis of asymptotic state estimators for linear dynamic systems. The first recurrent neural network is composed of two layers for computing output gain matrices with desired poles. The second recurrent neural network is composed of four layers for computing output gain matrices with desired poles and minimal norm. The proposed recurrent neural networks are shown to be capable of synthesizing linear control systems in real time. The operating characteristics of the recurrent neural networks and closed-loop control systems are demonstrated by use of many illustrative examples.

I. INTRODUCTION

Neural networks are composed of massively connected simple neurons. Resembling more or less their biological counterparts in structure, artificial neural networks are representational and computational models that process information in a parallel distributed fashion. One of the major application areas of neural networks is dynamic system identification and control. In recent years, neural networks for modeling, monitoring, and controlling dynamic systems have been explored by numerous researchers. Results of these investigations have been reported extensively in recent literature (e.g., [1–14]. See [15] and [16] for literature surveys on control applications of neural networks. Most of the proposed neural network approaches to control applications use neural networks (typically feedforward neural networks) as black-box representations of plants and/or controllers trained via supervised learning. These approaches are certainly justifiable for control of nonlinear systems, given that nonlinear control theory is far from complete. For control of linear dynamic systems, however, the use of neural networks as representational models may not be necessary because existing linear control theory provides many effective design methods for synthesis of linear controllers. Instead of replacing existing design methodology, neural networks can be used as computational models to play a supportive role for synthesizing and tuning linear control systems, which is more acceptable to practical control engineers. Because of the parallel distributed nature of neural computation [17, 18], neural networks can be a viable tool for synthesizing linear control systems in real time [19–22].

Pole assignment (placement) is a common method for synthesis of feedback control systems. When all of the state variables of a system are completely controllable and measurable, the closed-loop poles of the system (the roots of characteristic equation) can be placed at the desired locations on the complex plane with state feedback through appropriate gains [23]. Whereas the transient behavior of a feedback control system is largely determined by its closed-loop poles, pole placement is a very effective state-space approach for designing feedback control systems, especially for multivariable systems.

In some specific applications, some states in the linear system may not be available for feedback, because they are not measurable or such a measurement is too slow, too costly, or some other reason. In this case, a state estimator (observer) has to be used to estimate the unavailable states. A state estimator (observer) estimates the state variables of a dynamic system based on the measurements of the output and input (control) variables. For linear dynamic systems, the state estimator design task can be reduced to finding an output gain matrix. The output matrices in most linear state estimators are time-invariant. For time-varying dynamic systems, state estimators with time-invariant output matrices cannot follow the variation of system parameters, hence real-time gain updating of the output matrices of state estimators is necessary.

This chapter presents two two-layer and two four-layer recurrent neural networks for synthesizing and tuning linear feedback control systems in real time. By solving two coupled linear matrix equations with time-varying parameters, the two-layer recurrent neural network for feedback controller synthesis is able to synthesize and tune a linear state feedback controller with prescribed closed-loop poles. By solving a quadratic optimization problem with time-varying parameters, the four-layer recurrent neural network for controller synthesis is able to synthesize and tune a linear state feedback controller with the minimum norm of feedback gain matrix as well as prescribed closed-loop poles. In analogy, by solving two linear matrix equations with time-varying parameters, the two-layer recurrent neural network for state observer synthesis is able to synthesize and tune an asymptotic state estimator with prescribed poles. By solving a quadratic optimization problem with time-varying parameters, the four-layer recurrent neural network for observer synthesis is able to synthesize and tune an asymptotic state estimator with the prescribed eigenvalues and the minimum norm of output gain matrices.

II. BACKGROUND INFORMATION

Consider a continuous-time linear dynamic system:

$$\dot{x}(t) = Ax(t) + Bu(t), \qquad x(0) = x_0, \tag{1}$$
$$y(t) = Cx(t), \tag{2}$$

where $x \in \Re^n$ is the state vector, $u \in \Re^m$ is the control vector, $y \in \Re^p$ is the output vector, and $A \in \Re^{n \times n}$, $B \in \Re^{n \times m}$, and $C \in \Re^{p \times n}$ are known coefficient matrices associated with $x(t)$, $u(t)$, and $y(t)$, respectively.

If the linear system described in Eqs. (1) and (2) is completely state controllable, then a linear state feedback control law

$$u(t) = r(t) + Kx(t) \tag{3}$$

can be applied to control the state and output of the system, where $r \in \Re^m$ is a reference input vector and $K \in \Re^{m \times n}$ is a state feedback gain matrix. The closed-loop system is in the form

$$\dot{x}(t) = (A + BK)x(t) + Br(t), \qquad x(0) = x_0. \qquad (4)$$

According to the linear system theory, matrix K may be chosen by using different design strategies, such as the linear-quadratic optimal control method or pole assignment method, depending on the design requirements. In this chapter, we will focus on the pole assignment approach to obtain the feedback gain matrix K.

In the control literature, many authors presented a variety of numerical algorithms to compute matrix K by solving the matrix equations [24–27]

$$AZ - Z\Lambda = -BG, \qquad (5)$$

$$KZ = G, \qquad (6)$$

for a fixed $\Lambda \in \mathfrak{R}^{n \times n}$ and almost any $G \in \mathfrak{R}^{m \times n}$, where $Z \in \mathfrak{R}^{n \times n}$ is a matrix of instrumental variables. Equation (5) is known as the Sylvester equation [24, 26, 28]. It has been proven that if Λ is cyclic (i.e., its characteristic polynomial is equal to its minimal polynomial), has prescribed eigenvalues, A and Λ have no common eigenvalue [i.e., $\sigma(A) \cap \sigma(\Lambda) = \emptyset$], and (Λ, G) is observable, then (i) the unique solution Z of Eq. (5) is almost surely nonsingular with respect to parameter G [i.e., rank$(Z) = n$] and (ii) the spectrum of $(A + BK)$ equals that of Λ [i.e., $\sigma(A + BK) = \sigma(\Lambda)$]. The uniqueness and nonsingularity of Z are very important properties. For an arbitrary initial condition of Z, an algorithm can converge to the unique solution of nonsingular Z. Therefore, a unique solution of K can be obtained.

The usual procedure for computing K is as follows:

1. Choose $\Lambda \in \mathfrak{R}^{n \times n}$ such that Λ is cyclic with the desired spectrum and $\sigma(A) \cap \sigma(\Lambda) = \emptyset$, pick G so that (Λ, G) is an observable pair, and solve the Sylvester equation (5) for Z.
2. Solve the linear algebraic equation (6) for K; that is, $K = GZ^{-1}$.

Suppose that the linear system described in Eqs. (1) and (2) is completely state observable. Then the dynamic equation of an asymptotic (Luenberger) state estimator is in the form

$$\dot{\hat{x}}(t) = (A - LC)\hat{x}(t) + Ly(t) + Bu(t), \qquad \hat{x}(0) = \hat{x}_0, \qquad (7)$$

where $\hat{x}(t) \in \mathfrak{R}^n$ is the estimated state vector and $L \in \mathfrak{R}^{n \times p}$ is an output gain matrix.

According to linear system theory, the output gain matrix L can be determined by using a pole assignment method so that $A - LC$ has the desired poles (preferably with more negative real parts than those of A) [25]. As the dual problem of computing a state feedback matrix in designing a linear control system via pole assignment, the output gain matrix L can be obtained by solving two sets of linear matrix equations in the same way as obtaining the feedback gain matrix L^T in the dual system

$$\dot{z}(t) = A^T z(t) + C^T v(t), \qquad z(0) = z_0, \tag{8}$$
$$v(t) = L^T z(t), \tag{9}$$

where $z \in \Re^n$ is the dual state vector, $v \in \Re^p$ is the dual control vector, and the superscript T denotes the transpose operator. The feedback gain matrix L^T in the above dual system can be obtained by solving the following matrix equations [24] for a fixed $\Lambda \in \Re^{n \times n}$ and almost any $G \in \Re^{n \times p}$ such that $A^T - C^T L^T$ has the same eigenvalues as Λ^T,

$$A^T Z^T - Z^T \Lambda^T = -C^T G^T, \tag{10}$$
$$L^T Z^T = -G^T, \tag{11}$$

where $Z \in \Re^{n \times n}$ is a matrix of instrumental variables similar to that in Eq. (5). Rewriting Eqs. (10) and (11), we have

$$ZA - \Lambda Z = -GC, \tag{12}$$
$$ZL = -G. \tag{13}$$

Given that eigenvalues of Λ^T and $A^T - C^T L^T$ are, respectively, the same as those of Λ and $A - LC$, the output matrix of the asymptotic state observer can be obtained by solving Eqs. (12) and (13). It has been proven that if Λ is cyclic, has prescribed eigenvalues, A and Λ have no common eigenvalue [i.e., $\sigma(A) \cap \sigma(\Lambda) = \emptyset$], and (Λ^T, G^T) is observable, then (i) the unique solution Z of Eq. (12) is almost surely nonsingular with respect to a given G [i.e., rank$(Z) = n$] and (ii) the spectrum of $(A - LC)$ equals that of Λ [i.e., $\sigma(A - LC) = \sigma(\Lambda)$] [24].

In analogy, given (A, B, C), the usual procedure for computing L is as follows:

1. Choose $\Lambda \in \Re^{n \times n}$ such that Λ is cyclic and has the desired spectrum $\sigma(\Lambda)$, and $\sigma(A) \cap \sigma(\Lambda) = \emptyset$; Then choose G so that (Λ^T, G^T) is an observable pair and solve the Sylvester equation (12) for Z.
2. Solve the linear algebraic equation (13) for L; that is, $L = -Z^{-1}G$.

III. PROBLEM FORMULATION

The feedback gain matrix obtained by solving Eqs. (5) and (6) subject to the (Λ, G) observability constraint is not unique, but depends on the selection of G. In other words, a different G may result in a different K with the same desired closed-loop poles. To optimize the performance of the feedback control system, the minimization of an objective function such as the norm of the feedback gain matrix or closed-loop system matrix is desirable.

The selection of an objective function depends on the design objective of a specific control system. If the design objective is to minimize the average magnitude of control signals while placing closed-loop poles to desired positions, then $\|K\|$ should be used, where $\|\cdot\|$ is a predetermined matrix norm. If the design objective is to minimize the average magnitude of state variables while placing closed-loop poles to desired positions, then $\|A + BK\|$ should be used.

Among the different matrix norms, the Frobenius norm, defined as $\|K\|_F \overset{\Delta}{=} \sqrt{\text{trace}\{K^T K\}} = \sqrt{\sum_i \sum_j k_{ij}^2}$, is obviously the most appropriate and convenient to use because of its differentiability. The pole assignment design problem then can be formulated as the quadratic minimization problem

$$\text{minimize} \quad \|K\|_F^2 \text{ or } \|A + BK\|_F^2, \tag{14}$$

$$\text{subject to} \quad AZ - Z\Lambda + BG = 0, \tag{15}$$

$$KZ - G = 0, \tag{16}$$

$$\text{rank}\left[G^T \middle| \Lambda^T G^T \middle| \cdots \middle| (\Lambda^T)^{n-1} G^T \middle| \right] = n. \tag{17}$$

The first two constraints [Eqs. (15) and (16)] are simply the restated form of Eqs. (5) and (6). The third constraint [Eq. (17)] is the observability constraint on the (Λ, G) pair.

Whereas Eq. (17) is not in an explicit form, a reformulation of this constraint is necessary. First, consider the case where all the desired poles are real. Notice that Λ represents the spectrum of a closed-loop system and is normally a diagonal matrix; hence $\Lambda = \Lambda^T$ and Λ^k is still a diagonal matrix for any integer k if all the desired eigenvalues are real. Let the desired closed-loop poles be $\{\lambda_i, i = 1, 2, \ldots, n\}$, where $\lambda_i < 0$ $(i = 1, 2, \ldots, n)$ required by the asymptotic stability of the closed-loop system. If

$$\Lambda = \begin{pmatrix} \lambda_1 & 0 & \cdots & 0 \\ 0 & \lambda_2 & \cdots & 0 \\ \vdots & \vdots & \ddots & \vdots \\ 0 & 0 & \cdots & \lambda_n \end{pmatrix},$$

then

$$\Lambda^k = \begin{pmatrix} \lambda_1^k & 0 & \cdots & 0 \\ 0 & \lambda_2^k & \cdots & 0 \\ \vdots & \vdots & \ddots & \vdots \\ 0 & 0 & \cdots & \lambda_n^k \end{pmatrix}.$$

Thus

$$[G^T | \Lambda^T G^T | \cdots | (\Lambda^T)^{n-1} G^T]$$

$$= \begin{pmatrix} g_{11} & g_{21} & \cdots & g_{m1} & g_{11}\lambda_1 & g_{21}\lambda_1 & \cdots \\ g_{12} & g_{22} & \cdots & g_{m2} & g_{12}\lambda_2 & g_{22}\lambda_2 & \cdots \\ \vdots & \vdots & \ddots & \vdots & \vdots & \vdots & \ddots \\ g_{1n} & g_{2n} & \cdots & g_{mn} & g_{1n}\lambda_n & g_{2n}\lambda_n & \cdots \end{pmatrix}$$

$$\begin{matrix} g_{m1}\lambda_1 & \cdots & g_{11}\lambda_1^{n-1} & g_{21}\lambda_1^{n-1} & \cdots & g_{m1}\lambda_1^{n-1} \\ g_{m2}\lambda_2 & \cdots & g_{12}\lambda_2^{n-1} & g_{22}\lambda_2^{n-1} & \cdots & g_{m2}\lambda_2^{n-1} \\ \vdots & \ddots & \vdots & \vdots & \ddots & \vdots \\ g_{mn}\lambda_n & \cdots & g_{1n}\lambda_n^{n-1} & g_{2n}\lambda_n^{n-1} & \cdots & g_{mn}\lambda_n^{n-1} \end{matrix} \Bigg).$$

1. Suppose that the desired poles λ_i $(i = 1, 2, \ldots, n)$ are distinct. Whereas $\forall i$, $\lambda_i < 0$ by the requirement of asymptotic stability, it is not difficult to see that rank$[G^T | \Lambda^T G^T | \cdots | (\Lambda^T)^{n-1} G^T] = n$ if and only if $\forall j$ $(j = 1, 2, \ldots, n)$, $\exists i$ such that $g_{ij} \neq 0$. In other words, (Λ, G) is observable if and only if there is no column of zero element(s) in G. Therefore, if all the desired closed-loop poles are real and distinct, then the observability constraint [Eq. (17)] is equivalent to finding h_j $(0 < h_j < \infty)$ such that

$$h_j \sum_{i=1}^{m} g_{ij}^2 = 1, \qquad j = 1, 2, \ldots, n. \tag{18}$$

Equation (18) can be written in a vector form as

$$\tilde{G}h = \theta, \tag{19}$$

where $h = [h_1, h_2, \ldots, h_n]^T$, $\tilde{G} \triangleq \mathrm{diag}\{\sum_{i=1}^{m} g_{ij}^2, j = 1, 2, \ldots, n\}$, and $\theta \triangleq [1, 1, \ldots, 1]^T$ as an n vector.

2. Suppose that k desired poles out of n are equal $(2 \leq k \leq m)$. Without loss of generality, let $\lambda_1 = \lambda_2 = \cdots = \lambda_k$. It also can be found that (Λ, G) is observable if and only if $g_1, g_2, \ldots, g_k$, the first k columns in G, are linearly independent and $\forall j$ $(j = k + 1, k + 2, \ldots, n)$, $\exists i$ such that $g_{ij} \neq 0$.

3. Assume $m \leq n$. If there are more than m equal poles, then (Λ, G) is obviously not observable, because g_i is an m vector.

Now let us consider a state estimator with complex poles. In this case Λ contains the complex conjugate pairs of the poles. For simplicity, we first consider the case with only two conjugate poles: $\alpha_1 \pm \beta_1 i$, where $\alpha_1 < 0$ and $\beta_1 \neq 0$. Let

$$\Lambda = \begin{pmatrix} \alpha_1 & \beta_1 \\ -\beta_1 & \alpha_1 \end{pmatrix},$$

then

$$\Lambda^2 = \begin{pmatrix} \alpha_1^2 - \beta_1^2 & 2\alpha_1\beta_1 \\ -2\alpha_1\beta_1 & \alpha_1^2 - \beta_1^2 \end{pmatrix}.$$

Defining $\alpha_2 = \alpha_1^2 - \beta_1^2$ and $\beta_2 = 2\alpha_1\beta_1$, we have

$$\Lambda^2 = \begin{pmatrix} \alpha_2 & \beta_1 \\ -\beta_1 & \alpha_2 \end{pmatrix}.$$

Thus

$$\Lambda^k = \begin{pmatrix} \alpha_k & \beta_k \\ -\beta_k & \alpha_k \end{pmatrix},$$

where $\alpha_k = \alpha_{k-1}^2 - \beta_{k-1}^2$ and $\beta_k = 2\alpha_{k-1}\beta_{k-1}$. Because $\det(\Lambda^k) = \alpha_k^2 + \beta_k^2 \neq 0$, rank$(\Lambda^k) = 2$. Furthermore, because

$$\det \begin{pmatrix} \alpha_k & \beta_k \\ \alpha_{k+1} & \beta_{k+1} \end{pmatrix} = \det \begin{pmatrix} \alpha_k & \beta_k \\ \alpha_k^2 - \beta_2^2 & 2\alpha_k\beta_k \end{pmatrix} = \beta_k(\alpha_k^2 + \beta_k^2) \neq 0$$

for $k = 1, 2, \ldots, n$, then

$$\text{rank} \begin{pmatrix} \alpha_k & \beta_k \\ \alpha_{k+1} & \beta_{k+1} \end{pmatrix} = 2.$$

In the worst case, there is only one nonzero element in G, that is, $G' \triangleq [g_1, 0]^T$ or $G'' \triangleq [0, g_2]^T$, where $g_1 \neq 0$ and $g_2 \neq 0$. Then

$$[(G')^T | \Lambda^T (G')^T | \cdots | (\Lambda^T)^{n-1}(G')^T] = \begin{pmatrix} g_1 & g_1\alpha_1 & g_1\alpha_2 & \cdots & g_1\alpha_{n-1} \\ 0 & g_1\beta_1 & g_1\beta_2 & \cdots & g_1\beta_{n-1} \end{pmatrix}$$

$$\sim \begin{pmatrix} 1 & \alpha_1 & \alpha_2 & \cdots & \alpha_{n-1} \\ 0 & \beta_1 & \beta_2 & \cdots & \beta_{n-1} \end{pmatrix},$$

which is of full row rank, and

$$[(G'')^T | \Lambda^T (G'')^T | \cdots | (\Lambda^T)^{n-1}(G'')^T] = \begin{pmatrix} 0 & g_2\beta_1 & g_2\beta_2 & \cdots & g_2\beta_{n-1} \\ g_2 & g_1\alpha_1 & g_2\alpha_2 & \cdots & g_2\alpha_{n-1} \end{pmatrix}$$

$$\sim \begin{pmatrix} 0 & \beta_1 & \beta_2 & \cdots & \beta_{n-1} \\ 1 & \alpha_1 & \alpha_2 & \cdots & \alpha_{n-1} \end{pmatrix},$$

which also has full row rank.

Hence (Λ^T, G^T) is observable if and only if there is at least one nonzero element in G. The constraint in Eq. (17) can be reformulated as finding $h \in \Re$ $(0 < h < \infty)$ such that

$$h \sum_{i=1}^{n} \sum_{j=1}^{p} g_{ij}^2 = h\|G\|_F^2 = 1. \tag{20}$$

For the case with k equal to complex poles, $2 \leq k \leq \lceil m/2 \rceil$, (Λ, G) is observable if and only if the rows of G associated with α_k and β_k, respectively, are linearly independent.

In practice, Eqs. (18) or (19) also can be used to assign distinct complex conjugate pairs of closed-loop poles.

Similar to the feedback gain matrix, the output gain matrix obtained by solving Eqs. (12) and (13) subject to the (Λ^T, G^T) observability constraint depends on the selection of G. In other words, a different G may result in a different L with the same desired poles. To optimize the performance of an asymptotic state estimator, the minimization of an objective function such as the norm of the output matrix or system matrix is desirable.

The selection of an objective function depends on the design objective of a specific system. If the design objective is to minimize the average magnitude of the feedback signal to a state estimator from $y(t)$ to avoid saturation while placing the poles of the asymptotic state estimator to desired positions, then $\|L\|$ should be used, where $\| \cdot \|$ is a predetermined matrix norm. If the design objective is to minimize the average magnitude of estimated state variables to reduce sensitivity while placing the poles to desired positions, then $\|A - LC\|$ should be used.

Similar to the norm of feedback gain matrix, the Frobenius norm, defined as $\|L\|_F \triangleq \sqrt{\text{trace}\{L^T L\}} = \sqrt{\sum_i \sum_j \ell_{ij}^2}$, is obviously the most appropriate and convenient to use in this context. The observer design problem then can be formulated as the quadratic minimization problem

$$\text{minimize} \quad \|L\|_F^2 \quad \text{or} \quad \|A - LC\|_F^2, \tag{21}$$

$$\text{subject to} \quad ZA - \Lambda Z + GC = 0, \tag{22}$$

$$ZL + G = 0, \tag{23}$$

$$\text{rank}\left[G | \Lambda G | \cdots | \Lambda^{n-1} G \right] = n. \tag{24}$$

The first two constraints [Eqs. (22) and (23)] are simply the restated form of Eqs. (12) and (13). The third constraint [Eq. (24)] is the observability constraint on the (Λ^T, G^T) pair. Because Eq. (24) is not in an explicit form, a reformulation of this constraint is necessary.

Notice that $[G | \Lambda G | \cdots | \Lambda^{n-1} G]$ is an $n \times np$ matrix. Equation (24) holds if and only if there exists a unique $n \times np$ matrix F such that

$$F[G | \Lambda G | \cdots | \Lambda^{n-1} G]^T = I, \tag{25}$$

where I is an $n \times n$ identity matrix. Let $F \triangleq [F_1 | F_2 | \cdots | F_n]$ and $F_i \in \Re^{n \times p}$ $(i = 1, 2, \ldots, n)$. Then one approach to reformulating Eq. (25) is as follows: Find unique F_i $(i = 1, 2, \ldots, n)$ such that

$$F_1 G^T + F_2 G^T \Lambda^T + \cdots + F_n G^T (\Lambda^T)^{n-1} - I = 0. \tag{26}$$

Consider the cases where all the desired poles are real. Notice that Λ represents the spectrum of a state estimator and is normally a diagonal matrix; hence, $\Lambda = \Lambda^T$ and Λ^k is still a diagonal matrix for any integer k if all the desired eigenvalues are real. Let the desired poles of a state estimator be $\{\lambda_i, i = 1, 2, \ldots, n\}$, where $\lambda_i < 0$ $(i = 1, 2, \ldots, n)$ as required by the asymptotic stability of the state estimator. If

$$
\Lambda = \Lambda^T = \begin{pmatrix} \lambda_1 & 0 & \cdots & 0 \\ 0 & \lambda_2 & \cdots & 0 \\ \vdots & \vdots & \ddots & \vdots \\ 0 & 0 & \cdots & \lambda_n \end{pmatrix},
$$

then

$$
\Lambda^k = \left(\Lambda^T\right)^k = \begin{pmatrix} \lambda_1^k & 0 & \cdots & 0 \\ 0 & \lambda_2^k & \cdots & 0 \\ \vdots & \vdots & \ddots & \vdots \\ 0 & 0 & \cdots & \lambda_n^k \end{pmatrix},
$$

and

$$
G^T\left(\Lambda^T\right)^k = \begin{pmatrix} g_{11}\lambda_1^k & g_{21}\lambda_2^k & \cdots & g_{n1}\lambda_n^k \\ g_{12}\lambda_1^k & g_{22}\lambda_2^k & \cdots & g_{n2}\lambda_n^k \\ \vdots & \vdots & \ddots & \vdots \\ g_{1p}\lambda_1^k & g_{2p}\lambda_2^k & \cdots & g_{np}\lambda_n^k \end{pmatrix}.
$$

Thus

$$
\begin{pmatrix} G^T \\ G^T\Lambda^T \\ \vdots \\ G^T(\Lambda^T)^{n-1} \end{pmatrix} = \begin{pmatrix} g_{11} & g_{21} & \cdots & g_{n1} \\ g_{12} & g_{22} & \cdots & g_{n2} \\ \vdots & \vdots & \ddots & \vdots \\ g_{1p} & g_{2p} & \cdots & g_{np} \\ g_{11}\lambda_1 & g_{21}\lambda_2 & \cdots & g_{n1}\lambda_n \\ g_{12}\lambda_1 & g_{22}\lambda_2 & \cdots & g_{n2}\lambda_n \\ \vdots & \vdots & \ddots & \vdots \\ g_{1p}\lambda_1 & g_{2p}\lambda_2 & \cdots & g_{np}\lambda_n \\ \vdots & \vdots & \vdots & \vdots \\ g_{11}\lambda_1^{n-1} & g_{21}^{n-1}\lambda_2^{n-1} & \cdots & g_{n1}\lambda_n^{n-1} \\ g_{12}\lambda_1^{n-1} & g_{22}^{n-1}\lambda_2^{n-1} & \cdots & g_{n2}\lambda_n^{n-1} \\ \vdots & \vdots & \ddots & \vdots \\ g_{1p}\lambda_1^{n-1} & g_{2p}^{n-1}\lambda_2^{n-1} & \cdots & g_{np}\lambda_n^{n-1} \end{pmatrix}.
$$

1. Suppose that the desired poles λ_i $(i = 1, 2, \ldots, n)$ are distinctive. Because $\forall i, \lambda_i < 0$ by the requirement of asymptotic stability, it is not difficult to see that $\text{rank}[G|\Lambda G| \cdots |\Lambda^{n-1} G] = n$ if and only if $\forall i$ $(i = 1, 2, \ldots, n)$, $\exists j$ such that $g_{ij} \neq 0$. In other words, (Λ^T, G^T) is observable if and only if there is no row of zero element(s) in G. Therefore, if all the desired poles of the state estimator are real and distinctive, then the observability constraint [Eq. (24)] is equivalent to finding h_i $(0 < h_i < \infty)$ such that

$$h_i \sum_{j=1}^{p} g_{ij}^2 = 1, \qquad i = 1, 2, \ldots, n. \tag{27}$$

Equation (27) also can be written in a vector form as

$$\widetilde{G} h = \theta. \tag{28}$$

2. Suppose that k desired poles out of n are equal $(2 \leq k \leq p)$. Without loss of generality, let $\lambda_1 = \lambda_2 = \cdots = \lambda_k$. It also can be found that (Λ^T, G^T) is observable if and only if $g_1, g_2, \ldots, g_k$, the first k rows in G, are linearly independent, and $\forall j$ $(j = k+1, k+2, \ldots, n)$, $\exists i$ such that $g_{ij} \neq 0$. Thus Eq. (17) can be reformulated as finding α_i $(i = 1, 2, \ldots, k)$ and h_i $(i = k+1, k+2, \ldots, n)$ such that

$$\sum_{i=1}^{k} \alpha_i g_i = 0, \tag{29}$$

$$\sum_{i=1}^{k} \alpha_i^2 = 0, \tag{30}$$

$$h_i \sum_{j=1}^{p} g_{ij}^2 = h_i g_i g_i^T = h_i \|g_i\|_2^2 = 1, \qquad i = k+1, k+2, \ldots, n. \tag{31}$$

3. Since g_i is a p vector, if there are more than p equal poles, then (Λ^T, G^T) is obviously not observable.

The case with complex poles of conjugate pairs can be dealt with in a manner similar to the computation of K.

IV. NEURAL NETWORKS FOR CONTROLLER SYNTHESIS

In recent years, in parallel to the neural network approaches to control, neural networks have been proposed for solving a wide variety of linear algebra problems. For example, nonlinear and linear recurrent neural networks were proposed for matrix inversion [29–31]. Feedforward and recurrent neural networks also were proposed for solving systems of linear algebraic equations [32–34], linear matrix equations [31, 35], and a variety of other matrix algebra problems [36, 37]. The results of these investigations have laid the basis for pole assignment via the Sylvester equation using neural networks.

The first recurrent neural network for pole placement is composed of two inter-related layers. One layer of the recurrent neural networks (hidden layer) computes Z and the other (output layer) computes K. For simplicity of the presentation, the same notations, K and K, are used to represent the state variables of the two-layer recurrent neural network as well as the corresponding decision variables of the matrix equations. Based on the problem formulation discussed in the preceding two sections, an energy function can be defined as

$$E\big[K(t), Z(t)\big] \overset{\Delta}{=} E_{n\times n}\big[AZ(t) - Z(t)\Lambda + BG(t)\big] + E_{m\times n}\big[K(t)Z(t) - G(t)\big],$$
(32)

where $E_{p\times q} \overset{\Delta}{=} \sum_{i=1}^{p} \sum_{j=1}^{q} e_{ij}$ and $e_{ij} \colon \Re \to \Re$ is a bounded convex function.

To define a gradient flow by setting the time derivatives of the variables directly proportional to the negative partial derivative of the defined energy function with respect to the variables, the dynamic equations of the recurrent neural network for pole placement can be derived as

$$\frac{dZ(t)}{dt} = -\mu_z \big\{ A^T F_{n\times n}\big[AZ(t) - Z\Lambda + BG\big]$$
$$\qquad - F_{n\times n}\big[AZ(t) - Z\Lambda + BG\big]\Lambda^T \big\},$$
(33)

$$\frac{dK(t)}{dt} = -\mu_k F_{m\times n}\big[K(t)Z(t) - G\big]Z(t)^T,$$
(34)

where μ_z and μ_k are positive scaling constants (design parameters), $Z(t)$ is an $n \times n$ activation state matrix of the hidden layer corresponding to Z, $K(t)$ is an $n \times m$ activation state matrix of the output layer corresponding to the feedback gain matrix K and $F_{p\times q}$ is a $p \times q$ matrix of nondecreasing activation functions defined as $f_{ij} = de_{ij}(u)/du$.

By defining hidden state matrices $U(t)$ and $V(t)$, the dynamic equations of the recurrent neural network can be decomposed as

$$\frac{dZ(t)}{dt} = -\mu_z \big[A^T U(t) - U(t)\Lambda^T \big],$$
(35)

$$U(t) = F_{n\times n}\big[AZ(t) - Z\Lambda + BG\big],$$
(36)

$$\frac{dK(t)}{dt} = -\mu_k V(t)Z(t)^T,$$
(37)

$$V(t) = F_{m\times n}\big[K(t)Z(t) - G\big].$$
(38)

The hidden layer of the recurrent neural network for computing Z consists of two bidirectionally connected sublayers, and each sublayer consists of an $n \times n$ array of neurons. The neuron (i, j) in the output (hidden) sublayer is connected with the neurons of the ith row and jth column in the hidden (output) sublayer only. The connection weights from the hidden neurons (i, k) and (k, j) to the output neuron (i, j) in the sublayers are defined as $\mu_z \lambda_{jk}$ and $-\mu_z a_{ik}$, respectively. The connection weights from the output neurons (i, k) and (k, j) to the hidden neuron (i, j) are defined as $-\lambda_{kj}$ and a_{ki}, respectively. The biasing threshold (bias) matrix for the neurons in the hidden sublayer is defined as BG. There are no biases for the neurons in the other sublayer.

The output layer of the recurrent neural network for computing K also consists of two sublayers. Each sublayer in the layer consists of an $n \times m$ array of neurons. The time-varying connection weight matrix from the hidden sublayer to the output sublayer is defined as $-\mu_k Z(t)^T$ and the time-varying connection weight matrix from the output sublayer to the hidden sublayer is defined as $Z(t)$. There is no lateral connection among neurons in each sublayer. The biasing threshold (constant input) matrix in the hidden sublayer is defined as $-G$ and there are no biases for the neurons in the output sublayer. Figure 1 illustrates the dynamic pole placement process.

Equations (37) and (38) show that $k_{ij}(t)$ is connected with $v_{1j}(t), v_{2j}(t), \ldots, v_{nj}(t)$ only and $v_{ij}(t)$ is connected with $k_{1j}(t), k_{2j}(t), \ldots, k_{nj}(t)$ only. As pointed out in the previous studies [30, 31] this pattern of connectivity shows that the output layer can actually be decomposed into n independent subnetworks. Each subnetwork represents one row vector of $K(t)$. Equations (37) and (38) also indicate that the connection weight matrices are identical for each subnetwork. Because of the identical connection weight matrices for every subnetwork, the output layer also can be realized by a single subnetwork with time-sharing threshold vectors. In each time slot, the subnetwork biased by the corresponding threshold vector generates one row vector of the inverse matrix. Therefore, the spatial complexity of the output layer can be reduced by a factor of n.

In conventional approaches to pole assignment via the Sylvester equation, Z and K are solved successively through off-line computation. In many real-time control applications, the dynamics of the plants are time-varying. For a time-varying system, the system parameters vary over time and the closed-loop poles vary accordingly. With off-line computed feedback gains, the design criteria, even the system stability, cannot be guaranteed unless the time-varying parameters are known exactly beforehand. To ensure system stability and other design criteria, the autotuning of feedback gains is not only desirable, but also necessary.

Adaptive control systems based on system identification are an effective approach for control of time-varying systems. In such applications, the time-varying nature of the plants entails on-line computation of controller parameters and hence complicates the computation tasks. The proposed neural network approach com-

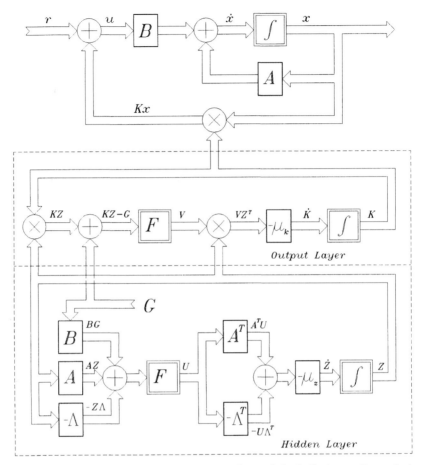

Figure 1 Architecture of the two-layer recurrent neural network for feedback controller synthesis.

putes Z and K simultaneously through on-line computation to solve Eqs. (5) and (6). Thus, the state feedback control law becomes

$$u(t) = r(t) + K(t)x(t), \qquad K(0) = K_0, \tag{39}$$

where $K(t)$ is the time-varying state feedback gain matrix. A salient feature of the proposed neural network approach is that the feedback gain matrix $K(t)$ is not constant, but time-varying. As will be shown in the ensuing simulation results, there is little difference in transient responses between the time-varying and constant feedback gain matrices if proper design parameters are chosen. The on-line computation of K is also essential in an adaptive control system for tuning the gain-scheduled controller, where the system parameters are slowly time-varying.

The two-layer recurrent neural network for solving the Sylvester equation is tailored from the generic recurrent neural network that has proven capability to solve linear matrix equations [31]. Therefore, the stable value of $K(t)$ is equal to K; that is,

$$\overline{K} \stackrel{\triangle}{=} \lim_{t \to \infty} K(t) = K. \tag{40}$$

The previous study [31] indicated that any nondecreasing activation functions in F can be used. Typical examples of the activation functions are linear function, quadratic function, and trigonometric function. The relationship between the nondecreasing activation functions and a convex energy function also has been revealed [31].

The dynamic pole assignment process is virtually a multiple-time-scale system. The plant and controller are in a time scale. The output layer is in another time scale smaller than that of the plant and controller; that is, the short-term memory $K(t)$ of the output layer is the long-term memory (feedback gain matrix K) of the controller [the short-term memory of the feedback control system is its state vector $x(t)$]. The hidden layer is in a time scale smaller than that of the output layer; that is, the short-term memory $Z(t)$ of the hidden layer is the long-term memory (connection weight matrix) of the output layer. This time-scale concept is not a requirement. It is just a way to explain the solution process in analogy to biological neural networks.

Whereas the convergence rate of the recurrent neural networks increases as the scaling constants μ_z and μ_k increase, the convergence of the recurrent neural networks can be expedited by selecting sufficiently large scaling constants. The impact of values of μ_z and μ_k will be demonstrated by means of illustrative examples in the ensuing section.

Recently, neural networks have been proposed for solving a wide variety of convex optimization problems (e.g., [18, 38–43]). The results of these investigations laid the basis for real-time pole assignment and norm minimization in the synthesis of linear feedback control systems using neural networks through on-line solutions to the quadratic minimization problem described in Eqs. (14)–(17).

The proposed recurrent neural network for pole assignment and norm minimization is composed of four interrelated layers of neurons: an output layer representing K, two hidden layers representing Z and G, and an input layer for h. For simplicity of the presentation, the same notations, K, Z, G, and h are used to represent the state variables of the neural network as well as the corresponding decision variables of the quadratic minimization problem. Based on the problem formulation discussed in the preceding section, an energy function can be defined. For example, if all the desired closed-loop poles are real and distinct, then the energy function can be defined as

$$
\begin{aligned}
E\big[K(t), Z(t), G(t), h(t)\big] &\triangleq T(t)\|K(t)\|_F^2 \left[\text{or} T(t)\|A + BK(t)\|_F^2\right] \\
&\quad + E_{n\times n}\big[AZ(t) - Z(t)\Lambda + BG(t)\big] \\
&\quad + E_{m\times n}\big[K(t)Z(t) - G(t)\big] \\
&\quad + E_{n\times 1}\big[\widetilde{G}(t)h(t) - \theta\big],
\end{aligned}
\tag{41}
$$

where $K(t)$ is an $m \times n$ activation state matrix of the output neurons corresponding to the feedback gain matrix K, $Z(t)$ is an $n \times n$ activation state matrix of the hidden neurons corresponding to Z, $G(t)$ is an $m \times n$ activation state matrix of the hidden neurons corresponding to G, $h(t)$ is an n-dimensional activation state vector of the hidden neurons corresponding to h,

$$
\widetilde{G}(t) \triangleq \operatorname{diag}\left\{\sum_{i=1}^{m} g_{ij}(t)^2,\ j = 1, 2, \ldots, n\right\}, \qquad E_{p\times q} \triangleq \sum_{i=1}^{p}\sum_{j=1}^{q} e_{ij},
$$

$e_{ij}: \Re \to \Re$ is a bounded convex function, and $T(t)$ is a scalar temperature parameter. The first term of the energy function corresponds to the objective function in Eq. (14). The second, third, and fourth terms are associated with the constraints in Eqs. (15), (16), and (19), respectively. An example of $E_{p\times q}(\cdot)$ is the squared Frobenius norm $\|\cdot\|_F^2$. The role of $T(t)$ is to provide an "annealing effect" for the minimization of an objective function, as discussed at full length in [38]. A choice of $T(t)$ in this application is

$$
\begin{aligned}
T(t) &\triangleq \|AZ(t) - Z(t)\Lambda + BG(t)\|_F^2 \\
&\quad + \|K(t)Z(t) - G(t)\|_F^2 + \|\widetilde{G}(t)h(t) - \theta\|_2^2.
\end{aligned}
\tag{42}
$$

Note that $T(t) \geq 0$ and $T(t) = 0$ if and only if $K(t)$, $Z(t)$, $G(t)$, and $h(t)$ constitute a feasible solution to the quadratic minimization problem described by Eqs. (14)–(17).

By setting the time derivatives of the variables to be directly proportional to the negative partial derivative of the defined energy function with respect to the variables, the dynamic equations of the recurrent neural network for minimizing $\|K\|_F^2$ and assigning real and distinct poles can be described in matrix form as

$$
\frac{dK(t)}{dt} = -\mu_k F_{m\times n}\big[K(t)Z(t) - G(t)\big]Z(t)^T - \mu_t T(t)K(t),
\tag{43}
$$

$$
\begin{aligned}
\frac{dZ(t)}{dt} = -\mu_z\big\{&A^T F_{n\times n}\big[AZ(t) - Z(t)\Lambda + BG(t)\big] + F_{n\times n}\big[AZ(t) \\
&- Z(t)\Lambda + BG(t)\big]\Lambda^T + K^T(t)F_{m\times n}\big[K(t)Z(t) - G(t)\big]\big\},
\end{aligned}
\tag{44}
$$

$$\frac{dG(t)}{dt} = -\mu_g \{ B^T F_{n \times n} [AZ(t) - Z(t)\Lambda + BG(t)] + F_{m \times n} [K(t)Z(t) - G(t)]$$
$$+ G(t)H(t)F_{n \times n} [\widetilde{G}(t)H(t) - I] \}, \tag{45}$$

$$\frac{dh(t)}{dt} = -\mu_h F_{n \times 1} [\widetilde{G}(t)h(t) - \theta], \tag{46}$$

where μ_t, μ_k, μ_z, μ_g, and μ_h are positive scaling constants (design parameters), $H(t) \triangleq \mathrm{diag}\{h_1(t), h_2(t), \ldots, h_n(t)\}$, and $F_{p \times q}$ is a $p \times q$ matrix of nondecreasing activation functions. To minimize $\|A + BK\|_F^2$, Eq. (43) should be replaced by

$$\frac{dK(t)}{dt} = -\mu_k F_{m \times n} [K(t)Z(t) - G(t)]Z(t)^T - \mu_t T(t)B^T [A + BK(t)]. \tag{47}$$

Figure 2 shows a block diagram of the four-layer recurrent neural network for the assignment of real and distinct closed-loop poles. The proposed recurrent neural network for the synthesis of minimum-norm linear feedback control systems is tailored from the deterministic annealing neural network that has proven capability to solve convex programming problems [38]. According to the previous results [38], if the temperature parameter is slowly and monotone decreasing with respect to time and the activation function is monotone nondecreasing with respect to its input, then the stable value of $K(t)$ is equal to K and minimizes the Frobenius norm; that is, $\overline{K} \triangleq \lim_{t \to \infty} K(t) = K$ and $\overline{K} = \arg\min\{\|K\|_F \text{ or } \|A + BK\|_F$, subject to Eqs. (15), (16), and (17)}. The relation between a bounded convex energy function and a nondecreasing activation function has been revealed [39]. Specifically, $f_{ij}(y_{ij}) = de_{ij}/dy_{ij}$. To name two examples: if $E_{p \times q}(Y) = \|Y\|_F^2$, then $f_{ij}(y_{ij}) = 2y_{ij}$; if $e_{ij}(y_{ij}) = y_{ij} \arctan(y_{ij}) - \ln\sqrt{1 + y_{ij}^2}$, then $f_{ij}(y_{ij}) = \arctan(y_{ij})$, where $F_{p \times q}(Y) = [f_{ij}(y_{ij})]$ and $Y = [y_{ij}] \in \Re^{p \times q}$. A general guideline for selecting the design parameters is $\mu_k \leq \mu_z \leq \mu_g \leq \mu_h$. The magnitudes of the eigenvalues (hence the convergence rate) of the multilayer recurrent neural network increase as the values of the design parameters increase, as will be shown in the ensuing illustrative example. The energy function and dynamic equations of the multilayer recurrent neural network for assigning equal closed-loop poles and/or complex poles can be similarly derived. As mentioned in the preceding section, the recurrent neural network described in Eqs. (43)–(47) also can be used to synthesize control systems with distinct complex closed-loop poles, as will be shown in an illustrative example later.

Although the deterministic annealing neural network possesses the properties of asymptotic stability and optimality for solving static convex programming problems [38] this does not necessarily translate into the stability and convergence of any closed-loop adaptive control system with on-line pole assign-

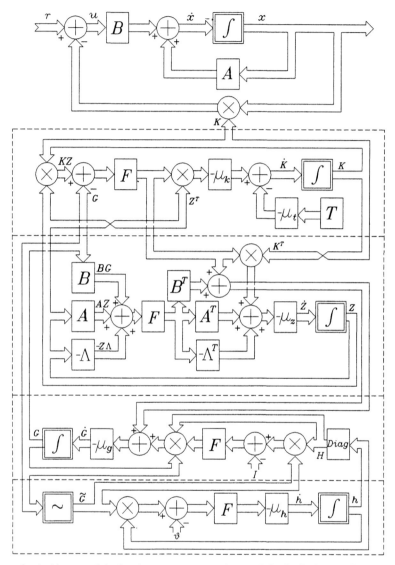

Figure 2 Architecture of the four-layer recurrent neural network for feedback controller synthesis. Reprinted from *Automatica*, J. Wang and G. Wu, "A multilayer recurrent neural network for synthesis of minimum-norm linear feedback control systems via pole placement," 32:435–442, Copyright 1996, with kind permission from Elsevier Science Ltd, The Boulevard, Langford Lane, Kidlington OX5 1GB, UK.

ment. Nevertheless, simulation results highlighted in the ensuing section show that good performance can be achieved using the proposed approach in most instances.

V. NEURAL NETWORKS FOR OBSERVER SYNTHESIS

In conventional approaches to state estimator synthesis for linear dynamic systems via pole assignment, the output gain matrix L is usually obtained through off-line computation. In many real-time applications, the system dynamics are time-varying. In such applications, the time-varying nature of the plants entails on-line state estimation and hence complicates the computation.

The recent results on recurrent neural networks for solving algebra problems have paved the way for real-time synthesis of state estimators for linear dynamic systems using neural networks through on-line solutions to the linear matrix equations described in Eqs. (12) and (13) and the quadratic minimization problem described in Eqs. (21)–(24). The proposed multilayer recurrent neural networks for synthesis of linear state estimators are tailored from the recurrent neural networks that have been proven to be capable of solving linear matrix equations and convex programming problems [38, 39].

The neural network-based asymptotic state estimators presented in this chapter are applicable for full-order observers. The proposed neural network approach, with some modifications, can be extended for synthesizing reduced-order state observers.

The first proposed recurrent neural network for synthesizing an asymptotic state estimator with desired poles is composed of two interrelated layers of neurons: an output layer representing L and an input layer for Z. For simplicity of presentation, the same notations, L and Z, are used to represent the state variables of the neural networks as well as the corresponding solutions of the linear matrix equations. Based on the problem formulation discussed in the preceding section, an energy function can be defined. For example, an energy function can be defined as

$$E\big[L(t), Z(t)\big] \stackrel{\Delta}{=} E_{n \times n}\big[Z(t)A - \Lambda Z(t) + GC\big] + E_{n \times p}\big[Z(t)L(t) + G\big], \quad (48)$$

where $L(t)$ is an $n \times p$ activation state matrix of the output neurons corresponding to the output gain matrix L and $Z(t)$ is an $n \times n$ activation state matrix of the hidden neurons corresponding to Z. The first term of the energy function corresponds to Eq. (12) and the second term is associated with Eq. (13).

By letting the time derivatives of the variables be directly proportional to the negative partial derivative of the defined energy function with respect to the variables, the dynamic equations of the two-layer recurrent neural network for computing L with desired poles can be described in the matrix form

$$\frac{dL(t)}{dt} = -\mu_l Z(t)^T F_{n \times p} [Z(t)L(t) + G], \tag{49}$$

$$\begin{aligned}\frac{dZ(t)}{dt} = -\mu_z \{ &F_{n \times n}[Z(t)A - \Lambda Z(t) + GC]A^T \\ &- \Lambda^T F_{n \times n}[Z(t)A - \Lambda Z(t) + GC] \\ &+ F_{n \times n}[Z(t)L(t) + G]L(t)^T \}, \end{aligned} \tag{50}$$

where μ_l and μ_z are positive scaling constants (design parameters). Figure 3 delineates the configuration of the two-layer recurrent neural network for synthesizing an asymptotic state estimator for a linear dynamic system.

The second proposed recurrent neural network for synthesizing an asymptotic state estimator with desired poles and minimal norm is composed of four interrelated layers of neurons: an output layer representing L, two hidden layers representing Z and G, and an input layer for h. For simplicity of presentation, the same notations, G and h, are used to represent the state variables of the neural networks as well as the corresponding decision variables of the quadratic minimization problem. Based on the problem formulation discussed in the preceding section, an energy function can be defined. For example, if all the poles of the state estimator are complex and distinctive, then an energy function can be defined as

$$\begin{aligned}E[L(t), Z(t), G(t), h(t)] \triangleq &T(t)\|L(t)\|_F^2 + E_{n \times n}[Z(t)A - \Lambda Z(t) + G(t)C] \\ &+ E_{n \times p}[Z(t)L(t) + G(t)] \\ &+ E_{n \times 1}[h(t)\|G(t)\|_F^2 - 1], \end{aligned} \tag{51}$$

where $G(t)$ is an $n \times p$ activation state matrix of the neurons corresponding to G, $h(t)$ is a scalar activation state of the hidden neuron corresponding to h, and $T(t)$ is a nonnegative and decreasing temperature parameter.

The first term of the energy function corresponds to the objective function in Eq. (21). The second, third, and fourth terms are associated with the constraints in Eqs. (22), (23), and (24), respectively. The role of the temperature parameter is described in [38]. According to the previous results [38] if the temperature parameter is monotonically decreasing with respect to time and the activation function is monotonically nondecreasing with respect to its input, then the stable value of $L(t)$ is equal to L and minimizes the Frobenius norm; that is, $\bar{L} \triangleq \lim_{t \to \infty} L(t) = L$ and $\bar{L} = \arg\min\{\|L\|_F$ or $\|A - LC\|_F$ subject to Eqs. (22), (23), and (24)}. In this chapter, we define

$$\begin{aligned}T(t) \triangleq &\|Z(t)A - \Lambda Z(t) + G(t)C\|_F^2 + \|Z(t)L(t) + G(t)\|_F^2 \\ &+ [h(t)\|G(t)\|_F^2 - 1]^2. \end{aligned} \tag{52}$$

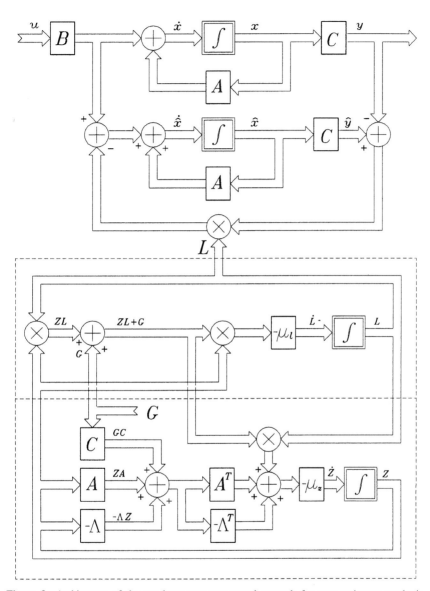

Figure 3 Architecture of the two-layer recurrent neural network for state estimator synthesis. Reprinted with permission from J. Wang and G. Wu, *Internat. J. Systems Sci.* 26:1205–1222, 1995 (© 1995 Taylor & Francis Ltd.).

Note that $T(t) \geq 0$ and $T(t) = 0$ if and only if $L(t)$, $Z(t)$, $G(t)$, and $h(t)$ constitute a feasible solution to the quadratic minimization problem described by Eqs. (21)–(24).

By letting the time derivatives of the variables be directly proportional to the negative partial derivative of the defined energy function with respect to the variables, the dynamic equations of the four-layer recurrent neural network for minimizing $\|L\|_F^2$ and assigning complex and distinct poles can be described in matrix form as

$$\frac{dL(t)}{dt} = -\mu_l Z(t)^T F_{n \times p}[Z(t)L(t) + G(t)] - \mu_t T(t)L(t), \qquad (53)$$

$$\begin{aligned}
\frac{dZ(t)}{dt} = &-\mu_z \{ F_{n \times n}[Z(t)A - \Lambda Z(t) + G(t)C]A^T \\
&- \Lambda^T F_{n \times n}[Z(t)A - \Lambda Z(t) + G(t)C] \\
&+ F_{n \times p}[Z(t)L(t) + G(t)]L(t)^T \},
\end{aligned} \qquad (54)$$

$$\begin{aligned}
\frac{dG(t)}{dt} = &-\mu_g \{ F_{n \times n}[Z(t)A - \Lambda Z(t) + G(t)C]C^T \\
&+ F_{n \times p}[Z(t)L(t) + G(t)] \\
&+ F_{n \times 1}[h(t)\|G(t)\|_F^2 - 1]G(t) \},
\end{aligned} \qquad (55)$$

$$\frac{dh(t)}{dt} = -\mu_h F_{n \times 1}[h(t)\|G(t)\|_F^2 - 1]h(t), \qquad (56)$$

where μ_l, μ_t, μ_z, μ_g, and μ_h are positive scaling constants (design parameters). To minimize $\|A - LC\|_F^2$, Eq. (53) should be replaced by

$$\frac{dL(t)}{dt} = -\mu_l Z(t)^T F_{n \times p}[Z(t)L(t) + G(t)] + \mu_t T(t)[A - L(t)C]C^T. \quad (57)$$

Figure 4 delineates the configuration of the four-layer recurrent neural network for synthesizing an asymptotic state estimator for a linear dynamic system.

The dynamic synthesis process is also a multiple-time-scale system. The linear dynamic system and state estimator are in a time scale. $L(t)$ is in another time scale smaller than that of the linear system and state estimator, and $Z(t)$ is in a time scale smaller than that of $L(t)$. For the four-layer recurrent neural network, $G(t)$ is in a time scale smaller than that of $Z(t)$ and $h(t)$ is in a time scale smaller than that of $G(t)$; that is, the short-term memory $L(t)$ is the long-term memory (output gain matrix L) of the state estimator [the short-term memory of the feedback system is its state vector $x(t)$], $L(t)$ is relatively stable compared with $Z(t)$, $Z(t)$ is relatively stable compared with $G(t)$, and $G(t)$ is relatively stable compared with $h(t)$. Whereas the convergence rate of the recurrent neural network increases as the scaling constants increase, the multiple time scale can be set by

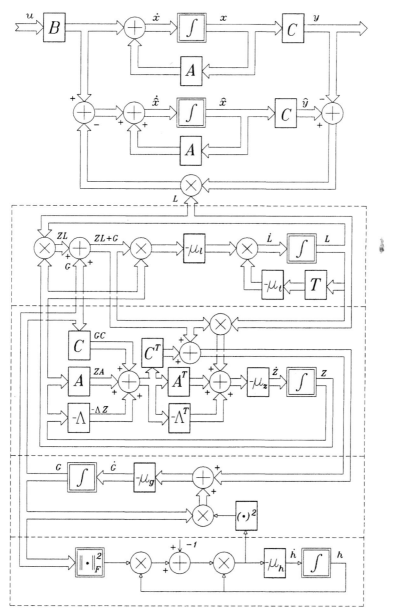

Figure 4 Architecture of the four-layer recurrent neural network for state estimator synthesis. Reprinted with permission from J. Wang and G. Wu, *Internat. J. Systems Sci.* 26:1205–1222, 1995 (© 1995 Taylor & Francis Ltd.).

selecting $\mu_l \le \mu_z \le \mu_g \le \mu_h$. Furthermore, the convergence of the recurrent neural network can be expedited by selecting sufficiently large scaling constants.

VI. ILLUSTRATIVE EXAMPLES

In this section, we discuss the results of simulation by means of eight illustrative examples using self-developed simulators coded in C language based on the fourth-order Runge–Kutta method.

EXAMPLE 1.   Consider an open-loop unstable system with coefficient matrices

$$A = \begin{pmatrix} 0 & 1 \\ -3.25 & 3 \end{pmatrix}, \qquad B = \begin{pmatrix} 1 & 2.5 \\ 2 & -1 \end{pmatrix}.$$

The two open-loop poles of the system are $1.5 \pm i$. Suppose that the desired closed-loop poles are $-2 \pm 0.5i$. Let

$$G = \begin{pmatrix} 0.25 & 2 \\ 1.5 & -10 \end{pmatrix}, \qquad \Lambda = \begin{pmatrix} -2 & 0.5 \\ -0.5 & -2 \end{pmatrix}.$$

Then the solutions to Eqs. (5) and (6) are

$$Z = \begin{pmatrix} -3.2110 & 9.2124 \\ -2.1841 & 2.9697 \end{pmatrix}, \qquad K = \begin{pmatrix} 0.4828 & -0.8243 \\ -1.6425 & 1.7280 \end{pmatrix}.$$

Therefore, the closed-loop system becomes $\dot{x}(t) = (A + BK)x(t) + Br(t)$. Applying the final value theorem, we have

$$x(\infty) = \lim_{s \to 0} s\left[sI - (A + BK)\right]^{-1} BR(s)$$
$$= -(A + BK)^{-1} Br(\infty),$$

where $R(s)$ is the Laplace transform of $r(t)$ and I is the identity matrix. In the simulation, the reference input $r(t)$ is a unit step vector. Hence $R(s) = (1/s \quad 1/s)^T$, $r(\infty) = \lim_{s \to 0} s R(s) = (1 \quad 1)^T$, and $x(\infty) = (1.3679 \quad 0.3240)^T$.

The activation function f_{ij} in $F_{n \times n}$ used in this example is the trigonometric function $f_{ij}(y) = \arctan(y)$ and f_{ij} in $F_{m \times n}$ is the linear function $f_{ij}(y) = y$. Without loss of generality, $Z(0) = 0$ and $K(0) = 0$. The initial closed-loop poles thus are equal to the open-loop ones. Figure 5a and b shows the transients of the two-layer recurrent neural network where $(\mu_k, \mu_z) = (10000, 10000)$. From Fig. 5, we can see that $Z(t)$ approaches its stable value rapidly (less than 40 μs), whereas $K(t)$ takes less than 80 μs to converge to K. Figure 5c illustrates the transient responses of the feedback control system. Figure 6 illustrates the root locus (trajectories of the closed-loop poles) of the feedback control systems in the dynamic pole placement process. It shows that the closed-loop poles move from

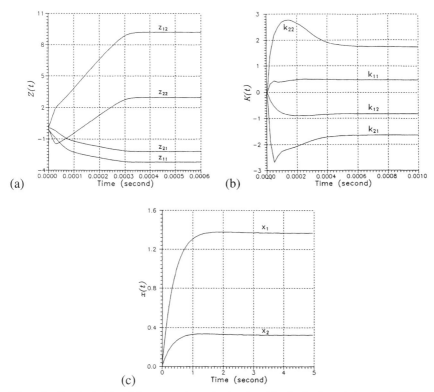

Figure 5 Transient states of the recurrent neural network and the feedback control system in Example 1. Reprinted with permission from J. Wang and G. Wu, *Internat. J. Systems Sci.* 26:2369–2382, 1995 (© 1995 Taylor & Francis Ltd.).

the open-loop poles [because $K(0) = 0$] to the negative real axis first, form an arc symmetric to the negative real axis on the left half of the complex plane, then converge to the negative real axis again, and finally reach the desired locations.

EXAMPLE 2. Consider the example in the literature [26] to demonstrate the operating characteristics of the proposed neural network approach. Given

$$A = \begin{pmatrix} 1 & 1 \\ 0 & 2 \end{pmatrix}, \qquad B = \begin{pmatrix} 1 & 1 \\ 1 & -1 \end{pmatrix}.$$

The open-loop system has poles at 1 and 2. Let the desired closed-loop poles be -1 and -2,

$$G = \begin{pmatrix} 1 & 3 \\ 2 & 4 \end{pmatrix}, \qquad \Lambda = \begin{pmatrix} -1 & 0 \\ 0 & -2 \end{pmatrix}.$$

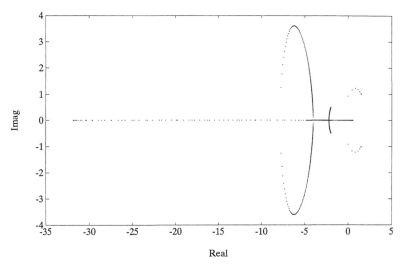

Figure 6 Trajectories of closed-loop poles of the feedback control system on the complex plane in Example 1. Reprinted with permission from J. Wang and G. Wu, *Internat. J. Systems Sci.* 26:2369–2382, 1995 (© 1995 Taylor & Francis Ltd.).

Then

$$Z = \begin{pmatrix} -1.6667 & -2.4167 \\ 0.3333 & 0.2500 \end{pmatrix}, \qquad K = \begin{pmatrix} -1.9286 & -6.6429 \\ -2.1429 & -4.7143 \end{pmatrix}.$$

The reference input $r(t)$ is again a unit step vector. Whereas $r(\infty) = (1 \quad 1)^T$, $x(\infty) = -(A + BK)^{-1}Br(\infty) = (-0.0713 \quad 0.2143)^T$.

The activation function f_{ij} in $F_{n \times n}$ used in the example is also the trigonometric function $f_{ij}(y) = \arctan(y)$ and f_{ij} in $F_{m \times n}$ is the linear function $f_{ij}(y) = y$. With $Z(0) = 0$ and $K(0) = 0$, two groups of design parameters, $(\mu_k, \mu_z) \in \{(1000, 100), (100, 10)\}$, were used. Figure 7 illustrates the transient states of the hidden layer and the output layer, respectively. They show that the convergence rate of $Z(t)$ is much higher than that of $K(t)$. Two different traces of $Z(t)$ and $K(t)$ were obtained as shown in Fig. 7. The ones with larger μ_k and μ_z approach stable value faster. Figure 7e and f illustrates the transient responses of the states $x(t)$ of the closed-loop control system based on constant K and time-varying $K(t)$ with different μ_z and μ_k. The trace marked with solid boxes is associated with the precomputed constant K. The other two traces are associated with two real-time computed values of $K(t)$. The trace with open circles has a larger value of μ_k and μ_z ($\mu_k = 1000, \mu_z = 100$) than the trace without any marks ($\mu_k = 100, \mu_z = 10$). From Fig. 7e and f, we can see that there is

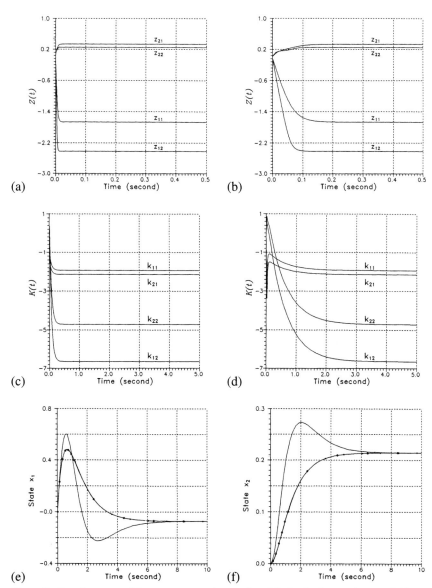

Figure 7 Transient states of the recurrent neural network and the feedback control system in Example 2. Reprinted with permission from J. Wang and G. Wu, *Internat. J. Systems Sci.* 26:2369–2382, 1995 (© 1995 Taylor & Francis Ltd.).

almost no difference between the transient responses based on off-line computed K (marked with solid boxes) and those using the neural networks (marked with circles) when $\mu_k = 1000$, $\mu_z = 100$. Figure 8 illustrates the transients of the closed-loop poles. The closed-loop poles initially swing abruptly. Figure 8 indicates that the closed-loop poles are placed to the desired locations in about 0.1 s.

EXAMPLE 3.   Consider an open-loop unstable system:

$$A = \begin{pmatrix} 1 & -1 & 0 \\ 1 & 1 & 0 \\ 2 & 1 & -1 \end{pmatrix}, \qquad B = \begin{pmatrix} 0 & 0 \\ 1.5 & 1 \\ -1 & 1 \end{pmatrix}.$$

The eigenvalues of A are -1 and $1 \pm i$ [i.e., $\sigma(A) = \{-1, 1+i, 1-i\}$]. Suppose the desired closed-loop poles are $\{-1.5, -2.5, -3.0\}$. Let

$$G = \begin{pmatrix} 1 & 1 & -1 \\ 0 & 1.5 & 1 \end{pmatrix}, \qquad \Lambda = \begin{pmatrix} -1.5 & 0 & 0 \\ 0 & -2.5 & 0 \\ 0 & 0 & -3.0 \end{pmatrix}.$$

Then the constant feedback gain matrix is

$$K = \begin{pmatrix} 35.5483 & -10.9846 & 0.6922 \\ -44.5297 & 10.1899 & -1.0208 \end{pmatrix}.$$

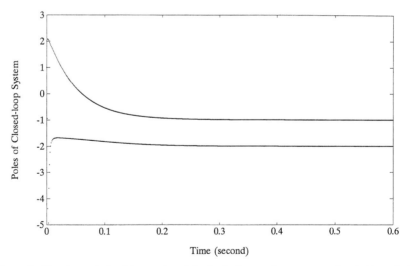

Figure 8   Transients of closed-loop poles of the feedback control system over time in Example 2. Reprinted with permission from J. Wang and G. Wu, *Internat. J. Systems Sci.* 26:2369–2382, 1995 (© 1995 Taylor & Francis Ltd.).

The reference input $r(t)$ is still a unit step vector. Whereas $r(\infty) = (1 \quad 1)^T$, according to the final value theorem, the steady state vector is $x(\infty) = \lim_{s\to 0} s[sI - (A + BK)]^{-1}BR(s) = -(A + BK)^{-1}Br(\infty) = (-0.6029 - 0.6029 \times 12.4232)^T$. Like the first two examples $[Z(0) = 0$ and $K(0) = 0]$, the activation function f_{ij} in $F_{n\times n}$ is the trigonometric function $f_{ij}(y) = \arctan(y)$ and f_{ij} in $F_{m\times n}$ is the linear function $f_{ij}(y) = y$. Figure 9a illustrates the transient states of the output layer. It shows that the activation state matrix of the output layer (the time-varying gain matrix) $K(t)$ approaches K rapidly and $\overline{K} = K$. The validity of the approach can be checked from Fig. 9. From Fig. 9a and b, the effect of μ_k and μ_z on the convergence of the feedback gain matrix can be seen clearly. With a larger value of μ_k and μ_z, $K(t)$ approaches the stable value $\overline{K}$ faster. Figure 9c, d and e illustrates the transient responses of the closed-loop control system. There are three traces in each subplot, which correspond to three different simulation settings. The trace marked with solid boxes is associated with the precomputed constant K. The other two traces are associated with two real-time computed values of $K(t)$. The trace with open circles has a larger value of μ_k and μ_z ($\mu_k = 10000$, $\mu_z = 1000$) than the trace without any marks ($\mu_k = 1000$, $\mu_z = 10$). Figure 9c, d, and e shows that the transient responses of the feedback control systems based on constant K and time-varying $K(t)$ are almost identical when $\mu_k = 10000$ and $\mu_z = 1000$. From Fig. 9c, d, and e, the effect of the magnitude of the scaling constants on the system transient responses can be seen clearly. Figure 10 illustrates the root locus (trajectories of the closed-loop poles) on the complex plane. It shows that the closed-loop poles start from the open-loop poles, move to the real axis, and then converge to desired locations as time evolves.

EXAMPLE 4. Consider an open-loop unstable system with the coefficient matrices

$$A = \begin{pmatrix} 0.5 & 1 \\ -1 & 2 \end{pmatrix}, \qquad B = \begin{pmatrix} 3 & 2.5 \\ 1.5 & -1 \end{pmatrix}.$$

The open-loop poles of the system are $1.25 \pm 0.6614i$. Suppose that the desired closed-loop poles are $-1 \pm 1.5i$.

If $\|K\|_F^2$ is chosen as the objective function, then the optimal solution to this minimum-norm pole assignment problem is

$$h = \begin{pmatrix} 10.482182 \\ 2.238136 \end{pmatrix}, \qquad G = \begin{pmatrix} -0.300639 & -0.422034 \\ -0.070834 & 0.307130 \end{pmatrix},$$

$$Z = \begin{pmatrix} 0.336659 & 0.371874 \\ 0.16202 & 0.445452 \end{pmatrix}, \qquad K = \begin{pmatrix} -0.882888 & -0.210371 \\ -0.253777 & 0.901337 \end{pmatrix},$$

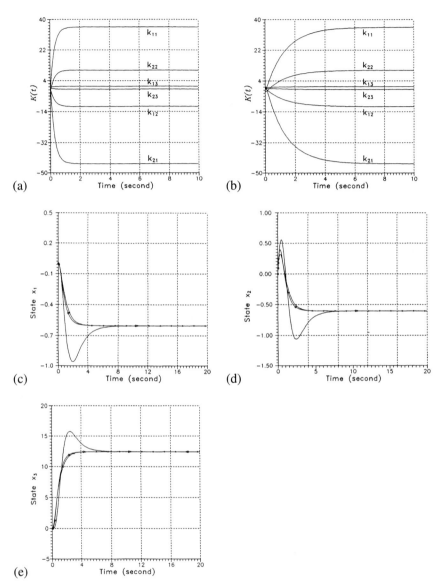

Figure 9 Transient states of the recurrent neural network and the feedback control system in Example 3. Reprinted with permission from J. Wang and G. Wu, *Internat. J. Systems Sci.* 26:2369–2382, 1995 (© 1995 Taylor & Francis Ltd.).

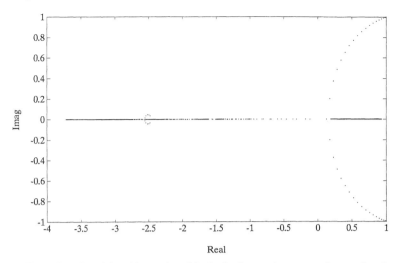

Figure 10 Trajectories of closed-loop poles of the feedback control system on the complex plane in Example 3. Reprinted with permission from J. Wang and G. Wu, *Internat. J. Systems Sci.* 26:2369–2382, 1995 (© 1995 Taylor & Francis Ltd.).

and corresponding norms are $\|K\|_F = 1.304055$ and $\|A + BK\|_F = 4.418396$. Let the arbitrarily chosen initial state matrices

$$h(0) = \begin{pmatrix} 1 \\ 1 \end{pmatrix}, \qquad G(0) = \begin{pmatrix} 1 & 0 \\ 0 & 1 \end{pmatrix},$$

$$Z(0) = \begin{pmatrix} 1 & 1 \\ 1 & 1 \end{pmatrix}, \qquad K(0) = \begin{pmatrix} 1 & 0 \\ 0 & 1 \end{pmatrix},$$

and $f_{ij}(y) = \arctan(y)$, and let the design parameters $\{\mu_t, \mu_k, \mu_z, \mu_g, \mu_h\} = \{300, 1000, 1000, 1000, 1000\}$. Then the neural network simulator yields the solution

$$\bar{h} = \begin{pmatrix} 1.899651 \\ 3.038066 \end{pmatrix}, \qquad \overline{G} = \begin{pmatrix} -0.712091 & -0.293395 \\ 0.139065 & 0.493028 \end{pmatrix},$$

$$\overline{Z} = \begin{pmatrix} 0.772235 & 0.165968 \\ 0.381305 & 0.557015 \end{pmatrix}, \qquad \overline{K} = \begin{pmatrix} -0.776239 & -0.295439 \\ -0.301291 & 0.974897 \end{pmatrix},$$

with the corresponding norms $\|\overline{K}\|_F = 1.315687$ and $\|A + B\overline{K}\|_F = 4.121085$. The difference between $\|K\|_F$ and $\|\overline{K}\|_F$ is small (about 0.0116), although the elements in matrices K and $\overline{K}$ are not identical. A better solution with a norm closer to the minimal norm can be achieved by adjusting the design parameters. The first four subplots in the left column of Fig. 11 show the transient behavior

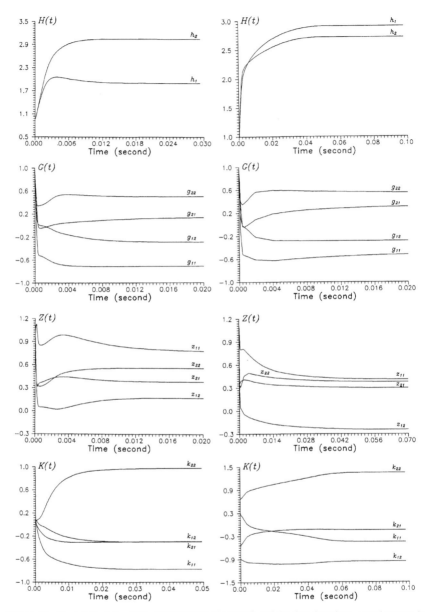

Figure 11 Transient states of the recurrent neural network and the time-invariant control system in Example 4. Reprinted from *Automatica*, J. Wang and G. Wu, "A multilayer recurrent neural network for synthesis of minimum-norm linear feedback control systems via pole placement," 32:435–442, Copyright 1996, with kind permission from Elsevier Science Ltd, The Boulevard, Langford Lane, Kidlington OX5 1GB, UK.

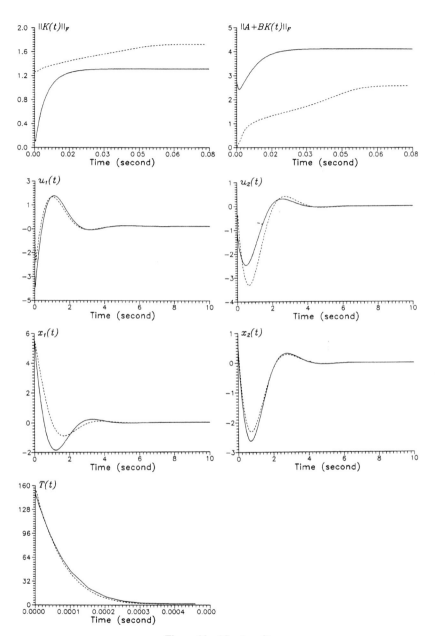

Figure 11 (*Continued*)

of state matrices $h(t)$, $G(t)$, $Z(t)$, and $K(t)$. We can see that $h(t)$, $G(t)$, and $Z(t)$ converge rapidly (less than 20 μs), whereas $K(t)$ takes less than 50 μs to stabilize.

If $\|A + BK\|_F^2$ is used as the objective function, then the optimal solution to this problem is

$$h = \begin{pmatrix} 4.217226 \\ 1.717112 \end{pmatrix}, \quad G = \begin{pmatrix} -0.373037 & -0.218248 \\ 0.312996 & 0.466614 \end{pmatrix},$$

$$Z = \begin{pmatrix} 0.304370 & -0.239938 \\ 0.239973 & 0.304669 \end{pmatrix}, \quad K = \begin{pmatrix} -0.407679 & -1.037412 \\ -0.110540 & 1.444495 \end{pmatrix},$$

with the norms $\|A + BK\|_F = 2.549499$ and $\|K\|_F = 1.827898$. By using the developed neural network simulator and choosing the same initial values of matrices $K(t)$, $G(t)$, $Z(t)$, and $h(t)$ and the same design parameters as before except $\mu_t = 500$, the neural network simulator generates the solution

$$\bar{h} = \begin{pmatrix} 2.914593 \\ 2.722218 \end{pmatrix}, \quad \overline{G} = \begin{pmatrix} -0.473538 & -0.257028 \\ 0.344765 & 0.548894 \end{pmatrix},$$

$$\overline{Z} = \begin{pmatrix} 0.413815 & -0.240831 \\ 0.299223 & 0.380813 \end{pmatrix}, \quad \overline{K} = \begin{pmatrix} -0.450342 & -0.959750 \\ -0.143479 & 1.350631 \end{pmatrix},$$

and $\|A + B\overline{K}\|_F = 2.583999$ and $\|\overline{K}\|_F = 1.722997$. The difference between $\|A + B\tilde{K}\|_F$ and $\|A + BK\|_F$ is about 0.0345. The first four subplots in the right column of Fig. 11 show the state matrices $h(t)$, $G(t)$, $Z(t)$, and $K(t)$ take less than 100 μs to converge. In Fig. 11, the Frobenius norms of $K(t)$ and $A + BK(t)$, control signals, and state variables also are illustrated. Notice that the initial value of $x(0)$ is 0, and the reference input is assumed to be an impulse vector; that is, $r(t) = [\delta(t), \delta(t)]$. The solid lines correspond to the case in which $\|K\|_F^2$ is used as the objective function, whereas the dotted lines denote the case in which $\|A + BK\|_F^2$ is used as the objective function. We can see that the former has smaller control signal magnitude, whereas the latter has smaller state variable magnitude. In Fig. 2, the transient behavior of the temperature parameter also is shown. Figure 12 illustrates the root locus (trajectories of the closed-loop poles) of the feedback control system and their transient behavior in the dynamic pole placement process, where $\|K\|_F^2$ is chosen as the objective function. It shows that the closed-loop poles start from $\{0.429973, 4.070027\}$ [because $K(0) = I$], meet on the real axis, and then move into the complex plane toward the desired positions. It also shows that the closed-loop poles reach the desired positions in about 20 μs.

Now assume that the element a_{22} in A changes slowly and periodically; that is, $a_{22}(t) = 2 + \sin(0.1\pi t)$. To follow the change closely, we choose the design parameters as $\{\mu_t, \mu_k, \mu_z, \mu_g, \mu_h\} = \{3000, 10000, 10000, 10000, 10000\}$. Other simulation conditions are the same as before. Figure 13 illustrates the tran-

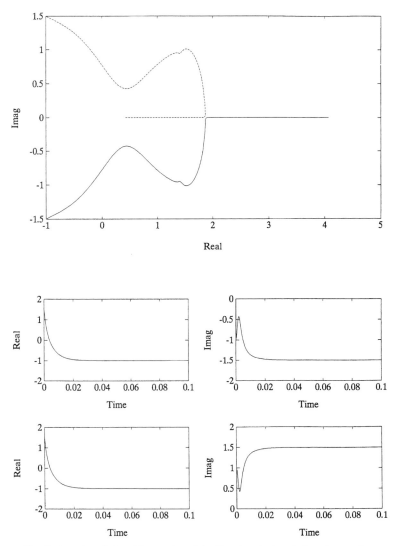

Figure 12 Trajectories of closed-loop poles of the time-invariant control system on the complex plane in Example 4. Reprinted from *Automatica*, J. Wang and G. Wu, "A multilayer recurrent neural network for synthesis of minimum-norm linear feedback control systems via pole placement," 32:435–442, Copyright 1996, with kind permission from Elsevier Science Ltd, The Boulevard, Langford Lane, Kidlington OX5 1GB, UK.

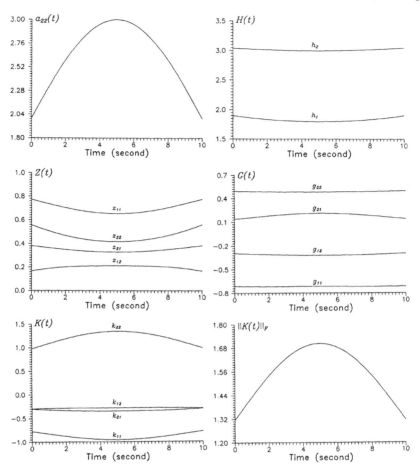

Figure 13 Transient states of the recurrent neural network and the time-varying control system in Example 4. Reprinted from *Automatica*, J. Wang and G. Wu, "A multilayer recurrent neural network for synthesis of minimum-norm linear feedback control systems via pole placement," 32:435–442, Copyright 1996, with kind permission from Elsevier Science Ltd, The Boulevard, Langford Lane, Kidlington OX5 1GB, UK.

sient behavior of the neural network and control system, where the dotted lines again are associated with the feedback control with constant optimal gain matrix which does respond to the change of A. It is interesting to note that all the state variables in the neural network behave with the same pattern as $a_{22}(t)$. Figure 14 illustrates the transient behavior of the closed-loop poles. Comparing Fig. 14 with Fig. 12, we can see that the patterns of convergence are identical. The trajectories of closed-loop poles on the complex plane are thus also identical. Because larger

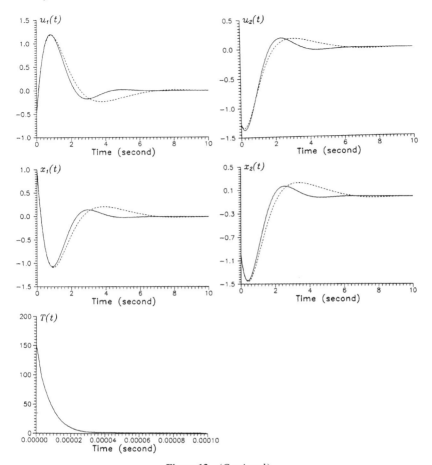

Figure 13 (*Continued*)

scaling constants are used in this example, convergence time is shorter. Table I lists the off-line computed minimal norm $\|K\|_F$ and the on-line generated norm $\|K(t)\|_F$ using the proposed neural network at several sample time instants, where $Q(t) \overset{\Delta}{=} (\|\bar{K}\|_F - \|K(t)\|_F)/\|K\|_F \times 100$ (%) is a performance index (relative error). It shows that the relative error of norm of the gain matrix is always less than 1.6%.

EXAMPLE 5. Consider another open-loop unstable system:

$$A = \begin{pmatrix} 2 & -1 \\ 3 & 0 \end{pmatrix}, \qquad B = \begin{pmatrix} 6 \\ -2 \end{pmatrix}.$$

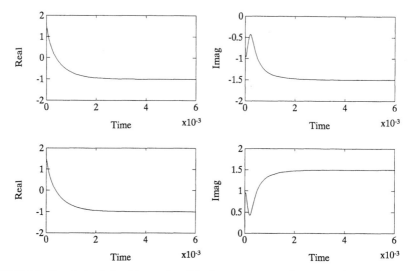

Figure 14 Transients of closed-loop poles of the time-varying control system on the complex plane in Example 4. Reprinted from *Automatica*, J. Wang and G. Wu, "A multilayer recurrent neural network for synthesis of minimum-norm linear feedback control systems via pole placement," 32:435–442, Copyright 1996, with kind permission from Elsevier Science Ltd, The Boulevard, Langford Lane, Kidlington OX5 1GB, UK.

The open-loop poles of the system are $1 \pm \sqrt{2}i$. Suppose that the desired closed-loop poles are $-1 \pm 1i$ and $\|K\|_F^2$ is chosen as the objective function. The optimal solution is

$$h = \begin{pmatrix} 2.595072 \\ 3.000106 \end{pmatrix}, \quad Z = \begin{pmatrix} -0.863324 & -0.407617 \\ 0.726984 & 3.104514 \end{pmatrix},$$
$$G = (0.620762 \quad 0.577340), \quad K = (-0.632353 \quad 0.102941),$$

with $\|K\|_F = 0.640677$. The reference input $r(t)$ in this example is again a unit

Table I

The Frobenius Norms of Feedback Gain Matrix at Sample Time Instants

t (s)	0.81	1.67	2.70	5.00	7.30	8.33	9.20	10.00
$a_{22}(t)$	2.25	2.50	2.75	3.00	2.75	2.50	2.25	2.00
$\|K\|_F$	1.40	1.49	1.59	1.68	1.59	1.49	1.40	1.30
$\|K(t)\|_F$	1.41	1.51	1.60	1.70	1.60	1.51	1.41	1.32
$Q(t)$ (%)	0.71	1.34	0.63	1.19	0.63	1.34	0.71	1.54

step vector. Using the developed neural network simulator, choosing the initial values

$$h(0) = \begin{pmatrix} 1 \\ 1 \end{pmatrix}, \qquad G(0) = (0 \quad 0),$$

$$Z(0) = \begin{pmatrix} 1 & 0 \\ 0 & 1 \end{pmatrix}, \qquad K(0) = (0 \quad 0),$$

and the design parameters $\{\mu_k, \mu_t, \mu_z, \mu_g, \mu_h\} = \{200, 1000, 1000, 1000, 1000\}$, the simulated neural network generates the solution

$$\bar{h} = \begin{pmatrix} 8.859943 \\ 7.202431 \end{pmatrix}, \qquad \overline{Z} = \begin{pmatrix} -0.497774 & -0.285162 \\ 0.267261 & 1.867978 \end{pmatrix},$$
$$\overline{G} = (0.335958 \quad 0.372615), \qquad \overline{K} = (-0.632353 \quad 0.102941),$$

with $\|\overline{K}\|_F = 0.640677$. It is worth mentioning that matrices $\overline{K}$ and K are identical even though matrices $\bar{h}$, $\overline{Z}$, and $\overline{G}$ are different from h, Z, and G.

If a_{12} in matrix A jumps from -1 to 0 at time equal to 4 s, the eigenvalues of the closed-loop system change to $\{0.806850, -2.068497\}$; hence, the system becomes unstable. The optimal solution then is

$$h = \begin{pmatrix} 3.188950 \\ 4.158206 \end{pmatrix}, \qquad Z = \begin{pmatrix} -0.713735 & -1.218704 \\ -0.687865 & 3.949039 \end{pmatrix},$$
$$G = (0.559985 \quad 0.490396), \qquad K = (-0.696972 \quad -0.090909),$$

with $\|K\|_F = 0.702876$. Using the preceding stable values as the initial values of the matrices h, G, Z, and K, and the same design parameters as before, the initial value of $x(t)$ as 0, and the input as a step function, the on-line neural network solution to the problem is obtained as

$$\bar{h} = \begin{pmatrix} 10.007151 \\ 11.040638 \end{pmatrix}, \qquad \overline{Z} = \begin{pmatrix} -0.388433 & -0.731390 \\ -0.499276 & 2.296805 \end{pmatrix},$$
$$\overline{G} = (0.316115 \quad 0.300956), \qquad \overline{K} = (-0.696972 \quad -0.090909),$$

and $\|\overline{K}\|_F = 0.702876$. Figure 15 shows the transient behavior of the neural network and control system where the dotted line curves are associated with the constant optimal feedback gain matrix. Figure 16 shows that the control and state variables of the control system using the constant K and time-varying $K(t)$ are almost identical before the coefficient in A changes. Because feedback control with a constant gain matrix K cannot react to the change of A at time greater than 4 s, the control signals and the state variables become unstable as shown in Fig. 15. Figure 16 illustrates the root locus of the feedback control system in the dynamic pole placement process. It shows that the closed-loop poles move from the open-loop poles [because $K(0) = 0$] to the real axis first and then move into

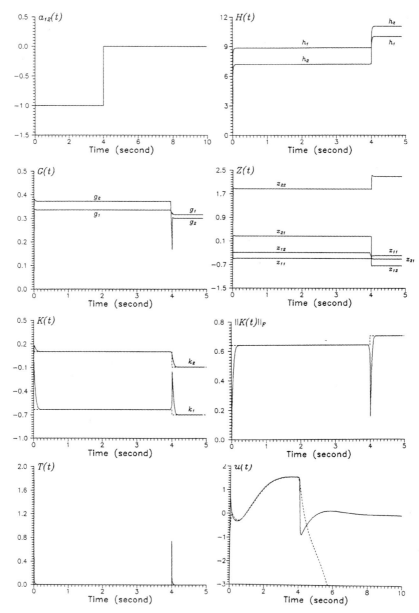

Figure 15 Transient states of the recurrent neural network and the feedback control system in Example 5.

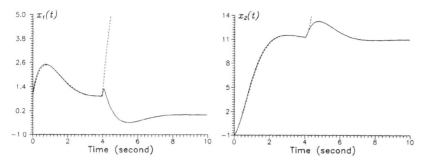

Figure 15 (*Continued*)

the complex plane toward the desired positions. When one element in A changes suddenly, the closed-loop poles jump to $\{0.806850, -2.068497\}$, then move into the complex plane again, and approach the desired positions. Figure 16 also illustrates the transient behavior of the closed-loop poles. It shows that the response time to the abrupt change is very short.

EXAMPLE 6. Consider a linear system with the coefficient matrices

$$A = \begin{pmatrix} -1 & 1 & 2 \\ 0 & -3 & -2 \\ -0.1 & -1 & -2 \end{pmatrix}, \quad B = \begin{pmatrix} 1 & 0 \\ 0 & 3 \\ 2 & 1 \end{pmatrix}, \quad C = \begin{pmatrix} 0 & 1 & 0 \\ 1 & 0 & 0 \\ 0 & 0 & 1 \end{pmatrix}.$$

The poles of the linear system are -3.9552 and $-1.0224 \pm 0.2592i$. Suppose that the desired poles of the asymptotic state estimator are $-10, -10 \pm 2i$; that is,

$$\Lambda = \begin{pmatrix} -10 & 2 & 0 \\ -2 & -10 & 0 \\ 0 & 0 & -10 \end{pmatrix}.$$

Let

$$G = \begin{pmatrix} 1 & -2 & 0 \\ 0 & 0 & -1 \\ 1 & 0 & 0 \end{pmatrix}.$$

Then the solutions to Eqs. (12) and (13) are

$$Z = \begin{pmatrix} 0.211624 & -0.158669 & -0.050602 \\ -0.045162 & 0.075769 & 0.167883 \\ -0.000410 & -0.148071 & -0.036915 \end{pmatrix},$$

$$L = \begin{pmatrix} 0.161173 & 9.583133 & 0.355202 \\ 7.597003 & -0.754098 & -1.701226 \\ -3.385327 & 2.918289 & 6.819862 \end{pmatrix}.$$

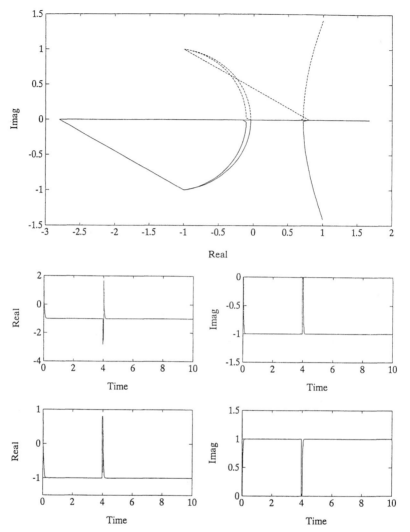

Figure 16 Trajectories of closed-loop poles of the feedback control system on the complex plane in Example 5.

Let the initial state matrices $Z(0) = 0$, $L(0) = 0$, $x(0) = \hat{x}(0) = 0$, and $u(t)$ be the step function defined as $u(t) = 0$ if $t < 0$ or $u(t) = 1$ if $t \geq 0$, let the activation function f_{ij} be the linear function $f_{ij}(y) = y$, and let the design parameters $\mu_l = \mu_z = 1000$. The simulated two-layer recurrent neural network yields the identical solution as Z and L [i.e., $Z(\infty) = Z$ and $L(\infty) = L$] and the corresponding asymptotic state estimator yields the exact asymptotic estimate of the state [i.e., $\hat{x}(\infty) = x(\infty) = (3.636379, 0.181758, 1.227333)^T$]. Figure 17 depicts the transients of the states of the recurrent neural network, the asymptotic state estimator, and the linear system. In the last three subplots of Fig. 17, the solid lines correspond to the estimated states using the two-layer recurrent neural network and the dashed lines correspond to the true states. Figure 18 depicts the poles of the asymptotic state estimator on the complex plane. It shows that the

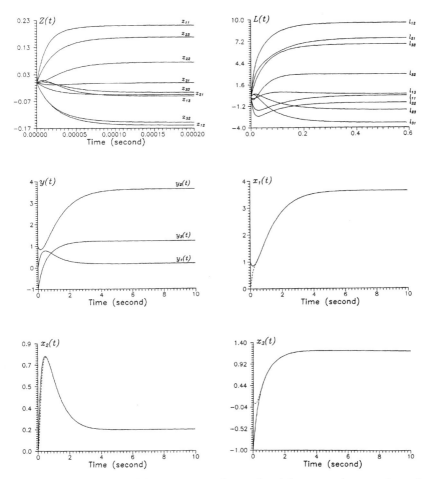

Figure 17 Transient states of the recurrent neural network and the asymptotic state estimator in Example 6. Reprinted with permission from J. Wang and G. Wu, *Internat. J. Systems Sci.* 26:1205–1222, 1995 (© 1995 Taylor & Francis Ltd.).

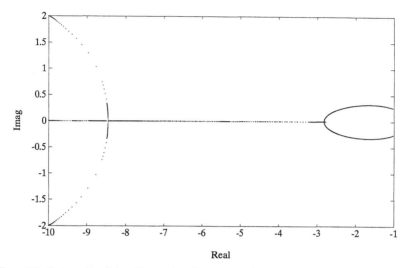

Figure 18 Trajectories of closed-loop poles of the asymptotic state estimator on the complex plane in Example 6. Reprinted with permission from J. Wang and G. Wu, *Internat. J. Systems Sci.* 26:1205–1222, 1995 (© 1995 Taylor & Francis Ltd.).

poles of the asymptotic state estimator move from the poles of the linear systems [because $L(0) = 0$] to the desired ones.

EXAMPLE 7. Consider another linear system,

$$A = \begin{pmatrix} -1.3 & 2.1 \\ 0.1 & -1.2 \end{pmatrix}, \qquad B = \begin{pmatrix} 1 & 1 \\ 1 & -1 \end{pmatrix}, \qquad C = \begin{pmatrix} 1 & 2 - 3s(t-4) \\ 3 & 1 \end{pmatrix},$$

where $s(t)$ is a step function. Note that c_{12} changes from 2 to -1 when $t = 4$. The poles of the linear system are -1.7110 and -0.7890. Suppose that the two desired poles of the state estimator are -9; that is,

$$\Lambda = \begin{pmatrix} -9 & 0 \\ 0 & -9 \end{pmatrix}.$$

Let

$$G = \begin{pmatrix} 4 & 2 \\ 1 & 5 \end{pmatrix}.$$

Then the solutions to Eqs. (12) and (13) when $t < 4$ are

$$Z' = \begin{pmatrix} -1.286550 & -0.935673 \\ -2.073517 & -0.339181 \end{pmatrix}, \qquad L' = \begin{pmatrix} -2.80 & 2.66 \\ 4.66 & -1.52 \end{pmatrix}.$$

Using the developed neural network simulator, choosing the initial values $Z(0) = L(0) = 0$ and $x(0) = \hat{x}(0) = 0$, the control $u(t) = [\sin(\pi t), 1.5\sin(0.3\pi t + \pi/3)]$, the activation function f_{ij} to be the linear function $f_{ij}(y) = y$, and the design parameters $\mu_l = \mu_z = 1000$, the simulated two-layer neural network generates the identical solution as Z' and L' when $t \to 4$; that is, $\lim_{t \to 4} Z(t) = Z'$ and $\lim_{t \to 4} L(t) = L'$.

Whereas c_{12} in matrix C changes from 2 to -1 when $t = 4$, the eigenvalues of the state estimator change to $\{0.806850, -2.068497\}$; hence, the state estimator becomes unstable. The solution based on the changed C then is

$$Z'' = \begin{pmatrix} -1.306600 & 0.608187 \\ -2.078530 & 0.046784 \end{pmatrix}, \qquad L'' = \begin{pmatrix} 0.350 & 2.450 \\ -5.825 & 1.925 \end{pmatrix}.$$

Using the preceding stable values as the values of the matrices $Z(4)$ and $L(4)$, and the same design parameters as before, the on-line neural network solutions to Eqs. (12) and (13) are obtained as the same as Z'' and L''. Figure 19 shows the transient behavior of the two-layer neural network and the corresponding linear state observer. The third and fourth subplots depict the estimated states using the precomputed constant L (the dashed lines) and using the time-varying $L(t)$ generated by the recurrent neural network (the solid lines). Figure 19 shows that the control and state variables of the linear system using the constant L and time-varying $L(t)$ are almost identical before the coefficient in C changes. Because state estimation with a constant gain matrix L cannot react to the change of C when $t > 4$, the estimated state variables based on constant L become unstable as shown in Fig. 19. The last three subplots of Fig. 19 illustrate the transients of the poles of the synthesized state estimator (the last three subplots are zoomed to illustrate the transient behavior), where the solid and dashed lines represent two different poles. Figure 19 shows that the abrupt change of c_{22} causes a displacement of one pole and the response time of the neural network to the pole displacement is very short.

EXAMPLE 8. Consider another linear system:

$$A = \begin{pmatrix} -1 & 1 \\ -3 & -4 \end{pmatrix}, \qquad B = \begin{pmatrix} 1 & 0 \\ 2 & -1 \end{pmatrix}, \qquad C = \begin{pmatrix} 2 & 1 \\ 0 & 2 - 4s(t-4) \end{pmatrix}.$$

Note that c_{22} in C drops from 2 to -2 when $t = 4$. The eigenvalues of A are $-2.5 \pm 0.866025i$. Let the eigenvalues of the state observer be $-10 \pm 4i$; that is,

$$\Lambda = \begin{pmatrix} -10 & -4 \\ 4 & -10 \end{pmatrix}.$$

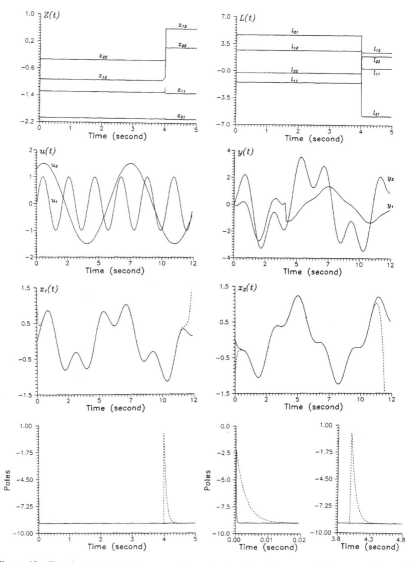

Figure 19 Transient states of the recurrent neural network and the asymptotic state estimator in Example 7. Reprinted with permission from J. Wang and G. Wu, *Internat. J. Systems Sci.* 26:1205–1222, 1995 (© 1995 Taylor & Francis Ltd.).

Let $u(t)$ be a step function, $G(0) = Z(0) = L(0) = I$, $h(0) = 0$, the activation function f_{ij} be the linear function $f_{ij}(y) = y$, and the design parameters be $\{\mu_t, \mu_l, \mu_z, \mu_g, \mu_h\} = \{1.3 \times 10^5, 10^6, 10^6, 10^6, 10^6\}$. Then the simulation result shows

$$h(4^-) = 107.182751, \qquad G(4^-) = \begin{pmatrix} -0.010622 & 0.033404 \\ 0.086982 & -0.023139 \end{pmatrix},$$

$$Z(4^-) = \begin{pmatrix} 0.008490 & -0.005723 \\ -0.018084 & -0.007585 \end{pmatrix},$$

$$L(4^-) = \begin{pmatrix} 3.444713 & -2.297795 \\ 3.254566 & 2.427751 \end{pmatrix},$$

$\|L(4^-)\|_F = 5.799318$, and $\hat{x}(4^-) = x(4^-) = (0.714278, -0.285688)^T$. A comparison of $\|L(4^-)\|_F = 5.799318$ with $\|L\|_F = 5.6680036$ shows that the four-layer neural network provides a near optimal solution.

When $t \geq 4$, the simulator yields the solutions

$$h(\infty) = 188.763275, \qquad G(\infty) = \begin{pmatrix} -0.017665 & -0.027485 \\ 0.063626 & 0.013486 \end{pmatrix},$$

$$Z(\infty) = \begin{pmatrix} 0.008355 & -0.003413 \\ -0.012525 & -0.006297 \end{pmatrix},$$

$$L(\infty) = \begin{pmatrix} 3.443513 & 2.296690 \\ 3.253164 & -2.425448 \end{pmatrix},$$

$\|L(\infty)\|_F = 5.796416$, and $\hat{x}(\infty) = x(\infty) = (0.714286, -0.285714)^T$. A comparison of $\|L(4^-)\|_F = 5.796416$ with $\|L\|_F = 5.6679946$ shows that the four-layer neural network again provides a near optimal solution. Figure 20 illustrates the transient behavior of the four-layer recurrent neural network and the asymptotic state estimator. In the last two subplots of Fig. 20, the solid lines correspond to the asymptotic state estimator synthesized using the four-layer recurrent neural network. The dashed lines in the seventh subplot represent the true states and the dashed lines in the eighth (last) subplot represent the estimated states using the constant L. To compare the performances of the state estimators based on the recurrent neural network and the constant L, the initial states of the state estimator using the constant L was purposely chosen to be the same as the true states, but the initial states of the state estimator synthesized by the neural network were chosen as zero. As shown in the last subplot, the estimated states based on the constant L become unstable after the abrupt change of c_{22}. Figure 21 illustrates the trajectories of the poles of the state estimator. It shows that the neural network can quickly adapt to the abrupt change of the output parameter.

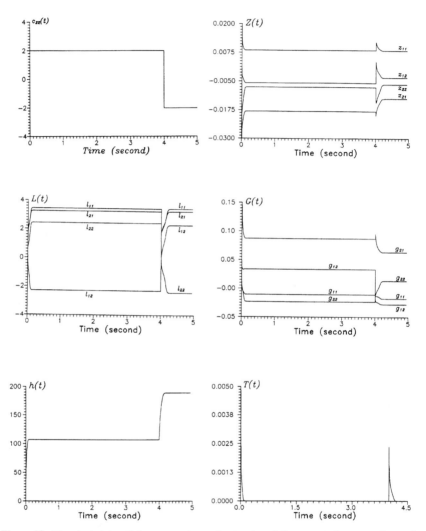

Figure 20 Transient states of the recurrent neural network and the asymptotic state estimator in Example 8. Reprinted with permission from J. Wang and G. Wu, *Internat. J. Systems Sci.* 26:1205–1222, 1995 (© 1995 Taylor & Francis Ltd.).

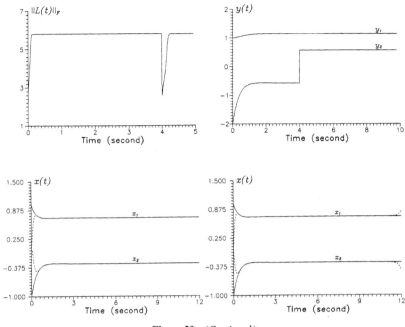

Figure 20   (*Continued*)

VII. CONCLUDING REMARKS

The proposed multilayer recurrent neural networks have been shown to be capable of synthesizing linear control systems via pole placement in real time. Compared with the traditional pole assignment methods, the proposed neural network approach is advantageous because on-line synthesis and autotuning are performed by using the recurrent neural networks. Compared with other neural network approaches to synthesis of linear control systems, the proposed recurrent neural network paradigm is autonomous for pole placement and/or norm minimization without the need for training. These features are especially desirable for the control of time-varying multivariable systems. Because of the inherently parallel distributed nature of neural computation, the proposed recurrent neural networks can be used for many real-time control applications. Further investigation can be directed to applications of the deterministic annealing neural network for pole assignment in robust and nonlinear control system synthesis. The computational power of neural networks makes them desirable tools for real-time synthesis of some nonlinear systems via on-line local pole assignment.

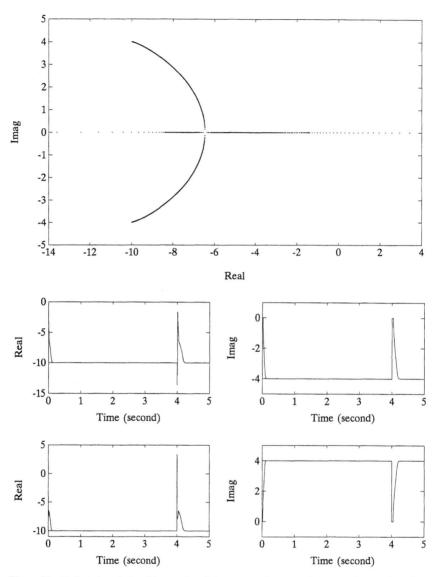

Figure 21 Trajectories of closed-loop poles of the asymptotic state estimator on the complex plane in Example 8. Reprinted with permission from J. Wang and G. Wu, *Internat. J. Systems Sci.* 26:1205–1222, 1995 (© 1995 Taylor & Francis Ltd.).

REFERENCES

[1] K. S. Narendra and K. Parthasarathy. Identification and control of dynamical systems using neural networks. *IEEE Trans. Neural Networks*1:4–26, 1990.

[2] P. J. Werbos. Neurocontrol and related techniques. In *Handbook of Neural Computing Applications* (A. J. Maren, C. T. Harston, and R. M. Pap, Eds.). Academic Press, New York, 1990.

[3] N. Bhat and T. J. McAvoy. Use of neural nets for dynamic modeling and control of chemical systems. *Comput. Chem. Eng.* 14:573–583, 1990.

[4] W. T. Miller, III, R. S. Sutton, and P. J. Werbos, Eds. *Neural Networks for Control*. MIT Press, Cambridge, MA, 1990.

[5] D. H. Nguyen and B. Widrow. Neural networks for self-learning control systems. *Internat. J. Control* 54:1439–1451, 1991.

[6] S. S. Kumar and A. Guez. ART based adaptive pole assignment for neurocontrollers. *Neural Networks* 4:319–335, 1991.

[7] J. Saint-Donnat, N. Bhat, and T. J. McAvoy. Neural net based model predictive control. *Internat. J. Control* 54:1453–1468, 1991.

[8] S. Chen and S. A. Billings. Neural networks for nonlinear dynamic system modeling and identification. *Internat. J. Control* 56:263–289, 1992.

[9] Y.-H. Pao, S. M. Philips, and D. J. Sobajic. Neural-net computing and the intelligent control of systems. *Internat. J. Control* 56:263–289, 1992.

[10] E. Tzirkel-Hancock and F. Fallside. Stable control of nonlinear systems using neural networks. *Internat. J. Robust Nonlinear Control* 2:63–86, 1992.

[11] K. Watanabe, T. Fukuda, and S. G. Tzafestas. An adaptive control for CARMA systems using linear neural networks. *Internat. J. Control* 56:483–497, 1992.

[12] A. U. Levin and K. S. Narendra. Control of nonlinear dynamic systems using neural networks: controllability and stabilization. *IEEE Trans. Neural Networks* 4:192–206, 1993.

[13] J. G. Kuschewski, S. Hui, and S. H. Żak. Application of feedforward neural networks to dynamical system identification and control. *IEEE Trans. Control Systems Technol.*1:37–49, 1993.

[14] C. J. Goh. On the nonlinear optimal regulator problem. *Automatica* 29:751–756, 1993.

[15] K. J. Hunt, D. Sbarbaro, R. Zbikowski, and P. J. Gawthrop. Neural networks for control systems — A survey. *Automatica* 28:1083–1112, 1992.

[16] S. N. Balakrishnan and R. D. Weil. Neurocontrol: a literature survey. *Math. Comput. Modeling* 23:101-117, 1996.

[17] J. J. Hopfield and D. W. Tank. Neural computation of decisions in optimization problems. *Biol. Cybernetics* 52:141–152, 1985.

[18] D. W. Tank and J. J. Hopfield. Simple neural optimization networks: an A/D converter, signal decision circuit, and a linear programming circuit. *IEEE Trans. Circuits Systems* 33:533–541, 1986.

[19] M. Sznaier and M. J. Damborg. An analog "neural network" based suboptimal controller for constrained discrete-time linear systems. *Automatica* 28:139–144, 1992.

[20] J. Wang and G. Wu. Multilayer recurrent neural networks for on-line synthesis of asymptotic state estimators for linear dynamic systems. *Internat. J. Systems Sci.* 26:1205–1222, 1995.

[21] J. Wang and G. Wu. Recurrent neural networks for synthesizing linear control systems via pole placement. *Internat. J. Systems Sci.* 26:2369–2382, 1995.

[22] J. Wang and G. Wu. A multilayer recurrent neural network for on-line synthesis of linear control systems via pole assignment. *Automatica* 32:435–442, 1996.

[23] W. M. Wonham. On pole assignment in multi-input, controllable linear systems. *IEEE Trans. Automat. Control* 12:660–665, 1967.

[24] S. P. Bhattacharyya and E. de Souza. Pole assignment via Sylvester's equation. *Systems Control Lett.* 1:261–263, 1982.

[25] C.-T. Chen. *Linear System Theory and Design*. Holt, Rinehart, Winston, New York, 1984.

[26] L. H. Keel, J. A. Fleming, and S. P. Bhattacharyya. Minimum norm pole assignment via Sylvester equation. In *Linear Algebra and Its Role in Systems Theory* (R. A. Brualdi, D. H. Carlson, B. N. Datta, C. R. Johnson, and R. J. Plemmons, Eds.), pp. 265–272. American Mathematical Society, Providence, RI, 1985.

[27] G.-R. Duan. Solutions of the equation $AV + BW = VF$ and their application to eigenstructure assignment in linear systems. *IEEE Trans. Automat. Control* 38:276–280, 1993.

[28] R. V. Patel, A. J. Laub, and P. M. Van Mooren, Eds. *Numerical Linear Algebra Techniques for Systems and Control*. IEEE Press, New York, 1994.

[29] F. L. Luo and B. Zheng. Neural network approach to computing matrix inversion. *Appl. Math. Computation* 47:109–120, 1992.

[30] J. Wang. A recurrent neural network for real-time matrix inversion. *Appl. Math. Computation* 55:89–100, 1993.

[31] J. Wang. Recurrent neural networks for solving linear matrix equations. *Comput. Math. Appl.* 26:23–34, 1993.

[32] A. Cichocki and R. Unbehauen. Neural networks for solving systems of linear equations and related problems. *IEEE Trans. Circuits Systems* 39:124–138, 1992.

[33] J. Wang. Electronic realization of a recurrent neural network for solving simultaneous linear equations. *Electron. Lett.* 28:493–495, 1992.

[34] J. Wang and H. Li. Solving simultaneous linear equations using recurrent neural networks. *Inform. Sci.* 76:255–278, 1993.

[35] L. X. Wang and J. M. Mendel. Three-dimensional structured network for matrix equation solving. *IEEE Trans. Comput.* 40:1337–1345, 1991.

[36] L. X. Wang and J. M. Mendel. Parallel structured networks for solving a wide variety of matrix algebra problems. *J. Parallel Distributed Comput.* 14:236–247, 1992.

[37] J. Wang and G. Wu. Recurrent neural networks for LU decomposition and Cholesky factorization. *Math. Comput. Modeling* 18:1–8, 1993.

[38] J. Wang. A deterministic annealing neural network for convex programming. *Neural Networks* 7:629–641, 1994.

[39] J. Wang. Analysis and design of a recurrent neural network for linear programming. *IEEE Trans. Circuits Systems I Fund. Theory Appl.* 40:613–618, 1993.

[40] M. P. Kennedy and L. O. Chua. Neural networks for nonlinear programming. *IEEE Trans. Circuits Systems* 35:554–562, 1988.

[41] C. Y. Maa and M. A. Shanblatt. Linear and quadratic programming neural network analysis. *IEEE Trans. Neural Networks* 3:580–594, 1992.

[42] J. Wang. On the asymptotic properties of recurrent neural networks for optimization. *Internat. J. Pattern Recognition Artificial Intelligence* 5:581–601, 1991.

[43] J. Wang and V. Chankong. Recurrent neural networks for linear programming: analysis and design principles. *Comput. Oper. Res.* 19:297–311, 1992.

Direct and Indirect Techniques to Control Unknown Nonlinear Dynamical Systems Using Dynamical Neural Networks

George A. Rovithakis
Department of Electronic and
Computer Engineering
Technical University of Crete
73100 Chania, Crete, Greece

Manolis A. Christodoulou
Department of Electronic and
Computer Engineering
Technical University of Crete
73100 Chania, Crete, Greece

I. INTRODUCTION

Humans have two principal objectives in the scientific study of their environment: to understand and to control. The two goals reinforce each other, because deeper understanding permits firmer control, and, on the other hand, systematic application of scientific theories inevitably generates new problems which require further investigation, and so on.

It might be assumed that a fine-grained descriptive theory of terrestrial phenomena would be required before an adequate theory of control could be constructed. In actuality this is not the case, and, indeed, circumstances themselves force us into situations where we must exert regulatory and corrective influences without complete knowledge of basic causes and effects. In connection with the design of experiments, space travel, economics, and the study of cancer, we encounter processes which are not fully understood, yet design and control decisions

127

are required. It is easy to see that in the treatment of complex processes, attempts at complete understanding at a basic level may consume so much time and so large a quantity of resources as to impede us in more immediate goals of control.

Artificial neural networks have been studied for many years with the hope of achieving humanlike performance in solving certain problems in speech and image processing. There has been a recent resurgence in the field of neural networks due to the introduction of new network topologies, training algorithms, and very large scale integration (VLSI) implementation techniques. The potential benefits of neural networks such as parallel distributed processing, high computation rates, fault tolerance, and adaptive capability, have lured researchers from other fields such as controls, robotics, etc., to seek solutions to their complicated problems, mostly through empirical studies with unfortunately insufficient theoretical justification. Bridging the gap between theory and applications was mentioned by Hunt *et al.* [1], where the need to answer crucial questions like stability, convergence, and robustness, as well as the relation between neural and adaptive control, was pointed out.

The key relationship between neural and adaptive control arises from the fact that neural networks can approximate arbitrarily well static and dynamic highly nonlinear systems. Thus one can substitute an unknown system with a neural network model, which is of known structure but contains a number of unknown parameters, synaptic weights, plus a modeling error term. The unknown parameters may appear both linearly or nonlinearly with respect to the network nonlinearities, thus transforming the original problem into a nonlinear robust adaptive control problem. Several answers to the problem of nonlinear adaptive control exist in the literature; typical examples can be found in [2–8]. A common assumption made in the previous works is linear parameterization. An attempt to relax this assumption and provide global adaptive output feedback control for a class of nonlinear systems determined by specific geometric conditions was given by Marino and Tomei [9]. However, to our knowledge, the robust nonlinear adaptive control problem is generally left as an open problem.

The use of dynamical neural networks for identification and control was first introduced by Narendra and Parthasarathy [10], who proposed dynamic backpropagation schemes, which are static backpropagation neural networks, connected either in series or in parallel with linear dynamical systems. However, their method requires a great deal of computational time and, furthermore, lacks theoretical verification that simulations provide. Sontag [11] showed that despite what might be expected from the well known representation theorems [12–14], single hidden layer nets are not sufficient for stabilization, whereas two hidden layer nets are, provided that its neurons contain threshold activation functions. Therefore, Sontag did not provide a practical stabilization algorithm, but rather explored the capabilities and limitations of alternative network architectures. Sanner and Slotine [15] incorporated Gaussian radial-basis-function neural networks with sliding mode control and linear feedback to formulate a direct adaptive tracking control architecture for a class of continuous time nonlinear dynamic systems. However, use of the sliding mode, which is a discontinuous control law, generally creates various problems, such as existence and uniqueness of solutions [16], introduction of chattering phenomena [17], and possibly excitation of high frequency unmodeled dynamics [18]. Polycarpou and Ioannou [19] employed Lyapunov stability theory to develop stable adaptive laws for identification and control of dynamical systems with unknown nonlinearities using various neural network architectures. Their control results were restricted to single-input–single-output (SISO) feedback linearizable systems.

In this chapter we discuss direct and indirect techniques for controlling unknown nonlinear dynamical systems. The material herein is mainly based on previous work of the authors on the subject [20, 21].

In the indirect control case, a dynamic neural network identifier performs online identification that provides information to the controller which is a dynamic state feedback designed to guarantee convergence of the control error to zero and robustness even when unmodeled dynamics are present and affect the system performance.

In the direct control case, we analyze how well the adaptive regulator will work when model mismatch exists. It is clearly shown that when we have complete matching at zero, we are able to prove convergence of the state to zero plus boundedness of all other signals in the closed loop. However, when the preceding assumption fails, our adaptive regulator can still assure uniform ultimate boundedness. We further examine the behavior of our adaptive regulator in the case where the true plant is of higher order than assumed; in other words, when unmodeled dynamics are present. However, we do not confine the unmodeled dynamics in the framework of singular perturbation theory, as we did in Section II. In this way, we relax the assumption made in Section II, while we expand the space in which the unmodel dynamics live. Observe that not all the plant states are assumed to be available for measurement; this obviously refers to the state of the unmodeled dynamics.

A. Notation

The following notation and definitions will be used extensively throughout the paper. I denotes the identity matrix and $|\cdot|$ denotes the usual Euclidean norm of a vector. In the case where y is a scalar, $|y|$ denotes its absolute value. If A is a matrix, then $\|A\|$ denotes the Frobenius matrix norm [22], defined as

$$\|A\|^2 = \sum_{ij} |a_{ij}|^2 = \mathrm{tr}\{A^T A\},$$

where tr$\{\cdot\}$ denotes the trace of a matrix. Now let $d(t)$ be a vector function of time. Then

$$\|d\|_2 \overset{\triangle}{=} \left(\int_0^\infty |d(\tau)|^2 \, d\tau \right)^{1/2}$$

and

$$\|d\|_\infty \overset{\triangle}{=} \sup_{t \geq 0} |d(t)|.$$

We will say that $d \in L_2$ when $\|d\|_2$ is finite. Similarly, we will say that $d \in L_\infty$ when $\|d\|_\infty$ is finite.

We also recall from [23] the following definition:

DEFINITION 1.   The solutions of $\dot{x} = f(x, t)$ are said to be uniformly ultimately bounded if there exist constants b and c and for every $\alpha \in (0, c)$ there is a constant $T(\alpha)$ such that

$$|x(t_0)| < \alpha \quad \Rightarrow \quad |x(t)| < b, \qquad \forall t > t_0 + T.$$

They are said to be globally uniformly ultimately bounded if the preceding condition holds for arbitrarily large α.

II. PROBLEM STATEMENT AND THE DYNAMIC NEURAL NETWORK MODEL

We consider affine in the control, nonlinear dynamical systems of the form

$$\dot{x} = f(x) + G(x)u, \tag{1}$$

where the state x, living in an n-dimensional smooth manifold $\mathcal{M}$, is assumed to be completely measured, the control u is in $\mathfrak{R}^n$, f is an unknown smooth vector field called the drift term, and G is a matrix where the columns are the unknown smooth controlled vector fields g_i, $i = 1, 2, \ldots, n$, $G = [g_1 \quad g_2 \quad \cdots \quad g_n]$.

The state regulation problem, which is discussed in the direct case, is known to reduce the state to zero from an arbitrary initial value by applying feedback control to the plant input. On the other hand, in the tracking problem, which is studied in the indirect case, we want the state of the unknown system to follow a given reference trajectory. However, the problem, as previously stated for the system Eq. (1), is very difficult or even impossible to solve because the vector fields f, g_i, $i = 1, 2, \ldots, n$, are assumed to be completely unknown. Therefore, it is obvious that to provide a solution to our problem, it is necessary to have a more accurate model for the unknown plant. For that purpose we apply dynamical neural networks.

Dynamic neural networks are recurrent, fully interconnected nets that contain dynamical elements in their neurons. Therefore, they are described by the set of differential equations

$$\dot{\hat{x}} = -A\hat{x} + WS(x) + W_{n+1}S'(x)u,$$

where $\hat{x} \in \mathcal{M}$, the inputs $u \in \mathfrak{R}^n$, W is an $n \times n$ matrix of adjustable synaptic weights, W_{n+1} is an $n \times n$ diagonal matrix of adjustable synaptic weights of the form $W_{n+1} = \text{diag}[w_{1,n+1} \quad w_{2,n+1} \quad \cdots \quad w_{n,n+1}]$, and A is an $n \times n$ matrix with positive eigenvalues which for simplicity can be taken as diagonal. Finally, $S(x)$ is an n-dimensional vector and $S'(x)$ is an $n \times n$ diagonal matrix, with elements $s(x_i)$ and $s'(x_i)$, respectively, both smooth (at least twice differentiable), monotone increasing functions which are usually represented by sigmoidals of the form

$$s(x_i) = \frac{k}{1 + e^{-lx_i}},$$

$$s'(x_i) = \frac{k}{1 + e^{-lx_i}} + \lambda,$$

for all $i = 1, 2, 3, \ldots, n$, where k and l are parameters representing the bound (k) and slope (l) of the sigmoid's curvature and $\lambda > 0$ is a strictly positive constant that shifts the sigmoid, such that $s'(x_i) > 0$ for all $i = 1, 2, \ldots, n$.

Due to the approximation capabilities of the dynamic neural networks, we can assume, with no loss of generality, that the unknown system Eq. (1) can be described completely by a dynamical neural network plus a modeling error term $\omega(x, u)$. In other words, there exist weight values W^* and W_{n+1}^* such that the system Eq. (1) can be written as

$$\dot{x} = -Ax + W^*S(x) + W_{n+1}^*S'(x)u + \omega(x, u). \tag{2}$$

Therefore, the state regulation problem is analyzed for the system Eq. (2) instead of Eq. (1). Whereas W^* and W_{n+1}^* are unknown, our solution consists of designing a control law $u(W, W_{n+1}, x)$ and appropriate update laws for W and W_{n+1} to guarantee convergence of the state to zero and, in some cases that are analyzed in the following sections, boundedness of x and of all signals in the closed loop.

Remark 1. Observe that the dynamic neural networks used herein employ only first order sigmoid functions (we do not have products of sigmoid functions up to arbitrary order) as entries in $S(x)$ and $S'(x)$. Moreover, the unknown system should have the same number of control inputs as the number of measurable states. Extensions to more general cases can be found in [24].

III. INDIRECT CONTROL

A. IDENTIFICATION

In this subsection we discuss the case where only parametric uncertainties are present. For more information concerning more general cases, the interested reader is referred to [20]. So let us assume that an exact model of the plant is available (i.e., we have no modeling error). The purpose of this section is to find a learning law that guarantees stability of the neural network plus convergence of its output and weights to a desired value. Whereas we have only parametric uncertainties, we can assume that there exist weight values W^* and W_{n+1}^* such that the system Eq. (1) is completely described by a neural network of the form

$$\dot{x} = Ax + W^* S(x) + W_{n+1}^* S'(x)u, \tag{3}$$

where all matrices are as defined earlier.

Define the error between the identifier states and the real system states as

$$\mathbf{e} = \hat{\mathbf{x}} - \mathbf{x}.$$

Then from Eqs. (2) and (3) we obtain the error equation

$$\dot{\mathbf{e}} = \mathbf{Ae} + \widetilde{\mathbf{W}}\mathbf{S(x)} + \widetilde{\mathbf{W}}_{n+1}\mathbf{S'(x)u}, \tag{4}$$

where

$$\widetilde{\mathbf{W}} = \mathbf{W} - \mathbf{W}^*,$$
$$\widetilde{\mathbf{W}}_{n+1} = \mathbf{W}_{n+1} - \mathbf{W}_{n+1}^*.$$

The Lyapunov synthesis method is used to derive stable adaptive laws. Therefore consider the Lyapunov function candidate

$$\mathcal{V}\big(\mathbf{e}, \widetilde{\mathbf{W}}, \widetilde{\mathbf{W}}_{n+1}\big) = \tfrac{1}{2}\mathbf{e}^T \mathbf{Pe} + \tfrac{1}{2}\mathrm{tr}\big\{\widetilde{\mathbf{W}}^T \widetilde{\mathbf{W}}\big\} + \tfrac{1}{2}\mathrm{tr}\big\{\widetilde{\mathbf{W}}_{n+1}^T \widetilde{\mathbf{W}}_{n+1}\big\}, \tag{5}$$

where $\mathbf{P} > 0$ is chosen to satisfy the Lyapunov equation

$$\mathbf{PA} + \mathbf{A}^T \mathbf{P} = -\mathbf{I}.$$

Observe that because $\mathbf{A}$ is a diagonal matrix, $\mathbf{P}$ can be chosen to be a diagonal matrix too, simplifying, in this way, the calculations. Differentiating Eq. (5) along the solution of Eq. (4) we obtain

$$\dot{\mathcal{V}} = \tfrac{1}{2}\big(\dot{\mathbf{e}}^T \mathbf{Pe} + \mathbf{e}^T \mathbf{P}\dot{\mathbf{e}}\big) + \mathrm{tr}\big\{\dot{\widetilde{\mathbf{W}}}^T \widetilde{\mathbf{W}}\big\} + \mathrm{tr}\big\{\dot{\widetilde{\mathbf{W}}}_{n+1}^T \widetilde{\mathbf{W}}_{n+1}\big\}$$
$$= \tfrac{1}{2}\big(-\mathbf{e}^T \mathbf{e} + \mathbf{S}^T(\mathbf{x})\widetilde{\mathbf{W}}^T \mathbf{Pe} + \mathbf{u}^T \mathbf{S'(x)}\widetilde{\mathbf{W}}_{n+1}\mathbf{Pe} + \big(\mathbf{S}^T(\mathbf{x})\widetilde{\mathbf{W}}^T \mathbf{Pe}\big)^T$$
$$+ \big(\mathbf{u}^T \mathbf{S'(x)}\widetilde{\mathbf{W}}_{n+1}\mathbf{Pe}\big)^T\big) + \mathrm{tr}\big\{\dot{\widetilde{\mathbf{W}}}^T \widetilde{\mathbf{W}}\big\} + \mathrm{tr}\big\{\dot{\widetilde{\mathbf{W}}}_{n+1}^T \widetilde{\mathbf{W}}_{n+1}\big\}.$$

Now because $\mathbf{S}^T(\mathbf{x})\widetilde{\mathbf{W}}^T\mathbf{Pe}$ and $\mathbf{u}^T\mathbf{S}'(\mathbf{x})\widetilde{\mathbf{W}}_{n+1}\mathbf{Pe}$ are scalars,

$$\mathbf{S}^T(\mathbf{x})\widetilde{\mathbf{W}}^T\mathbf{Pe} = (\mathbf{S}^T(\mathbf{x})\widetilde{\mathbf{W}}^T\mathbf{Pe})^T,$$

$$\mathbf{u}^T\mathbf{S}'(\mathbf{x})\widetilde{\mathbf{W}}_{n+1}\mathbf{Pe} = \left(\mathbf{u}^T\mathbf{S}'(\mathbf{x})\widetilde{\mathbf{W}}_{n+1}\mathbf{Pe}\right)^T.$$

Therefore,

$$\dot{V} = -\tfrac{1}{2}\mathbf{e}^T\mathbf{e} + \mathbf{S}^T(\mathbf{x})\widetilde{\mathbf{W}}^T\mathbf{Pe} + \mathbf{u}^T\mathbf{S}'(\mathbf{x})\widetilde{\mathbf{W}}_{n+1}\mathbf{Pe}$$
$$+ \operatorname{tr}\{\dot{\widetilde{\mathbf{W}}}^T\widetilde{\mathbf{W}}\} + \operatorname{tr}\{\dot{\widetilde{\mathbf{W}}}_{n+1}^T\widetilde{\mathbf{W}}_{n+1}\}. \tag{6}$$

Hence, if we choose

$$\operatorname{tr}\{\dot{\widetilde{\mathbf{W}}}^T\widetilde{\mathbf{W}}\} = -\mathbf{S}^T(\mathbf{x})\widetilde{\mathbf{W}}^T\mathbf{Pe}, \tag{7}$$

$$\operatorname{tr}\{\dot{\widetilde{\mathbf{W}}}_{n+1}^T\widetilde{\mathbf{W}}_{n+1}\} = -\mathbf{u}^T\mathbf{S}'(\mathbf{x})\widetilde{\mathbf{W}}_{n+1}\mathbf{Pe}, \tag{8}$$

then Eq. (6) becomes

$$\dot{V} = -\tfrac{1}{2}\mathbf{e}^T\mathbf{e} \tag{9}$$

or

$$\dot{V} = -\tfrac{1}{2}\|\mathbf{e}\|^2 \le 0. \tag{10}$$

From Eqs. (7) and (8) we obtain learning laws in an element form as

$$\dot{w}_{ij} = -p_i s(x_j)e_i,$$
$$\dot{w}_{in+1} = -s'(x_i)p_i u_i e_i,$$

for all $i, j = 1, 2, 3, \ldots, n$.

Now we can prove the following theorem:

THEOREM 1.   *Consider the identification scheme Eq. (4). The learning law*

$$\dot{w}_{ij} = -p_i s(x_j)e_i,$$
$$\dot{w}_{in+1} = -s'(x_i)p_i u_i e_i,$$

for all $i, j = 1, 2, 3, \ldots, n$, guarantees the properties

- $\mathbf{e}, \hat{\mathbf{x}}, \widetilde{\mathbf{W}}, \widetilde{\mathbf{W}}_{n+1} \in L_\infty, \mathbf{e} \in L_2;$
- $\lim_{t\to\infty}\mathbf{e}(t) = 0, \lim_{t\to\infty}\widetilde{\mathbf{W}}(t) = 0, \lim_{t\to\infty}\widetilde{\mathbf{W}}_{n+1}(t) = 0.$

Proof.   We have shown that using the learning law

$$\dot{w}_{ij} = -p_i s(x_j)e_i,$$
$$\dot{w}_{in+1} = -s(x_i)p_i u_i e_i,$$

for all $i, j = 1, 2, 3, \ldots, n$,

$$\dot{V} = -\tfrac{1}{2}\|\mathbf{e}\|^2 \le 0.$$

Hence, $\mathcal{V} \in L_\infty$, which implies $\mathbf{e}$, $\widetilde{\mathbf{W}}$, $\widetilde{\mathbf{W}}_{n+1} \in L_\infty$. Furthermore, $\hat{\mathbf{x}} = \mathbf{e} + \mathbf{x}$ is also bounded. Whereas $\mathcal{V}$ is a nonincreasing function of time and bounded from below, the $\lim_{t \to \infty} \mathcal{V} = \mathcal{V}_\infty$ exists. Therefore by integrating $\dot{\mathcal{V}}$ from 0 to ∞ we have

$$\int_0^\infty \|\mathbf{e}\|^2 \, dt = 2[\mathcal{V}(0) - \mathcal{V}_\infty] < \infty,$$

which implies that $\mathbf{e} \in L_2$. By definition the sigmoid functions $s(x_i)$, $i = 1, 2, \ldots, n$ are bounded for all $\mathbf{x}$ and by assumption all inputs to the neural network are also bounded. Hence from Eq. (4) we have that $\dot{\mathbf{e}} \in L_\infty$. Whereas $\mathbf{e} \in L_2 \cap L_\infty$ and $\dot{\mathbf{e}} \in L_\infty$, using Barbalat's lemma [26], we conclude that $\lim_{t \to \infty} \mathbf{e}(t) = 0$. Now using the boundedness of $\mathbf{u}$, $\mathbf{S}(\mathbf{x})$, and $\mathbf{S}'(\mathbf{x})$, and the convergence of $\mathbf{e}(t)$ to zero, we have that $\dot{\mathbf{W}}$, $\dot{\mathbf{W}}_{n+1}$ also converge to zero. ∎

Remark 2. Under the assumptions of Theorem 1, we cannot conclude anything about the convergence of the weights to their optimal values. To guarantee convergence, $\mathbf{S}(\mathbf{x})$, $\mathbf{S}'(\mathbf{x})$, and $\mathbf{u}$ need to satisfy a persistency of excitation condition. A signal $\mathbf{z}(t) \in \Re^n$ is persistently exciting in $\Re^n$ if there exist positive constants β_0, β_1, and T such that

$$\beta_0 \mathbf{I} \le \int_t^{t+T} \mathbf{z}(\tau)\mathbf{z}^T(\tau) \, d\tau \le \beta_1 \mathbf{I}, \qquad \forall t \ge 0.$$

However, such a condition cannot be verified *a priori* because $\mathbf{S}(\mathbf{x})$ and $\mathbf{S}'(\mathbf{x})$ are nonlinear functions of the state $\mathbf{x}$.

B. CONTROL

In this section we investigate the tracking problem. The unknown nonlinear dynamical system is identified by a dynamic neural network and then it is driven to follow the response of a known system. The purpose of the identification stage is to provide adequate initial values for the control stage, therefore leading to better transient response of the error.

1. Parametric Uncertainty

In this section we assume that the unknown system can be modeled exactly by a dynamical neural network of the form

$$\dot{\mathbf{x}} = \mathbf{A}\mathbf{x} + \mathbf{W}^*\mathbf{S}(\mathbf{x}) + \mathbf{W}_{n+1}^* S'(\mathbf{x})\mathbf{u}, \tag{11}$$

where all matrices are as defined previously. Define the error between the identifier states and the real system states as

$$\mathbf{e} = \hat{\mathbf{x}} - \mathbf{x}.$$

Then from Eqs. (2) and (11) we obtain the error equation

$$\dot{\mathbf{e}} = \mathbf{Ae} + \widetilde{\mathbf{W}}\mathbf{S}(\mathbf{x}) + \widetilde{\mathbf{W}}_{n+1}\mathbf{S}'(\mathbf{x})\mathbf{u}, \tag{12}$$

where

$$\widetilde{\mathbf{W}} = \mathbf{W} - \mathbf{W}^*,$$
$$\widetilde{\mathbf{W}}_{n+1} = \mathbf{W}_{n+1} - \mathbf{W}^*_{n+1}.$$

In the tracking problem we want the real system states to follow the states of a reference model. Let the reference model be described by

$$\dot{\mathbf{x}}_m = \mathbf{A}_m\mathbf{x}_m + \mathbf{B}_m\mathbf{r},$$

where $\mathbf{x}_m \in \Re^n$ are the model states, $\mathbf{r} \in \Re^n$ are the model inputs, and $\mathbf{A}_m, \mathbf{B}_m$ are constant matrices of appropriate dimensions. Define the error between the identifier states and the model states as

$$\mathbf{e}_c = \hat{\mathbf{x}} - \mathbf{x}_m. \tag{13}$$

Differentiating Eq. (13) we obtain

$$\dot{\mathbf{e}}_c = \dot{\hat{\mathbf{x}}} - \dot{\mathbf{x}}_m$$

or

$$= \mathbf{A}\hat{\mathbf{x}} + \mathbf{W}\mathbf{S}(\mathbf{x}) + \mathbf{W}_{n+1}\mathbf{S}'(\mathbf{x})\mathbf{u} - \mathbf{A}_m\mathbf{x}_m - \mathbf{B}_m\mathbf{r}. \tag{14}$$

Taking

$$\mathbf{u} = -\left[\mathbf{B}\mathbf{W}_{n+1}\mathbf{S}'(\mathbf{x})\right]^{-1}\left[\mathbf{A}\mathbf{x}_m + \mathbf{W}\mathbf{S}(\mathbf{x}) - \mathbf{A}_m\mathbf{x}_m - \mathbf{B}_m\mathbf{r}\right] \tag{15}$$

and substituting into Eq. (14) we finally obtain

$$\dot{\mathbf{e}}_c = \mathbf{A}\mathbf{e}_c. \tag{16}$$

The Lyapunov synthesis method is again used to derive stable adaptive laws. Therefore, if we take the Lyapunov function candidate

$$\mathcal{V}\left(\mathbf{e}, \mathbf{e}_c, \widetilde{\mathbf{W}}, \widetilde{\mathbf{W}}_{n+1}\right) = \tfrac{1}{2}\mathbf{e}^T\mathbf{P}\mathbf{e} + \tfrac{1}{2}\mathbf{e}_c^T\mathbf{P}\mathbf{e}_c + \tfrac{1}{2}\mathrm{tr}\{\widetilde{\mathbf{W}}^T\widetilde{\mathbf{W}}\} + \tfrac{1}{2}\mathrm{tr}\{\widetilde{\mathbf{W}}^T_{n+1}\widetilde{\mathbf{W}}_{n+1}\},$$

where $\mathbf{P} > 0$ is chosen to satisfy the Lyapunov equation

$$\mathbf{PA} + \mathbf{A}^T\mathbf{P} = -\mathbf{I},$$

we obtain (following the same procedure as in Section III.A) that the learning laws

$$\dot{w}_{ij} = -p_i s(x_j) e_i,$$
$$\dot{w}_{in+1} = -s'(x_i) p_i u_i e_i,$$

for all $i, j = 1, 2, 3, \ldots, n$, make

$$\dot{V} = -\tfrac{1}{2} \|\mathbf{e}\|^2 - \tfrac{1}{2} \|\mathbf{e}_c\|^2 \le 0.$$

Furthermore, it is trivial to verify that the foregoing learning laws can be written in matrix form as

$$\dot{\mathbf{W}} = -\mathbf{EPS}_0,$$
$$\dot{\mathbf{W}}_{n+1} = -\mathbf{PS'UE},$$

where all matrices are defined as

$$\mathbf{P} = \mathrm{diag}[p_1, p_2, \ldots, p_n],$$
$$\mathbf{E} = \mathrm{diag}[e_1, e_2, \ldots, e_n],$$
$$\mathbf{U} = \mathrm{diag}[u_1, u_2, \ldots, u_n],$$

$$\mathbf{S}_0 = \begin{bmatrix} s(x_1) & \cdots & s(x_n) \\ \vdots & & \vdots \\ s(x_1) & \cdots & s(x_n) \end{bmatrix}.$$

To apply the control law Eq. (15), we have to assure the existence of $(\mathbf{W}_{n+1} \times \mathbf{S'(x)})^{-1}$. Since $\mathbf{W}_{n+1}$ and $\mathbf{S'(x)}$ are diagonal matrices and $s'(x_i) \ne 0$, $\forall i = 1, 2, \ldots, n$, all we need to establish is $w_{in+1}(t) \ne 0$, $\forall t \ge 0$, $\forall i = 1, 2, \ldots, n$. Hence $\mathbf{W}_{n+1}(t)$ is confined through the use of a projection algorithm [1, 17, 18] to the set $\mathcal{W} = \{\mathbf{W}_{n+1} : \|\widetilde{\mathbf{W}}_{n+1}\| \le w_m\}$, where w_m is a positive constant. Furthermore, $\widetilde{\mathbf{W}}_{n+1} = \mathbf{W}_{n+1} - \mathbf{W}_{n+1}^*$ and $\mathbf{W}_{n+1}^*$ contains the initial values of $\mathbf{W}_{n+1}$ that identification provides. In particular, the standard adaptive laws are modified to

$$\dot{\mathbf{W}} = -\mathbf{EPS}_0,$$

$$\dot{\mathbf{W}}_{n+1} = \begin{cases} -\mathbf{PS'UE}, & \text{if } \mathbf{W}_{n+1} \in \mathcal{W} \text{ or } \{\|\widetilde{\mathbf{W}}_{n+1}\| = w_m \\ & \text{and } \mathrm{tr}\{-\mathbf{PS'UE}\widetilde{\mathbf{W}}_{n+1}\} \le 0\}, \\[2ex] -\mathbf{PS'UE} + \mathrm{tr}\{\mathbf{PS'UE}\widetilde{\mathbf{W}}_{n+1}\} & \\ \quad \times \left(\dfrac{1 + \|\widetilde{\mathbf{W}}_{n+1}\|}{w_m}\right)^2 \widetilde{\mathbf{W}}_{n+1}, & \{\|\widetilde{\mathbf{W}}_{n+1}\| = w_m \text{ and} \\ & \mathrm{tr}\{-\mathbf{PS'UE}\widetilde{\mathbf{W}}_{n+1}\} > 0\}. \end{cases}$$

Therefore, if the initial weights are chosen such that $\|\widetilde{\mathbf{W}}(0)_{n+1}\| \leq w_m$, then we have that $\|\widetilde{\mathbf{W}}_{n+1}\| \leq w_m$ for all $t \geq 0$. This can be established readily by noting that whenever $\|\widetilde{\mathbf{W}}(t)_{n+1}\| = w_m$, then

$$\frac{d\|\widetilde{\mathbf{W}}(t)_{n+1}\|^2}{dt} \leq 0, \tag{17}$$

which implies that the weights $\mathbf{W}_{n+1}$ are directed toward the inside or the ball $\{\mathbf{W}_{n+1}: \|\widetilde{\mathbf{W}}_{n+1}\| \leq w_m\}$. A proof of the inequality Eq. (17), can be found in [20]. Now we can prove the following theorem:

THEOREM 2. *Consider the control scheme Eqs. (12), (15), and (16). The learning law*

$$\dot{\mathbf{W}} = -\mathbf{EPS}_0,$$

$$\dot{\mathbf{W}}_{n+1} = \begin{cases} -\mathbf{PS'UE}, & \text{if } \mathbf{W}_{n+1} \in \mathcal{W} \text{ or } \{\|\widetilde{\mathbf{W}}_{n+1}\| = w_m \\ & \text{and } \mathrm{tr}\{-\mathbf{PS'UE}\widetilde{\mathbf{W}}_{n+1}\} \leq 0\}, \\ -\mathbf{PS'UE} + \mathrm{tr}\{\mathbf{PS'UE}\widetilde{\mathbf{W}}_{n+1}\} \\ \quad \times \left(\dfrac{1+\|\widetilde{\mathbf{W}}_{n+1}\|}{w_m}\right)^2 \widetilde{\mathbf{W}}_{n+1}, & \{\|\widetilde{\mathbf{W}}_{n+1}\| = w_m \text{ and} \\ & \mathrm{tr}\{-\mathbf{PS'UE}\widetilde{\mathbf{W}}_{n+1}\} > 0\}, \end{cases}$$

guarantees the properties

- $\mathbf{e}, \mathbf{e}_c, \hat{\mathbf{x}}, \widetilde{\mathbf{W}}, \widetilde{\mathbf{W}}_{n+1} \in L_\infty, \mathbf{e}, \mathbf{e}_c \in L_2$;
- $\lim_{t\to\infty} \mathbf{e}(t) = 0$, $\lim_{t\to\infty} \mathbf{e}_c(t) = 0$;
- $\lim_{t\to\infty} \dot{\widetilde{\mathbf{W}}}(t) = 0$, $\lim_{t\to\infty} \dot{\widetilde{\mathbf{W}}}_{n+1}(t) = 0$.

Proof. With the previously mentioned adaptive laws,

$$\dot{V} = -\frac{1}{2}\|\mathbf{e}\|^2 - \frac{1}{2}\|\mathbf{e}_c\|^2 + I_n \mathrm{tr}\left\{\mathrm{tr}\{\mathbf{PS'UE}\widetilde{\mathbf{W}}_{n+1}\}\left(\frac{1+\|\widetilde{\mathbf{W}}_{n+1}\|}{w_m}\right)^2 \widetilde{\mathbf{W}}_{n+1}^T\widetilde{\mathbf{W}}_{n+1}\right\}$$

$$\leq -\frac{1}{2}\|\mathbf{e}\|^2 - \frac{1}{2}\|\mathbf{e}_c\|^2 + I_n \mathrm{tr}\{\mathbf{PS'UE}\widetilde{\mathbf{W}}_{n+1}\}\left(\frac{1+\|\widetilde{\mathbf{W}}_{n+1}\|}{w_m}\right)^2 \mathrm{tr}\{\widetilde{\mathbf{W}}_{n+1}^T\widetilde{\mathbf{W}}_{n+1}\}$$

$$\leq -\frac{1}{2}\|\mathbf{e}\|^2 - \frac{1}{2}\|\mathbf{e}_c\|^2 + I_n \mathrm{tr}\{\mathbf{PS'UE}\widetilde{\mathbf{W}}_{n+1}\}\left(\frac{1+w_m}{w_m}\right)^2 \|\widetilde{\mathbf{W}}_{n+1}\|^2 g$$

$$\leq -\frac{1}{2}\|\mathbf{e}\|^2 - \frac{1}{2}\|\mathbf{e}_c\|^2 + I_n \mathrm{tr}\{\mathbf{PS'UE}\widetilde{\mathbf{W}}_{n+1}\}(1+w_m)^2,$$

where I_n is an indicator function defined as $I_n = 1$ if the conditions $\|\widetilde{\mathbf{W}}_{n+1}\| = w_m$ and $\mathrm{tr}\{-\mathbf{PS'UE}\widetilde{\mathbf{W}}_{n+1}\} > 0$ are satisfied. Now because $\mathrm{tr}\{\mathbf{PS'UE}\widetilde{\mathbf{W}}_{n+1}\} < 0$, then $I_n \mathrm{tr}\{\mathbf{PS'UE\widetilde{W}}_{n+1}\} \times (1+w_m)^2 < 0$. Hence, $\dot{\mathcal{V}} \leq 0$. Therefore, the additional terms introduced by the projection can only make $\dot{\mathcal{V}}$ more negative. Because $\dot{\mathcal{V}}$ is negative semidefinite we have that $\mathcal{V} \in L_\infty$, which implies $\mathbf{e}, \mathbf{e}_c, \widetilde{\mathbf{W}}, \widetilde{\mathbf{W}}_{n+1} \in L_\infty$. Furthermore, $\hat{\mathbf{x}} = \mathbf{e} + \mathbf{x}$ is also bounded. Whereas $\mathcal{V}$ is a nonincreasing function of time and bounded from below, the $\lim_{t \to \infty} \mathcal{V} = \mathcal{V}_\infty$ exists. Therefore by integrating $\dot{\mathcal{V}}$ from 0 to ∞ we have

$$\int_0^\infty \tfrac{1}{2}\left(\|\mathbf{e}\|^2 + \|\mathbf{e}_c\|^2\right) dt - I_n(1+w_m)^2 \int_0^\infty \mathrm{tr}\{\mathbf{PS'UE\widetilde{W}}_{n+1}\} dt$$
$$\leq [\mathcal{V}(0) - \mathcal{V}_\infty] < \infty,$$

which implies that $\mathbf{e}, \mathbf{e}_c \in L_2$. By definition the sigmoid functions $\mathbf{S}(x), \mathbf{S}'(x)$ are bounded for all $\mathbf{x}$ and by assumption all inputs to the reference model are also bounded. Hence from Eq. (15) we have that $\mathbf{u}$ is bounded and from Eqs. (12) and (16), $\dot{\mathbf{e}}, \dot{\mathbf{e}}_c \in L_\infty$. Whereas $\mathbf{e}, \mathbf{e}_c \in L_2 \cap L_\infty$ and $\dot{\mathbf{e}}, \dot{\mathbf{e}}_c \in L_\infty$, using Barbalat's lemma [26], we conclude that $\lim_{t \to \infty} \mathbf{e}(t) = \lim_{t \to \infty} \mathbf{e}_c(t) = 0$. Now using the boundedness of $\mathbf{u}$, $\mathbf{S}(\mathbf{x})$, and $\mathbf{S}'(\mathbf{x})$, and the convergence of $\mathbf{e}(t)$ to zero, we have that $\dot{\mathbf{W}}, \dot{\mathbf{W}}_{n+1}$ also converge to zero.   ∎

Remark 3.   The preceding analysis implies that the projection modification guarantees boundedness of the weights without affecting the rest of the stability properties established in the absence of projection.

2. Parametric plus Dynamic Uncertainties

In this subsection we examine a more general case where parametric and dynamic uncertainties are present. To analyze the problem, the complete singular perturbation model, which can be found in [20], is used. Therefore, the control scheme is now described by the set of nonlinear differential equations

$$\dot{\mathbf{e}} = \mathbf{Ae} + \widetilde{\mathbf{W}}\mathbf{S}(\mathbf{x}) + \widetilde{\mathbf{W}}_{n+1}\mathbf{S}'(\mathbf{x})\mathbf{u} - \mathbf{F}(\mathbf{x}, \mathbf{W}, \mathbf{W}_{n+1})\boldsymbol{\eta},$$
$$\dot{\mathbf{e}}_c = \mathbf{Ae}_c,$$
$$\mu\dot{\boldsymbol{\eta}} = \mathbf{A}_0\boldsymbol{\eta} - \mu\mathbf{h}(\mathbf{e}, \widetilde{\mathbf{W}}, \widetilde{\mathbf{W}}_{n+1}, \boldsymbol{\eta}),$$
$$\mathbf{u} = -\left[\mathbf{W}_{n+1}\mathbf{S}'(\mathbf{x})\right]^{-1}[\mathbf{Ax}_m + \mathbf{WS}(\mathbf{x}) - \mathbf{A}_m\mathbf{x}_m - \mathbf{B}_m\mathbf{r}]. \qquad (18)$$

Before proceeding any further, we need the following lemma [20]:

LEMMA 1.   *It is true that* $\dot{\mathbf{h}}(\mathbf{e}, \widetilde{\mathbf{W}}, \widetilde{\mathbf{W}}_{n+1}, \boldsymbol{\eta}, \mathbf{u})$ *is bounded by*

$$\left\|\dot{\mathbf{h}}(\mathbf{e}, \widetilde{\mathbf{W}}, \widetilde{\mathbf{W}}_{n+1}, \boldsymbol{\eta}, \mathbf{u})\right\| \leq \rho_1\|\mathbf{e}\| + \rho_2\|\boldsymbol{\eta}\|$$

provided that the following inequalities hold:

$$\left\|\mathbf{h}_w\dot{\tilde{\mathbf{W}}}\right\| \le k_0\|\mathbf{e}\|,$$

$$\left\|\mathbf{h}_{w_{n+1}}\dot{\tilde{\mathbf{W}}}_{n+1}\right\| \le k_1\|\mathbf{e}\|,$$

$$\left\|\mathbf{h}_e\tilde{\mathbf{W}}_{n+1}\mathbf{S}'(\mathbf{x})\mathbf{u}\right\| \le k_2\|\mathbf{e}\|,$$

$$\left\|\mathbf{h}_e\tilde{\mathbf{W}}\mathbf{S}(\mathbf{x})\right\| \le k_3\|\mathbf{e}\|,$$

$$\left\|\mathbf{h}_e\mathbf{F}(\mathbf{x},\mathbf{W},\mathbf{W}_{n+1})\right\| \le \rho_2,$$

$$\left\|\mathbf{h}_e\mathbf{Ae}\right\| \le k_4\|\mathbf{e}\|,$$

and

$$\rho_1 = k_0 + k_1 + k_2 + k_3 + k_4.$$

We are now able to prove the following theorem [20]:

THEOREM 3. *The control scheme Eq. (18), is asymptotically stable for all*

$$\mu \in (0,\mu_0),$$

where $\mu_0 = \frac{1}{2}(1/(2\gamma_1\gamma_2+\gamma_3))$. Furthermore, the learning law

$$\dot{\mathbf{W}} = -\mathbf{EPS}_0,$$

$$\dot{\mathbf{W}}_{n+1} = \begin{cases} -\mathbf{PS}'\mathbf{UE}, & \text{if } \mathbf{W}_{n+1} \in \mathcal{W} \text{ or } \left\{\|\tilde{\mathbf{W}}_{n+1}\| = w_m \right. \\ & \left. \text{and } \mathrm{tr}\{-\mathbf{PS}'\mathbf{UE}\tilde{\mathbf{W}}_{n+1}\} \le 0\right\}, \\ -\mathbf{PS}'\mathbf{UE} + \mathrm{tr}\{\mathbf{PS}'\mathbf{UE}\tilde{\mathbf{W}}_{n+1}\} & \\ \quad \times \left(\dfrac{1+\|\tilde{\mathbf{W}}_{n+1}\|}{w_m}\right)^2 \tilde{\mathbf{W}}_{n+1}, & \{\|\tilde{\mathbf{W}}_{n+1}\| = w_m \text{ and} \\ & \mathrm{tr}\{-\mathbf{PS}'\mathbf{UE}\tilde{\mathbf{W}}_{n+1}\} > 0\}, \end{cases}$$

guarantees the properties

- $\mathbf{e}, \mathbf{e}_c, \eta, \hat{\mathbf{x}}, \tilde{\mathbf{W}}, \tilde{\mathbf{W}}_{n+1} \in L_\infty$, $\mathbf{e}, \mathbf{e}_c, \eta \in L_2$;
- $\lim_{t\to\infty}\mathbf{e}(t) = 0$, $\lim_{t\to\infty}\mathbf{e}_c(t) = 0$, $\lim_{t\to\infty}\eta(t) = 0$;
- $\lim_{t\to\infty}\dot{\tilde{\mathbf{W}}}(t) = 0$, $\lim_{t\to\infty}\dot{\tilde{\mathbf{W}}}_{n+1}(t) = 0$.

IV. DIRECT CONTROL

In this section we perform a robustness analysis when both modeling errors and unmodeled dynamics are present and affect the system performance. However, the unmodeled dynamics are not confined to the framework of singular perturbation theory; thus, the present section can be viewed not just as another method to control unknown systems, but as a considerable extension because it covers more general cases.

A. MODELING ERROR EFFECTS

Let us assume that the true plant is of known order n and can be modeled exactly by the dynamic neural network Eq. (11) plus a modeling error term $\omega(x, u)$:

$$\dot{x} = -Ax + W^*S(x) + W^*_{n+1}S'(x)u + \omega(x, u). \tag{19}$$

We choose a function $h(x)$ of class C^2 from $\mathcal{M}$ to $\mathfrak{R}^+$ whose derivative with respect to time is

$$\dot{h} = \frac{\partial h}{\partial x}\Big[-Ax + W^*S(x) + W^*_{n+1}S'(x)u + \omega(x, u)\Big]. \tag{20}$$

Equation (20) can also be written

$$\dot{h} + \frac{\partial h}{\partial x}Ax - \frac{\partial h}{\partial x}\omega(x, u) = \frac{\partial h}{\partial x}W^*S(x) + \frac{\partial h}{\partial x}W^*_{n+1}S'(x)u. \tag{21}$$

Define

$$v \stackrel{\text{def}}{=} \frac{\partial h}{\partial x}WS(x) + \frac{\partial h}{\partial x}W_{n+1}S'(x)u - \dot{h} - \frac{\partial h}{\partial x}Ax$$

and use the error filtering method

$$\dot{e} + re = v$$

$$= -\dot{h} + \frac{\partial h}{\partial x}\Big[-Ax + WS(x) + W_{n+1}S'(x)u\Big]. \tag{22}$$

Set

$$e \stackrel{\text{def}}{=} \eta - h. \tag{23}$$

Hence

$$\dot{\eta} + r\eta = rh + \frac{\partial h}{\partial x}\Big[-Ax + WS(x) + W_{n+1}S'(x)u\Big] \tag{24}$$

with the state $\eta \in \mathfrak{R}$ and r a strictly positive constant. Furthermore, if we choose

$$h(x) = \tfrac{1}{2}|x|^2,$$

then Eq. (24) finally becomes

$$\dot{\eta} = -r\eta + rh - x^T Ax + x^T W S(x) + x^T W_{n+1} S'(x)u. \tag{25}$$

Now consider the Lyapunovlike function

$$\mathcal{L} = \tfrac{1}{2}e^2 + \tfrac{1}{2}\mathrm{tr}\{\widetilde{W}^T \widetilde{W}\} + \tfrac{1}{2}\mathrm{tr}\{\widetilde{W}_{n+1}^T \widetilde{W}_{n+1}\}. \tag{26}$$

Taking the derivative of $\mathcal{L}$ with respect to time, Eq. (26) becomes

$$\dot{\mathcal{L}} = -re^2 + e\left[-\dot{h} - x^T Ax + x^T W S(x) + x^T W_{n+1} S'(x)u \right] \\ + \mathrm{tr}\{\dot{W}^T \widetilde{W}\} + \mathrm{tr}\{\dot{W}_{n+1}^T \widetilde{W}_{n+1}\}, \tag{27}$$

which after using Eq. (21) becomes

$$\dot{\mathcal{L}} = -re^2 + e\left[-x^T W^* S(x) - x^T W_{n+1}^* S'(x)u + x^T W S(x) \right. \\ \left. + x^T W_{n+1} S'(x)u - x^T \omega(x, u) \right] + \mathrm{tr}\{\dot{W}^T \widetilde{W}\} + \mathrm{tr}\{\dot{W}_{n+1}^T \widetilde{W}_{n+1}\}$$

or, equivalently,

$$\dot{\mathcal{L}} = -re^2 + ex^T \widetilde{W} S(x) + ex^T \widetilde{W}_{n+1} S'(x)u - ex^T \omega(x, u) \\ + \mathrm{tr}\{\dot{W}^T \widetilde{W}\} + \mathrm{tr}\{\dot{W}_{n+1}^T \widetilde{W}_{n+1}\}. \tag{28}$$

Furthermore, if we choose the learning laws

$$\dot{w}_{ij} = -ex_i s(x_j), \tag{29}$$

$$\dot{w}_{in+1} = -ex_i s'(x_i)u_i, \tag{30}$$

for all $i, j = 1, 2, \ldots, n$, and in matrix form as

$$\dot{W} = -ex S^T(x), \tag{31}$$

$$\dot{W}_{n+1} = -ex' S'(x)U, \tag{32}$$

where

$$x' = \mathrm{diag}[x_1, x_2, \ldots, x_n],$$
$$U = \mathrm{diag}[u_1, u_2, \ldots, u_n],$$

we obtain

$$\dot{\mathcal{L}} = -re^2 - ex^T \omega(x, u) \\ \leq -r|e|^2 + |e||x||\omega(x, u)|. \tag{33}$$

At this point we can distinguish two possible cases. The complete model matching at zero case and the modeling error at zero case.

1. Complete Model Matching at Zero Case

We make the following assumption:

Assumption A.1.   The modeling error term satisfies

$$|\omega(x, u)| \leq k_1'|x| + k_1''|u|,$$

where k_1' and k_1'' are known positive constants.

Employing A.1, inequality (33) becomes

$$\dot{\mathcal{L}} \leq -r|e|^2 + k_1'|e||x|^2 + k_1''|e||x||u|. \tag{34}$$

To continue, we need the following lemma:

LEMMA 2.   *The control law*

$$u = -[W_{n+1}S'(x)]^{-1}[WS(x) + v], \tag{35}$$

$$v = \tfrac{1}{2}rx - Ax, \tag{36}$$

where the synaptic weight estimates W and W_{n+1} are adjusted according to Eqs. (32) and (33), respectively, guarantees

- $\eta(t) \leq 0, \ \forall t \geq 0;$
- $\lim_{t \to \infty} \eta(t) = 0$ *exponentially fast,*

provided that $\eta(0) < 0$.

Proof.   Observe that if we use the control law Eqs. (35) and (36), Eq. (25) becomes

$$\dot{\eta} = -r\eta, \qquad \forall t \geq 0,\,^{.}$$

which is a homogeneous differential equation with solution

$$\eta(t) = \eta(0)e^{-rt}.$$

Hence, if $\eta(0)$, which represents the initial value of $\eta(t)$, is chosen negative, we obtain

$$\eta(t) \leq 0, \qquad \forall t \geq 0.$$

Moreover, $\eta(t)$ converges to zero exponentially fast.   ∎

Now let us make the following assumption:

Assumption A.2. $k_1'' = 0$.

Assumption A.2 tell us that at zero we have no modeling error in the controlled vector fields. Furthermore, observe that because of A.2,

$$\omega(x, u) \equiv \omega(x).$$

Employing A.2, inequality (33) becomes

$$\dot{\mathcal{L}} \leq -r|e|^2 + k_1'|e||x|^2. \tag{37}$$

It is true though that

$$h = \eta - e.$$

Hence, because $h \geq 0$ we have that

$$\eta(t) \geq e(t).$$

However,

$$\eta(t) \leq 0, \qquad \forall t \geq 0,$$

which implies

$$|\eta(t)| \leq |e(t)|, \qquad \forall t \geq 0. \tag{38}$$

From the foregoing analysis we have for Eq. (37)

$$\begin{aligned}
\dot{\mathcal{L}} &\leq -r|e|^2 + 2k_1'|e|(\eta - e) \\
&\leq -r|e|^2 + 2k_1'|e|(|\eta| + |e|) \\
&\leq -r|e|^2 + 4k_1'|e|^2 \\
&\leq -(r - 4k_1')|e|^2.
\end{aligned} \tag{39}$$

Therefore, if we choose

$$r \geq 4k_1', \tag{40}$$

inequality (39) becomes

$$\dot{\mathcal{L}} \leq 0. \tag{41}$$

Hence, we can prove the following theorem:

THEOREM 4.   *The update laws Eqs. (32) and (33) together with the control law Eqs. (35) and (36) guarantee the properties*

- $e, |x|, W, W_{n+1}, \eta, \dot{e} \in L_\infty, |e| \in L_2$;
- $\lim_{t \to \infty} e(t) = 0, \lim_{t \to \infty} |x(t)| = 0$;
- $\lim_{t \to \infty} \dot{W}(t) = 0, \lim_{t \to \infty} \dot{W}_{n+1}(t) = 0$.

Proof.   From Eq. (41) we have that $\mathcal{L} \in L_\infty$; hence, $e, \widetilde{W}, \widetilde{W}_{n+1} \in L_\infty$. Furthermore, $W = \widetilde{W} + W^* \in L_\infty$ and $W_{n+1} = \widetilde{W}_{n+1} + W_{n+1}^* \in L_\infty$. Whereas $e = \eta - h$ and $\eta \leq 0, \forall t \geq 0$ we have $\eta, h \in L_\infty$, which in turn implies $|x| \in L_\infty$. Moreover, because $\mathcal{L}$ is a monotone decreasing function of time and bounded from below, the $\lim_{t \to \infty} \mathcal{L}(t) = \mathcal{L}_\infty$ exists. Therefore by integrating $\dot{\mathcal{L}}$ from 0 to ∞ we have

$$\int_0^\infty |e|^2 \, dt \le \frac{1}{r - 4k_1'} [\mathcal{L}(0) - \mathcal{L}_\infty] < \infty,$$

which implies that $|e| \in L_2$. We also have that

$$\dot{e} = -re + x^T \widetilde{W} S(x) + x^T \widetilde{W}_{n+1} S'(x) u - x^T \omega(x).$$

Hence, $\dot{e} \in L_\infty$ because $u, |x| \in L_\infty$ the sigmoidals are bounded by definition, $\widetilde{W}, \widetilde{W}_{n+1} \in L_\infty$, and A.1 and A.2 hold. Therefore, because $e \in L_2 \cap L_\infty$ and $\dot{e} \in L_\infty$, applying Barbalat's lemma allows us to conclude that $\lim_{t \to \infty} e(t) = 0$. Now using the boundedness of u, $S(x)$, $S'(x)$, and x, and the convergence of $e(t)$ to zero, we obtain that $\dot{W}$, $\dot{W}_{n+1}$ also converge to zero. Furthermore, we know that $\eta(t)$ converges to zero exponentially fast. Hence whereas $e(t)$ also converges to zero, we have that

$$\lim_{t \to \infty} h(x(t)) = \lim_{t \to \infty} \eta(t) - \lim_{t \to \infty} e(t) = 0.$$

Thus

$$\lim_{t \to \infty} |x(t)| = 0. \qquad \blacksquare$$

Remark 4. From the preceding analysis we cannot conclude anything about the convergence of the synaptic weights W and W_{n+1} to their optimum values W^* and W_{n+1}^*, respectively. To guarantee convergence, $S(x)$ and $S'(x)u$ need to satisfy a persistency of excitation condition. A signal $z(t) \in \mathfrak{R}^n$ is persistently exciting in $\mathfrak{R}^n$ if there exist positive constants β_0, β_1, and T such that

$$\beta_0 I \le \int_t^{t+T} z(\tau) z^T(\tau) \, d\tau \le \beta_1 I, \qquad \forall t \ge 0.$$

However, such a condition cannot be verified *a priori* because $S(x)$ and $S'(x)u$ are nonlinear functions of the state x.

Now let us return to the status before making assumption A.2. We can verify easily that the same results can be derived provided that

$$|u| \le k_u |x|, \tag{42}$$

where k_u is a known positive constant. However, if we observe Eqs. (35) and (36), we can verify easily that Eq. (42) is valid, provided that W is uniformly bounded by a known positive constant w_m. Therefore, $W(t)$ is confined through the use of a projection algorithm to the set $\mathcal{W} = \{W: \|W\| \le w_m\}$. In particular, the standard update law is modified to

$$
\dot{W} =
\begin{cases}
-ex\,S^T(x), & \text{if } W \in \mathcal{W} \text{ or} \|W\| = w_m \\
 & \text{and tr}\{ex\,S^T(x)W\} \geq 0, \\
-ex\,S^T(x) + \text{tr}\{ex\,S^T(x)W\} \\
\quad \times \left(\dfrac{1 + \|W\|}{w_m}\right)^2 W, & \text{if } \|W\| = w_m \\
 & \text{and tr}\{ex\,S^T(x)W\} < 0.
\end{cases}
\tag{43}
$$

Therefore, if the initial weights are chosen such that $\|W(0)\| \leq w_m$, then we have $\|W\| \leq w_m$ for all $t \geq 0$. This can be established readily by noting that whenever $\|W\| = w_m$, then

$$
\frac{d}{dt}\left(\|W\|^2\right) \leq 0,
\tag{44}
$$

which implies that the weights W are directed toward the inside of the ball $\{W: \|W\| \leq w_m\}$. A proof of inequality (44) can be found in [21]. Furthermore we can prove the following lemma [21]:

LEMMA 3.   *Based on the adaptive law Eq. (43), the additional terms introduced in the expression for $\dot{\mathcal{L}}$ can only make $\dot{\mathcal{L}}$ more negative.*

Lemma 3 implies that the projection modification Eq. (43) guarantees boundedness of the weights without affecting the rest of the stability properties established in the absence of projection.

Now that we have established the validity of Eq. (42), we observe that if the design constant r is chosen such that

$$
r > 4(k_1' + k_1'' k_u),
\tag{45}
$$

then Eq. (41) is true and the results of Theorem 4 are still valid.

Remark 5.   The previous analysis reveals that the accuracy of the dynamic neural network model should be restricted only at the origin. In other words, when we have complete model matching at zero, our modified adaptive regulator can guarantee that the stability properties of the closed loop system do not alter. Furthermore, if we do not have modeling error in the controlled vector fields, that is, $k_1'' = 0$, there is no need to uniformly bound the Ws through the use of the projection algorithm (43), thus simplifying the implementation issue.

Remark 6.   Inequalities (40) and (45) show how the design constant r should be selected to guarantee convergence of the state x to zero, even in the presence of modeling error terms which are not uniformly bounded *a priori*, as assumption A.1 implies. The value of r becomes large as we allow for large model imperfections. However, r is implemented as a gain in the construction of $\dot{\eta}$ and for practical reasons it cannot take arbitrarily large values. Thus we are lead to a compromise between the value of r and the maximum allowable modeling error terms.

2. Modeling Error at Zero Case

In the previous subsection, we assumed that the modeling error term satisfied the condition

$$\left|\omega(x, u)\right| \leq k_1'|x| + k_1''|u|,$$

which implies that the modeling error becomes zero when $|x| = 0$ and we proved convergence of the state x to zero, plus boundedness of all signals in the closed loop. In this subsection, however, we examine the more general case which is described by the following assumption:

Assumption A.3. $\quad |\omega(x, u)| \leq k_0 + k_1'|x| + k_1''|u|.$

Therefore, we now allow a modeling error $k_0 \neq 0$ at zero. Furthermore, as stated in Section IV.A.1, we can find an *a priori* known constant $k_u > 0$, such that

$$|u| \leq k_u|x|,$$

thus making

$$\left|\omega(x, u)\right| \equiv |\omega(x)|$$

and A.3 equivalent to

$$\left|\omega(x)\right| \leq k_0 + k_1|x|, \tag{46}$$

where

$$k_1 = k_1' + k_1'' k_u \tag{47}$$

is a positive constant.

Employing inequality (46), inequality (33) becomes

$$\begin{aligned} \dot{\mathcal{L}} &\leq -r|e|^2 + |e||x|[k_0 + k_1|x|] \\ &\leq -r|e|^2 + k_1|e||x|^2 + k_0|e||x|. \end{aligned} \tag{48}$$

Again, using Lemma 2 and the fact that $|\eta(t)| \leq |e(t)|$ when $\eta(t) \leq 0$, $\forall t \geq 0$, Eq. (48) becomes

$$\dot{\mathcal{L}} \leq -r|e|^2 + 4k_1|e|^2 + k_0|e||x|. \tag{49}$$

However

$$\begin{aligned} \tfrac{1}{2}|x|^2 &\leq |e| + |\eta| \\ &\leq 2|e|. \end{aligned}$$

Hence

$$|x| \leq 2\sqrt{|e|}. \tag{50}$$

Substituting inequality (50) into inequality (49) we obtain

$$
\begin{aligned}
\dot{\mathcal{L}} &\leq -(r - 4k_1)|e|^2 + 2k_0|e|\sqrt{|e|} \\
&\leq \left[-(r - 4k_1)\sqrt{|e|} + 2k_0 \right]|e|\sqrt{|e|} \\
&\leq 0
\end{aligned} \tag{51}
$$

provided that

$$\sqrt{|e|} > \frac{2k_0}{r - 4k_1}$$

or, equivalently,

$$|e| > \frac{4k_0^2}{(r - 4k_1)^2} \tag{52}$$

with $r > 4k_1$. Inequality (52) together with inequality (50) demonstrates that the trajectories of $e(t)$ and $x(t)$ are uniformly ultimately bounded with respect to the arbitrarily small (since r can be chosen sufficiently large) sets

$$\mathcal{E} = \left\{ e(t): \left| e(t) \right| \leq \frac{4k_0^2}{(r - 4k_1)^2}, \ r > 4k_1 > 0 \right\}$$

and

$$\mathcal{X} = \left\{ x(t): \left| x(t) \right| \leq \frac{4k_0}{r - 4k_1}, \ r > 4k_1 > 0 \right\}.$$

Thus we have proven the following theorem:

THEOREM 5. *Consider the system Eq. (19) with the modeling error term satisfying inequality (46). Then the control law Eqs. (35) and (36) together with the update laws Eqs. (32) and (43) guarantee the uniform ultimate boundedness with respect to the sets*

$$\mathcal{E} = \left\{ e(t): \left| e(t) \right| \leq \frac{4k_0^2}{(r - 4k_1)^2}, \ r > 4k_1 > 0 \right\},$$

$$\mathcal{X} = \left\{ x(t): \left| x(t) \right| \leq \frac{4k_0}{r - 4k_1}, \ r > 4k_1 > 0 \right\}.$$

Furthermore,

$$\dot{e} = -re + x^T \widetilde{W} S(x) + x^T \widetilde{W}_{n+1} S'(x)u - x^T \omega(x). \tag{53}$$

Hence, because the boundedness of $\tilde{W}$ and $\tilde{W}_{n+1}$ is assured by the use of the projection algorithm and $\omega(x)$ due to Eq. (43) and Theorem 5, we conclude that $\dot{e} \in L_\infty$.

Remark 7.   The previous analysis reveals that in the case where we have a modeling error different from zero at $|x| = 0$, our adaptive regulator can guarantee at least uniform ultimate boundedness of all signals in the closed loop. In particular, Theorem 5 shows that if k_0 is sufficiently small or if the design constant r is chosen such that $r \gg 4k_1$, then $|x(t)|$ can be arbitrarily close to zero and in the limit as $r \to \infty$, actually becomes zero. However, as we stated in Remark 6, implementation issues constrain the maximum allowable value of r.

B. MODEL ORDER PROBLEMS

Let us assume that the true plant is of order $N \geq n$ and is therefore described by

$$\dot{x} = -Ax + W^* S(x) + W^*_{n+1} S'(x)u + \phi(x, x_{ud}),$$
$$\dot{x}_{ud} = B(x, x_{ud}), \tag{54}$$

where x_{ud} living in a p-dimensional manifold $\mathcal{M}_p$ is the state of the unmodeled dynamics and $\phi(\cdot)$, $B(\cdot)$ are unknown vector fields of x and x_{ud}. Obviously $p = N - n$.

Proceeding as previously, we choose a function $h(x)$ of class C^2 from $\mathcal{M}$ to $\mathfrak{R}^+$ whose derivative with respect to time is

$$\dot{h} = \frac{\partial h}{\partial x}\big[-Ax + W^* S(x) + W^*_{n+1} S'(x)u + \phi(x, x_{ud})\big], \tag{55}$$

which can also be written as

$$\dot{h} + \frac{\partial h}{\partial x}Ax - \frac{\partial h}{\partial x}\phi(x, x_{ud}) = \frac{\partial h}{\partial x}W^* S(x) + \frac{\partial h}{\partial x}W^*_{n+1} S'(x)u. \tag{56}$$

Define

$$v \stackrel{\text{def}}{=} \frac{\partial h}{\partial x}W S(x) + \frac{\partial h}{\partial x}W_{n+1} S'(x)u - \dot{h} - \frac{\partial h}{\partial x}Ax$$

and use the error filtering method

$$\dot{e} + re = v$$
$$= -\dot{h} + \frac{\partial h}{\partial x}\big[-Ax + W S(x) + W_{n+1} S'(x)u\big]. \tag{57}$$

Again set

$$e \stackrel{\text{def}}{=} \eta - h. \tag{58}$$

Therefore

$$\dot{\eta} + r\eta = rh + \frac{\partial h}{\partial x} \left[-Ax + WS(x) + W_{n+1}S'(x)u \right] \tag{59}$$

with the state $\eta \in \mathfrak{R}$ and r a strictly positive constant. Furthermore, if we choose

$$h(x) = \tfrac{1}{2}|x|^2,$$

then Eq. (59) finally becomes

$$\dot{\eta} = -r\eta + rh - x^T Ax + x^T WS(x) + x^T W_{n+1}S'(x)u. \tag{60}$$

Now consider the Lyapunovlike function

$$\mathcal{L} = \tfrac{1}{2}e^2 + \mathcal{V}(x_{\text{ud}}) + \tfrac{1}{2}\text{tr}\{\widetilde{W}^T \widetilde{W}\} + \tfrac{1}{2}\text{tr}\{\widetilde{W}_{n+1}^T \widetilde{W}_{n+1}\}, \tag{61}$$

where $\mathcal{V}(x_{\text{ud}})$ is a positive definite function of class C^1 from $\mathcal{M}_p$ to $\mathfrak{R}^+$. Taking the time derivative of Eq. (61) we obtain

$$\dot{\mathcal{L}} = -re^2 + e\left[-\dot{h} - x^T Ax + x^T WS(x) + x^T W_{n+1}S'(x)u \right] \\ + \dot{\mathcal{V}}(x_{\text{ud}}) + \text{tr}\{\dot{W}^T \widetilde{W}\} + \text{tr}\{\dot{W}_{n+1}^T \widetilde{W}_{n+1}\}, \tag{62}$$

which after using Eqs. (56) and (57) finally takes the form

$$\dot{\mathcal{L}} = -re^2 + ex^T \widetilde{W}S(x) + ex^T \widetilde{W}_{n+1}S'(x)u - ex^T \phi(x, x_{\text{ud}}) \\ + \dot{\mathcal{V}}(x_{\text{ud}}) + \text{tr}\{\dot{W}^T \widetilde{W}\} + \text{tr}\{\dot{W}_{n+1}^T \widetilde{W}_{n+1}\}. \tag{63}$$

Furthermore, if we choose the same update laws as Eqs. (32) and (43) we obtain

$$\dot{\mathcal{L}} = -re^2 - ex^T \phi(x, x_{\text{ud}}) + \dot{\mathcal{V}}(x_{\text{ud}}). \tag{64}$$

To proceed further we distinguish two different cases.

1. Uniform Asymptotic Stability in the Large Case

For completeness, we introduce from [25] the following definitions that are crucial to our discussion.

DEFINITION 2. The equilibrium point $x_{\text{ud}} = 0$ is said to be uniformly asymptotically stable in the large if the following statements hold.

- For every $M > 0$ and any $t_0 \in \mathfrak{R}^+$, there exists an $\bar{\alpha}(M) > 0$ such that $|x_{\text{ud}}(t; x_{\text{ud}}(0), t_0)| < M$ for all $t \geq t_0$ whenever $|x_{\text{ud}}(0)| < \bar{\alpha}(M)$.

- For every $\bar{\alpha} > 0$ and any $t_0 \in \mathfrak{R}^+$ there exists a $M(\bar{\alpha}) > 0$ such that $|x_{ud}(t; x_{ud}(0), t_0)| < M(\bar{\alpha})$ for all $t \geq t_0$ whenever $|x_{ud}(0)| < \bar{\alpha}$.
- For any $\bar{\alpha}$, any $M > 0$, and $t_0 \in \mathfrak{R}^+$ there exists $T(M, \bar{\alpha}) > 0$, independent of t_0 such that if $|x_{ud}(0)| < \bar{\alpha}$, then $|x_{ud}(t; x_{ud}(0), t_0)| < M$ for all $t \geq t_0 + T(M, \bar{\alpha})$.

For the state of the unmodeled dynamics we make the following assumption:

Assumption A.4. The origin $x_{ud} = 0$ of the unmodeled dynamics is uniformly stable in the large. More specifically, there is a C^1 function $\mathcal{V}(x_{ud})$ from $\mathcal{M}_p$ to $\mathfrak{R}^+$ and continuous, strictly increasing, scalar functions $\gamma_i(|x_{ud}|)$ from $\mathfrak{R}^+$ to $\mathfrak{R}^+$, $i = 1, 2, 3$, which satisfy

$$\gamma_i(0) = 0, \qquad i = 1, 2, 3,$$
$$\lim_{s \to \infty} \gamma_i(s) = \infty, \qquad i = 1, 2,$$

such that for $x_{ud} \in \mathcal{M}_p$,

$$\gamma_1(|x_{ud}|) \leq \mathcal{V}(x_{ud}) \leq \gamma_2(|x_{ud}|)$$

and

$$\frac{\partial \mathcal{V}}{\partial x_{ud}} B(x, x_{ud}) \leq -\gamma_3(|x_{ud}|). \tag{65}$$

Employing assumption A.4, inequality (64) becomes

$$\begin{aligned}
\dot{\mathcal{L}} &= -r|e|^2 - ex^T \phi(x, x_{ud}) + \frac{\partial \mathcal{V}}{\partial x_{ud}} B(x, x_{ud}) \\
&\leq -r|e|^2 - ex^T \phi(x, x_{ud}) - \gamma_3(|x_{ud}|) \\
&\leq -r|e|^2 - ex^T \phi(x, x_{ud}) \\
&\leq -r|e|^2 + |e||x||\phi(x, x_{ud})|.
\end{aligned} \tag{66}$$

To continue we consider the following cases:

Case 1. Assume that the unknown vector field $\phi(x, x_{ud})$ satisfies the condition

$$|\phi(x, x_{ud})| \leq k_\phi |x||\phi'(x_{ud})| \tag{67}$$

with $\phi'(x_{ud})$ an unknown vector field that depends only on x_{ud}. We further assume that $\phi'(x_{ud})$ is uniformly bounded by a constant θ. Hence

$$|\phi'(x_{ud})| \leq \theta. \tag{68}$$

Thus we have

$$\dot{\mathcal{L}} \leq -r|e|^2 + k_\phi \theta |e||x|^2. \tag{69}$$

Moreover, if we apply inequality (50), then inequality (69) becomes

$$\begin{aligned}
\dot{\mathcal{L}} &\le -r|e|^2 + 4k_\phi\theta|e|^2 \\
&\le -(r - 4k_\phi\theta)|e|^2 \\
&\le 0
\end{aligned} \tag{70}$$

provided that the design constant r is chosen such that

$$r > 4k_\phi\theta. \tag{71}$$

Thus we can prove the following theorem [21]:

THEOREM 6.   *Consider the closed loop system*

$$\begin{aligned}
\dot{x} &= -Ax + W^*S(x) + W^*_{n+1}S'(x)u + \phi(x, x_{ud}), \\
\dot{x}_{ud} &= B(x, x_{ud}), \\
\dot{\eta} &= -r\eta, \\
u &= -\big[W_{n+1}S'(x)\big]^{-1}\big[WS(x) + v\big], \\
v &= \tfrac{1}{2}rx - Ax, \\
h &= \tfrac{1}{2}|x|^2, \\
e &= \eta - h.
\end{aligned}$$

The update laws Eqs. (32) *and* (43) *guarantee the properties*

- $e, |x|, |x_{ud}|, \eta \in L_\infty, |e| \in L_2$;
- $\lim_{t\to\infty} e(t) = 0, \lim_{t\to\infty} |x(t)| = 0$;
- $\lim_{t\to\infty} \dot{W}(t) = 0, \lim_{t\to\infty} \dot{W}_{n+1}(t) = 0$,

provided that inequality (71) *holds.*

Case 2.   Now assume that the unknown vector field $\phi(x, x_{ud})$ satisfies the condition

$$\big|\phi(x, x_{ud})\big| \le \bar{\theta}. \tag{72}$$

In other words, $\phi(x, x_{ud})$ is assumed to be uniformly bounded by a constant. Thus inequality (66) becomes

$$\dot{\mathcal{L}} \le -r|e|^2 + \bar{\theta}|e||x|. \tag{73}$$

Following a similar procedure as in previous subsections we can prove the following theorem [21]:

THEOREM 7.   *Consider the system Eq.* (54) *with the unmodeled dynamics satisfying assumption A.4. Assume also that inequality* (72) *holds for the unknown vector field $\phi(x, x_{ud})$. Then the control law Eqs.* (35) *and* (36) *together with the* *update laws Eqs.* (32) *and* (43) *guarantee the uniform ultimate boundedness with respect to the sets*

$$\mathcal{E} = \left\{ e(t) : |e(t)| \leq \frac{4\bar{\theta}^2}{r^2}, \, r > 0 \right\},$$

$$\mathcal{X} = \left\{ x(t) : |x(t)| \leq \frac{4\bar{\theta}}{r}, \, r > 0 \right\}.$$

Furthermore,

$$\dot{e} = -re + x^T \widetilde{W} S(x) + x^T \widetilde{W}_{n+1} S'(x) u - x^T \phi(x, x_{\mathrm{ud}}). \tag{74}$$

Hence, because the boundedness of $\widetilde{W}$ and $\widetilde{W}_{n+1}$ is assured by the use of the projection algorithm and $\phi(x, x_{\mathrm{ud}})$ is uniformly bounded, we conclude that $\dot{e} \in L_\infty$.

The preceding analysis clearly shows the importance of the appropriate selection of the design constant r. If it is wisely chosen, our adaptive regulator can assure convergence of x to zero, or at least uniform ultimate boundedness of x and all other signals in the closed loop. Again, the more information we have about the unknown system, the better are the control results. The foregoing statement is mathematically translated as the constants $k_\phi, \theta, \bar{\theta}$ will be smaller, thus overcoming any implementation issues that may appear due to large value in r. Hence, it is desirable that the dynamic neural network match as accurately as possible the input–output behavior of the true but unknown nonlinear dynamical system.

Finally, observe that in the case where the unknown vector field $\phi(x, x_{\mathrm{ud}})$ is uniformly bounded by a constant, any positive value of r suffices to guarantee uniform boundedness of x.

2. Violation of the Uniform Asymptotic Stability in the Large Condition

In the present subsection we examine the effect of the unmodeled dynamics on the stability properties of the closed loop system when they violate the uniform asymptotic stability in the large condition, namely, inequality (65). Thus we assume that instead of inequality (65) we have

$$\frac{\partial V}{\partial x_{\mathrm{ud}}} B(x, x_{\mathrm{ud}}) \leq -\gamma_3 (|x_{\mathrm{ud}}|) + \rho |x|^2, \tag{75}$$

where ρ is a positive constant. Employing inequality (74) in inequality (68) we have

$$
\begin{aligned}
\dot{\mathcal{L}} &= -r|e|^2 - ex^T \phi(x, x_{\mathrm{ud}}) + \frac{\partial V}{\partial x_{\mathrm{ud}}} B(x, x_{\mathrm{ud}}) \\
&\leq -r|e|^2 - ex^T \phi(x, x_{\mathrm{ud}}) - \gamma_3 (|x_{\mathrm{ud}}|) + \rho |x|^2 \\
&\leq -r|e|^2 + |e||x||\phi(x, x_{\mathrm{ud}})| + \rho |x|^2.
\end{aligned} \tag{76}
$$

As in the previous subsection, we consider the following cases:

Case 1. Assume that the unknown vector field $\phi(x, x_{ud})$ satisfies inequalities (67) and (68). Then we can prove the following theorem [21]:

THEOREM 8. *Consider the system Eq. (54) with the unmodeled dynamics satisfying assumption (65). Assume also that inequalities (67) and (68) hold for the unknown vector field $\phi(x, x_{ud})$. Then the control law Eqs. (35) and (36) together with the update laws Eqs. (32) and (43) guarantee the uniform ultimate boundedness with respect to the sets*

$$\mathcal{E} = \left\{ e(t): \left| e(t) \right| \leq \frac{4\rho}{r - 4k_\phi\theta}, \ r > 0 \right\},$$

$$\mathcal{X} = \left\{ x(t): \left| x(t) \right| \leq 4\sqrt{\rho/(r - 4k_\phi\theta)}, \ r > 0 \right\}.$$

Furthermore,

$$\dot{e} = -re + x^T \widetilde{W} S(x) + x^T \widetilde{W}_{n+1} S'(x)u - x^T \phi(x, x_{ud}). \tag{77}$$

Hence, because the boundedness of $\widetilde{W}$ and $\widetilde{W}_{n+1}$ is assured by the use of the projection algorithm and $\phi(x, x_{ud})$ is bounded, we conclude that $\dot{e} \in L_\infty$.

Case 2. Now we assume that the unknown vector field $\phi(x, x_{ud})$ satisfies inequality (72). Then we can prove the following theorem [21]:

THEOREM 9. *Consider the system Eq. (54) with the unmodeled dynamics satisfying assumption (65). Assume further that inequality (72) holds for the unknown vector field $\phi(x, x_{ud})$. Then the control law Eqs. (35) and (36) together with the update laws Eqs. (32) and (43) guarantee the uniform ultimate boundedness with respect to the sets*

$$\mathcal{E} = \left\{ e(t): \left| e(t) \right| \leq \frac{(\bar{\theta} + \sqrt{\bar{\theta}^2 + 4\rho r})^2}{r^2}, \ r > 0 \right\},$$

$$\mathcal{X} = \left\{ x(t): \left| x(t) \right| \leq \frac{2(\bar{\theta} + \sqrt{\bar{\theta}^2 + 4\rho r})}{r}, \ r > 0 \right\}.$$

Furthermore,

$$\dot{e} = -re + x^T \widetilde{W} S(x) + x^T \widetilde{W}_{n+1} S'(x)u - x^T \phi(x, x_{ud}). \tag{78}$$

Hence, because the boundedness of $\widetilde{W}$ and $\widetilde{W}_{n+1}$ is assured by the use of the projection algorithm and $\phi(x, x_{ud})$ is bounded, we conclude that $\dot{e} \in L_\infty$.

Theorems 7 and 8 demonstrate that even in the case where the uniform asymptotic stability in the large condition is violated, our adaptive regulator still can assure the uniform ultimate boundedness of the state x and of all signals in the closed loop. Appropriate selection of the design constant r leads to better performance, provided it satisfies any implementation constraints. Again, the accuracy of our model (dynamical neural network) is a performance index of our adaptive regulator.

V. CONCLUSIONS

The purpose of this chapter is to present two techniques to adaptively control unknown nonlinear dynamical systems using dynamical neural networks. In the indirect control case, it is shown that even when unmodeled dynamics are present and affect the system performance, the control scheme can guarantee the convergence of the control error to zero, plus boundedness of all other signals in the closed loop, thereby confining the unmodeled dynamics into the framework of singular perturbation theory.

Considerable extensions to more general cases are provided in the direct case, because the restrictive singular perturbation modeling of the unmodeled dynamics is relaxed and a comprehensive and rigorous analysis on the effects of modeling errors, is also provided. In the worst case the control architecture guarantees a uniform ultimate boundedness property for the state and boundedness of all other signals in the closed loop.

REFERENCES

[1] K. J. Hunt, D. Sbarbaro, R. Zbikowski, and P. J. Gawthrop. Neural networks for control systems – A survey. *Automatica* 28:1083–1112, 1992.
[2] D. G. Taylor, P. V. Kokotovic, R. Marino, and I. Kanellakopoulos. Adaptive regulation of nonlinear systems with unmodeled dynamics. *IEEE Trans. Automat. Control* 34:405–412, 1989.
[3] I. Kanellakopoulos, P. V. Kokotovic, and R. Marino. An extended direct scheme for robust adaptive nonlinear control. *Automatica* 27:247–255, 1991.
[4] S. Sastry and A. Isidori. Adaptive control of linearizable systems. *IEEE Trans. Automat. Control* 34:1123–1131, 1989.
[5] I. Kanellakopoulos, P. V. Kokotovic, and A. S. Morse. Systematic design of adaptive controllers for feedback linearizable systems. *IEEE Trans. Automat. Control* 36:1241–1253, 1991.
[6] G. Campion and G. Bastin. Indirect adaptive state feedback control of linearly parametrized nonlinear systems. *Internat. J. Adaptive Control Signal Process.* 4:345–358, 1990.
[7] J.-B. Pomet and L. Praly. Adaptive nonlinear regulation: Estimation from the Lyapunov equation. *IEEE Trans. Automat. Control* 37:729–740, 1992.
[8] R. Marino and P. Tomei. Global adaptive output feedback control of nonlinear systems, Part I: Linear parameterization. *IEEE Trans. Automat. Control* 38:17–32, 1993.
[9] R. Marino and P. Tomei. Global adaptive output feedback control of nonlinear systems, Part II: Nonlinear parameterization. *IEEE Trans. Automat. Control* 38:33–48, 1993.
[10] K. S. Narendra and K. Parthasarathy. Identification and control of dynamical systems using neural networks. *IEEE Trans. Neural Networks* 1:4–27, 1990.
[11] E. D. Sontag. Feedback stabilization using two-hidden-layer nets. *IEEE Trans. Neural Networks* 3:981–990, 1992.

[12] G. Cybenco. Approximation by superpositions of a sigmoidal function. *Math. Control, Signals, Systems* 2:303–314, 1989.

[13] K. Funahashi. On the approximate realization of continuous mappings by neural networks. *Neural Networks* 2:183–192, 1989.

[14] K. M. Hornik, M. Stinchombe, and H. White. Multilayer feedforward networks are universal approximators. *Neural Networks* 2:359–366, 1989.

[15] R. M. Sanner and J.-J. Slotine. Gaussian networks for direct adaptive control. *IEEE Trans. Neural Networks* 3:837–863, 1992.

[16] M. M. Polycarpou and P. A. Ioannou. On the existence and uniqueness of solutions in adaptive control systems. *IEEE Trans. Automat. Control* 38:474–479, 1993.

[17] V. I. Utkin. *Sliding Modes and their Applications to Variable Structure Systems.* MIR, Moscow, 1978.

[18] K. D. Young, P. V. Kokotovic, and V. I. Utkin. A singular perturbation analysis of high gain feedback systems. *IEEE Trans. Automat. Control* AC-22:931–938, 1977.

[19] M. M. Polycarpou and P. A. Ioannou. Identification and control of nonlinear systems using neural network models: Design and stability analysis. Technical Report 91-09-01, Department Electrical Engineering and Systems, University of Southern California, Los Angeles, 1991.

[20] G. A. Rovithakis and M. A. Christodoulou. Adaptive control of unknown plants using dynamical neural networks. *IEEE Trans. Systems, Man, Cybernetics* 24:400–412, 1994.

[21] G. A. Rovithakis and M. A. Christodoulou. Direct adaptive regulation of unknown nonlinear dynamical systems via dynamic neural networks. *IEEE Trans. Systems, Man, Cybernetics* 25:1578–1595, 1995.

[22] G. H. Golub and C. F. Van Loan. *Matrix Computations*, 2nd ed. The John Hopkins Univ. Press, Baltimore, 1989.

[23] H. K. Khalil. *Nonlinear Systems.* Maxwell Macmillan International, Singapore, 1992.

[24] G. A. Rovithakis and M. A. Christodoulou. Neural adaptive regulation of unknown nonlinear dynamical systems. *IEEE Trans. Systems, Man, Cybernetics*, to appear.

[25] A. N. Michel amd R. K. Miller. *Qualitative Analysis of Large Scale Dynamic Systems.* Academic Press, New York, 1977.

[26] N. Rouche, P. Habets, and M. Laloy. *Stability Theory by Liapunov's Direct Method.* Springer-Verlag, New York, 1977.

A Receding Horizon Optimal Tracking Neurocontroller for Nonlinear Dynamic Systems

Young-Moon Park
Department of Electrical
Engineering
Seoul National University
Seoul 151-742, Korea

Myeon-Song Choi
Department of Electrical
Engineering
Myong Ji University
Yongin 449-728, Korea

Kwang Y. Lee
Department of Electrical
Engineering
Pennsylvania State
University
University Park,
Pennsylvania 16802

The receding horizon optimal control problem of nonlinear dynamic systems is considered using neural networks. A neural network training algorithm named generalized backpropagation-through-time (GBTT) is developed to deal with a cost function defined in a finite horizon. Multilayer neural networks are used to design an optimal tracking neurocontroller (OTNC) for discrete-time nonlinear dynamic systems with quadratic cost function. The OTNC is made up of two controllers: feedforward neurocontroller (FFNC) and feedback neurocontroller (FBNC). The FFNC controls the steady-state output of the plant, whereas the FBNC controls the transient-state output of the plant. The FFNC is designed using a novel inverse mapping concept with a neuro-identifier. The GBTT algorithm is used to minimize the general quadratic cost function for the FBNC training. The proposed methodology is useful as an off-line control method where the plant is first identified and then a controller is designed for it. Two case studies for a typical plant and a power system with nonlinear dynamics show good performance of the proposed OTNC.

Control and Dynamic Systems

I. INTRODUCTION

Traditional controller design involves complex mathematical analysis and yet, has many difficulties in controlling highly nonlinear plants. To overcome these difficulties, a number of new approaches which use neural networks for control have increased significantly in recent years. The use of neural networks' learning ability helps controller design to be flexible, especially when plant dynamics are complex and highly nonlinear. This is a distinct advantage over traditional control methods.

Poggio and Girosi [1] elucidated that the problem of learning between input and output spaces is equivalent to that of synthesizing an associative memory that retrieves appropriate output when the input is present and generalizes when a new input is applied. It is equivalent to the problem of estimating an input–output transformation using given input–output pairs as training sets. It also can be included in the classical framework of approximation theory.

Nguyen and Widrow [2] showed the possibility of using neural networks to control a plant with high nonlinearities. They exploited the neural networks' self-learning ability in the "truck-backer" problem. Chen [3] proposed a self-tuning adaptive controller with neural networks, and Iiguni and Sakai [4] constructed a neural network controller combined with a linear optimal controller to compensate for the uncertainties in model parameters. Recently, Ku and Lee [5] proposed an architecture of diagonal recurrent neural networks for identification and control, and applied it to a nuclear reactor control problem [6]. There are a number of other cases in which neural networks' learning ability is applied for plant control [7–11].

The use of neural networks in control has focused mostly on the model reference adaptive control (MRAC) problem [3, 5–11]. This chapter introduces a new class of control problems, namely, the optimal tracking problem, which minimizes a general quadratic cost function of tracking errors and control efforts. This results in a hybrid of *feedback* and *feedforward* neurocontrollers in parallel. The feedforward neurocontroller (FFNC) generates steady-state control input to keep the plant output at a given reference value, and the feedback neurocontroller (FBNC) generates the transient control input to stabilize error dynamics along the optimal path while minimizing the cost function. A novel inverse mapping concept is developed to design the FFNC using an *identificaiton neural network* (IDNN).

The use of the general quadratic cost function provides "optimal" performance with respect to trade-offs between the tracking error and control effort. Whereas the cost function is defined over a finite time interval, a *generalized backpropagation-through-time* (GBTT) algorithm is developed to train the feedback controller by extending Werbo's *backpropagation-through-time* (BTT) algorithm [10], which was originally developed for the cost function of the tracking error alone.

The control methodology in this chapter is useful as an off-line control method where the plant is first identified and then a controller is designed for it. This assumes that the plant can be identified without making the plant unstable. Repeatability of the control experiment also is assumed to tune the parameters of the neural networks. This is often the case when we design a supplementary controller for an existing control system in a power plant.

The organization of the chapter is as follows. Section II presents the formulation of the optimal tracking control problem and architecture for the *optimal tracking neurocontroller* (OTNC). Section III shows the development of neurocontrollers including their training algorithms. In Section IV, the proposed OTNC is implemented in a typical nonlinear plant and in power plant voltage control. Conclusions are drawn in Section V.

II. RECEDING HORIZON OPTIMAL TRACKING CONTROL PROBLEM FORMULATION

An optimal tracking control problem is first formulated in this section with an appropriate justification for the use of a general quadratic cost function. Motivation for the use of a feedforward controller is illustrated with the aid of linear optimal control theory. Finally, an architecture is given for the OTNC.

A. RECEDING HORIZON OPTIMAL TRACKING CONTROL PROBLEM OF A NONLINEAR SYSTEM

We consider a system in the form of the general nonlinear autoregressive moving average (NARMA) model

$$y_{(k+1)} = f(y_{(k)}, y_{(k-1)}, \ldots, y_{(k-n+1)}, u_{(k)}, u_{(k-1)}, \ldots, u_{(k-m+1)}), \quad (1)$$

where y and u, respectively, represent output and input variables, k represents time index, and n and m represent the respective output and input delay orders.

When the target output of a plant holds up for some time and varies from time to time, the control objectives can be defined as follows [12]:

1. Minimize the summation of the squares of regulating output error and the squares of input error in transient.
2. Reduce the steady-state error to zero.

The preceding control objectives can be achieved by minimizing the well-known quadratic cost function

$$J = \tfrac{1}{2} \sum_{k=1}^{N} \left(Q(y_{\text{ref}} - y_{(k+1)})^2 + R(u_{\text{ref}} - u_{(k)})^2 \right), \tag{2}$$

where y_{ref} is a reference output, u_{ref} is the steady-state input corresponding to y_{ref}, and Q and R are positive weighting factors. This quadratic cost function or the performance index not only forces the plant output to follow the reference, but also forces the plant input to be close to the steady-state value in maintaining the plant output at its reference value.

The instantaneous error alone as a performance index tends to exert large inputs and causes the plant to oscillate around the tracking reference. On the other hand, the quadratic performance index Eq. (2) limits the control energy to be expanded over some time interval while minimizing an accumulated error. For linear time-invariant dynamic systems, this performance index leads to a fixed feedback gain matrix as N approaches infinity.

Before solving the preceding control problem for a general nonlinear system, restriction of the model to a linear system will be helpful to gain insight for the need of both feedforward and feedback controls. A linear counterpart to the NARMA model Eq. (1) is the linear time-invariant system

$$\begin{aligned} x_{(k+1)} &= Ax_{(k)} + Bu_{(k)}, \\ y_{(k)} &= Cx_{(k)}, \end{aligned} \tag{3}$$

where x is an n-dimensional state vector and A, B, and C are constant matrices of appropriate dimension. When the output $y_{(k)}$ has the set point y_{ref} in steady state, then state equation (3) becomes

$$\begin{aligned} x_{\text{ref}} &= Ax_{\text{ref}} + Bu_{\text{ref}}, \\ y_{\text{ref}} &= Cx_{\text{ref}}, \end{aligned} \tag{4}$$

where x_{ref} is the state vector corresponding to y_{ref} in steady state [13]. By subtracting Eq. (4) from Eq. (3), and shifting the vectors as

$$u'_{(k)} = u_{(k)} - u_{\text{ref}}, \tag{5a}$$

$$x'_{(k)} = x_{(k)} - x_{\text{ref}}, \tag{5b}$$

$$y'_{(k)} = y_{(k)} - y_{\text{ref}}, \tag{5c}$$

the optimal tracking problem Eqs. (2) and (3) is converted to the optimal regulating problem with zero output in steady state,

$$\begin{aligned} x'_{(k+1)} &= Ax'_{(k)} + Bu'_{(k)}, \\ y'_{(k)} &= Cx'_{(k)}, \end{aligned} \tag{6}$$

with the quadratic cost function

$$J = \tfrac{1}{2} \sum_{k=1}^{N} \left(Q(y'_{(k+1)})^2 + R(u'_{(k+1)})^2 \right). \tag{7}$$

The control law for the optimal regulator problem defined by Eqs. (6) and (7) for $N = \infty$ is given as

$$u'_{(k)} = F x'_{(k)}, \tag{8}$$

where F is the optimal feedback gain matrix obtained by solving an algebraic matrix Riccati equation [12]. Thus from Eq. (5a), the optimal control for the original linear system, Eq. (3), is

$$u_{(k)} = F x'_{(k)} + u_{\text{ref}}. \tag{9}$$

This shows the important observation that the control input consists of two parts—feedforward and feedback,

$$u_{(k)} = u_{\text{fb}}\left(x'_{(k)}\right) + u_{\text{ff}}(y_{\text{ref}}), \tag{10}$$

where u_{fb} represents the feedback control and u_{ff} represents the feedforward control corresponding to the steady-state output y_{ref}. When the problem is limited to the case where u_{ref} exists for any y_{ref}, it is reasonable to assume that the control input for a *nonlinear* system can also be separated into feedforward and feedback parts. The role of the feedforward control is to keep the plant output close to the reference value in steady state. The role of the feedback control is to stabilize the tracking error dynamics during transient [14].

There have been a number of studies on solving the tracking control problem with feedforward and feedback controls [15, 16]. Meystel and Uzzaman [15] used an algorithmic procedure—*approximate inversion*—to find the feedforward control input for a given reference trajectory; however, although it guarantees admissible solutions, it leads to exhaustive binary search. Then they used a simple constant-gain proportional control for feedback control input. Jordan and Rumelhart [16] used an inverse learning model to find the feedforward control when the reference trajectory of plant output was given. However, it did not provide an optimal controller for the optimal tracking problem defined with the general quadratic performance index Eq. (2).

Although an explicit solution for the optimal tracking problem is available for linear systems, it is not possible, in general, for nonlinear systems. Iiguni and Sakai [4] tried to design a nonlinear regulator using neural networks by assuming the plant to be linear but with uncertain parameters. For linear systems, minimization of the quadratic performance index is possible, and training of the neurocontroller can be simplified by using the Ricatti equation. However, this approximation is not possible for nonlinear systems; hence, a method of minimizing the quadratic performance index is developed by generalizing the BTT algorithm.

B. ARCHITECTURE FOR AN OPTIMAL TRACKING NEUROCONTROLLER

Following the preceding observation, an optimal tracking neurocontroller (OTNC) is designed with two neurocontrollers to control a nonlinear plant that has a nonzero set point in steady state. A *feedforward* neurocontroller (FFNC) is constructed to generate feedforward control input corresponding to the set point, and is trained by the error backpropagation algorithm (BPA). A *feedback* neurocontroller (FBNC) is constructed to generate feedback control input and is trained by a generalized BTT (GBTT) algorithm to minimize the quadratic performance index.

An independent neural network called identification neural network (IDNN) is used when the FFNC and FBNC are in the training mode. This network is trained to emulate plant dynamics and to backpropagate an *equivalent error* or *generalized delta* [2] to the controllers under training. Figure 1 shows an architecture for the optimal tracking neurocontroller for a nonlinear plant. In the figure, the *tapped delay operator* Δ is defined as a delay mapping from a sequence of scalar input, $\{x_{(i)}\}$, to a vector output with an appropriate dimension defined as $\mathbf{x}_{(i-1)} = (x_{(i-1)}, x_{(i-2)}, \ldots, x_{(i-p)})$, where $p = n$ for the output variable y and $p = m - 1$ for the input variable u.

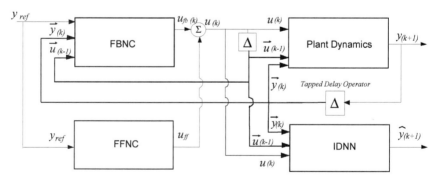

Figure 1 Block diagram for the optimal tracking neurocontroller.

III. DESIGN OF NEUROCONTROLLERS

A. STRUCTURE OF MULTILAYER FEEDFORWARD NEURAL NETWORKS

The proposed OTNC is made of three multilayer feedforward neural networks: identification neural network, feedforward neurocontroller, and feedback neuro-controller. The structure of multilayer feedforward neural networks is shown in Fig. 2, where I, O, x, and W represent input, output, state variable, and weight parameter matrix, respectively. Nl represents the number of layers in the neural networks and N_i is the number of nodes in the ith layer.

The structure of the multilayer feedforward neural network represents a nonlinear function with input and output. The characteristic of the nonlinear function is determined by the number of layers and nodes, or perceptrons, that have weight parameters and a nonlinear activation function. The organization of a perceptron is represented in Fig. 3, where a hyperbolic tangent function is used as the activation function. The weighting parameters of a perceptron are adjusted by the *equivalent error* with the error backpropagation algorithm in the training mode.

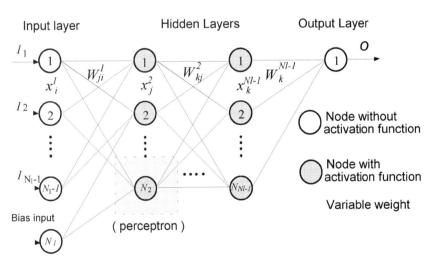

Figure 2 Structure of a multilayer feedforward neural network.

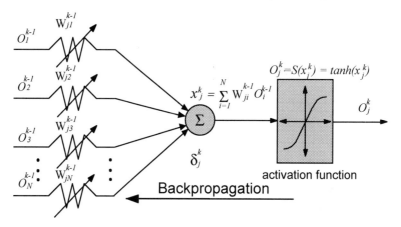

Figure 3   Structure of a perceptron.

B. IDENTIFICATION NEURAL NETWORK

The function of the training of IDNN corresponds to the identification of the plant dynamics. It is then used to backpropagate the equivalent error to the neurocontrollers. Training the IDNN can be regarded as an approximation process of a nonlinear function using input–output data sets [1].

A NARMA model Eq. (1) can be viewed as a nonlinear mapping from $(n+m)$-dimensional input space to a one-dimensional output space

$$y_{(k+1)} = f(\mathbf{I}_{i(k)}), \tag{11}$$

where $\mathbf{I}_{i(k)}$ is the identifier input vector defined as

$$\mathbf{I}_{i(k)} \overset{\Delta}{=} \{y_{(k)}, y_{(k-1)}, \ldots, y_{(k-n+1)}, u_{(k)}, u_{(k-1)}, \ldots, u_{(k-m+1)}\}.$$

IDNN can be viewed as a nonlinear function, F, to approximate f,

$$\hat{y}_{(k+1)} = F(\mathbf{I}_{i(k)}, \mathbf{W}_i), \tag{12}$$

where $\mathbf{W}_i$ is the weight parameter matrix in IDNN.

Then the training of IDNN is to adjust weight parameters so that IDNN can emulate the nonlinear function of the plant dynamics using the input–output training patterns obtained in the wide range operation under various conditions.

IDNN learns the way to generate the same output as the plant for the same input by using the error *backpropagation algorithm* (BPA). The block diagram for the training of IDNN is shown Fig. 4.

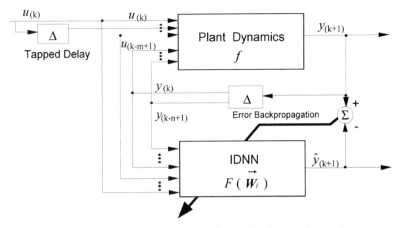

Figure 4   Block diagram for training the identification neural network.

The architecture of IDNN is the multilayer feedforward neural network shown in Fig. 2, where $\mathbf{I}_{i(k)}$ is applied as the input vector $\mathbf{I}$ and the output O represents the output estimate $\hat{y}_{(k+1)}$.

The general BPA can be used to reduce an *error function for identification* (EI) defined as an average of errors given for a training data set,

$$\text{EI} = \frac{1}{Nd} \sum_{i=1}^{Nd} \sum_{k=0}^{Nh-1} \frac{1}{2} (y_{(k+1)}^i - \hat{y}_{(k+1)}^i)^2, \tag{13}$$

where $y_{(k)}$ is the output of the plant, $\hat{y}_{(k)}$ is the output of IDNN at time step k, Nd is the number of training data sets, the superscript i represents that the data set is the ith training sample, and Nh is the number of the time horizon.

The training for the ith sample begins by generating sequences of IDNN outputs $\{\hat{y}_{(1)}^i, \hat{y}_{(2)}^i, \ldots, \hat{y}_{(Nh)}^i\}$ from the identification input vector defined as $\mathbf{I}_{i(k)} = \{y_{(k)}^i, y_{(k-1)}^i, \ldots, y_{(k-n+1)}^i, u_{(k)}^i, \ldots, u_{(k-m+1)}^i\}$ for $k = 0, 1, \ldots, Nh - 1$ using the given sequence of data set $\{y_{(-n+1)}^i, y_{(-n+2)}^i, \ldots, y_{(Nh)}^i, u_{(-m+1)}^i, u_{(-m+2)}^i, \ldots, u_{(Nh-1)}^i\}$. These variables provide the error terms $(y_{(k)}^i - \hat{y}_{(k)}^i)$ for $k = 1, 2, \ldots, Nh$. This process is repeated for other data sets. The BPA learning process for these sets of data is based on an output sensitivity of the identification performance index defined as

$$\delta_{y^i}^k \triangleq -\frac{\partial \text{EI}}{\partial \hat{y}_{(k)}^i} = \frac{1}{Nd} (y_{(k)}^i - \hat{y}_{(k)}^i), \qquad i = 1, 2, \ldots, Nd, \ k = 1, 2, \ldots, Nh. \tag{14}$$

Noting that the output $O_{(k)}$ of the IDNN, is an estimation of $\hat{y}_{(k+1)}$ at time step k, to use the BPA, the equivalent error on the output node IDNN for the ith training data set is defined as

$$\delta^k_{oi} = -\frac{\partial \mathbf{EI}}{\partial O^i_{(k)}} = -\frac{\partial \mathbf{EI}}{\partial \hat{y}^i_{(k+1)}} = \delta^{k+1}_{yi}, \qquad k = 0, 1, \ldots, Nh - 1. \tag{15}$$

This error is then used to compute an equivalent error for a node in an arbitrary layer by using the BPA.

The objective of the conventional BPA is to update the weight parameters by using equivalent errors computed from the equivalent error defined on the output node. An equivalent error on a node of a hidden layer is calculated by backpropagating the equivalent errors of the immediately previous layer. Then a weight parameter in the network is updated with equivalent errors. The general BPA corresponding to the network shown in Figs. 2 and 3 is summarized in subsequent text.

The equivalent errors for the ith data on the mth node in the layers are as follows.

The (Nl–1)th layer (preceding the output layer)

$$\delta^k_{(x^{Nl-1}_m)i} \triangleq -\frac{\partial \mathbf{EI}}{\partial (x^{Nl-1}_m)^i_k} = -\frac{\partial \mathbf{EI}}{\partial O^i_{(k)}} \cdot \frac{\partial O^i_{(k)}}{\partial (x^{Nl-1}_m)^i_k}$$

$$= \dot{s}\left((x^{Nl-1}_m)^i_k\right) \cdot \delta^k_{oi} \cdot W^{Nl-1}_m. \tag{16}$$

The (Nl–2)th layer

$$\delta^k_{(x^{Nl-2}_m)i} \triangleq -\frac{\partial \mathbf{EI}}{\partial (x^{Nl-2}_m)^i_k} = -\frac{\partial \mathbf{EI}}{\partial O^i_{(k)}} \cdot \sum_{j=1}^{N_{Nl-1}} \left[\frac{\partial O^i_{(k)}}{\partial (x^{Nl-1}_j)^i_k} \frac{\partial (x^{Nl-1}_j)^i_k}{\partial (x^{Nl-2}_m)^i_k} \right]$$

$$= \sum_{j=1}^{N_{Nl-1}} \left[-\frac{\partial \mathbf{EI}}{\partial O^i_{(k)}} \cdot \frac{\partial O^i_{(k)}}{\partial (x^{Nl-1}_j)^i_k} \frac{\partial (x^{Nl-1}_j)^i_k}{\partial (x^{Nl-2}_m)^i_k} \right]$$

$$= \dot{s}\left((x^{Nl-2}_m)^i_k\right) \cdot \sum_{j=1}^{N_{Nl-1}} \left[\delta^k_{(x^{Nl-1}_j)i} W^{Nl-2}_{mj} \right] \tag{17}$$

$$\vdots$$

The 1st layer (the input layer)

$$\delta^k_{(x^1_m)^i} \stackrel{\Delta}{=} -\frac{\partial \mathbf{EI}}{\partial (x^1_m)^i_k} = \sum_{j=1}^{N_2} \left[\delta^k_{(x^2_j)^i} W^1_{mj} \right]. \tag{18}$$

Then the weight update rules are

$$\mathbf{W}^{\text{new}} = \mathbf{W}^{\text{old}} + \Delta \mathbf{W}. \tag{19}$$

The (Nl−1)th layer (preceding the output layer)

$$\Delta W^{Nl-1}_m = -\alpha \frac{\partial \mathbf{EI}}{\partial (W^{Nl-1}_m)^{\text{old}}} = -\alpha \sum_{i=1}^{Nd} \sum_{k=1}^{Nh} \left[\frac{\partial \mathbf{EI}}{\partial O^i_{(k)}} \frac{\partial O^i_{(k)}}{\partial (W^{Nl-1}_m)^{\text{old}}} \right]$$

$$= \alpha \sum_{i=1}^{Nd} \sum_{k=1}^{Nh} \left[\delta^k_{0^i} \cdot s \left((x^{Nl-1}_m)^i_k \right) \right]. \tag{20}$$

The (Nl−2)th layer

$$\Delta W^{Nl-2}_{mj} = \alpha \sum_{i=1}^{Nd} \sum_{k=1}^{Nh} \left[\delta^k_{(x^{Nl-1}_j)^i} \cdot s \left((x^{Nl-2}_m)^i_k \right) \right], \quad j = 1, 2, \ldots, N_{Nl-2}, \tag{21}$$

$$\vdots$$

The 1st layer (the input layer)

$$\Delta W^1_{mj} = \alpha \sum_{i=1}^{Nd} \sum_{k=1}^{Nh} \left[\delta^k_{(x^2_j)^i} \cdot (x^1_m)^i_k \right], \quad j = 1, 2, \ldots, N_1, \tag{22}$$

where α is the learning rate and $s(\cdot)$ and $\dot{s}(\cdot)$ are the activation function and its derivative, respectively.

Through the learning process, the plant dynamic characteristics are stored in the weighting parameters of IDNN. The training is terminated when the average error between the plant and IDNN outputs converges to a small value. Then,

IDNN is presumed to have learned, approximately, the plant characteristics with converged weight parameters $\mathbf{W}_i^*$, that is,

$$y_{(k+1)} = f(\mathbf{I}_{i(k)}) \approx \hat{y}_{(k+1)} = F(\mathbf{I}_{i(k)}, \mathbf{W}_i^*). \tag{23}$$

C. FEEDFORWARD NEUROCONTROLLER

In designing a plant controller to follow an arbitrary reference output, it is necessary to keep the steady-state tracking error at zero. For this purpose, the FFNC is designed to generate a control input which will maintain the plant output to a given reference output in steady state. The FFNC is then required to learn the inverse dynamics of the plant in steady state. A novel approach is now proposed to develop the inverse mapping with the aid of the IDNN.

Note that the steady-state control input can be obtained by setting $y_{(k)} \equiv y_{\text{ref}}$ and $u_{(k)} \equiv u_{\text{ref}}$ for all k in the NARMA model Eq. (1), that is

$$y_{\text{ref}} = f(y_{\text{ref}}, y_{\text{ref}}, \ldots, y_{\text{ref}}, u_{\text{ref}}, u_{\text{ref}}, \ldots, u_{\text{ref}}) \tag{24}$$

or, equivalently,

$$u_{\text{ref}} = g(y_{\text{ref}}), \tag{25}$$

which is the inverse function of Eq. (24). The inverse is not unique, in general, and any one solution is sufficient for control purpose. However, noting that the control input is the control energy, the smallest solution of u_{ref} is preferred. The FFNC network G, as an inverse mapping of the plant in steady state, can be developed by using the IDNN F as shown in Fig. 5, that is,

$$\hat{y}_{\text{ref}} = F(y_{\text{ref}}, y_{\text{ref}}, \ldots, y_{\text{ref}}, u_{\text{ff}}, u_{\text{ff}}, \ldots, u_{\text{ff}}, \mathbf{W}), \tag{26}$$

$$u_{\text{ff}} = G(y_{\text{ref}}, \mathbf{W}), \tag{27}$$

where u_{ff} is the feedforward control input and $\hat{y}_{\text{ref}}$ is the output of the IDNN designed in Eq. (23). Training of the FFNC is to adjust its weight parameters so that the output of the IDNN $\hat{y}_{\text{ref}}$ approximates the given reference output y_{ref}, and when the training is finished u_{ff} approximates u_{ref}. Training the FFNC can be understood as an approximation process of the inverse dynamics for the plant in steady state. To train the FFNC an equivalent error is defined, which then is propagated back through the IDNN that has been trained already.

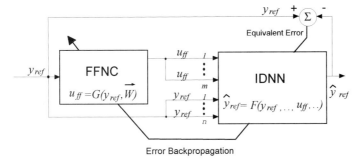

Figure 5 Block diagram for training the feedforward controller.

The objective of training the FFNC is to reduce the average error defined by

$$J = \frac{1}{N} \sum_{i=1}^{N} \frac{1}{2} \left(y_{\text{ref}}^i - \hat{y}_{\text{ref}}^i (u_{\text{ff}}^i) \right)^2, \tag{28}$$

where N is the number of samples in a training set group.

To update the weight parameters in the FFNC the equivalent error is propagated backward through the IDNN. The equivalent error on the output of the FFNC is defined as the negative sensitivity of the preceding performance index with respect to u_{ff}, which can be calculated from the equivalent error on the IDNN input nodes:

$$\delta_{u_{\text{ff}}}^i \triangleq -\frac{\partial J}{\partial u_{\text{ff}}^i} = -\frac{\partial J}{\partial \hat{y}_{\text{ref}}^i} \cdot \frac{\partial \hat{y}_{\text{ref}}^i}{\partial u_{\text{ff}}^i}. \tag{29}$$

Whereas u_{ff} is applied to the first m input nodes of the IDNN, that is, $(I_k)^i = u_{\text{ff}}^i$, $k = 1, 2, \ldots, m$, then,

$$\delta_{u_{\text{ff}}}^i = \sum_{k=1}^{m} -\frac{\partial J}{\partial \hat{y}_{\text{ref}}^i} \cdot \frac{\partial \hat{y}_{\text{ref}}^i}{\partial (I_k)^i}$$

$$= \sum_{k=1}^{m} \delta_{I_k}^i, \tag{30}$$

where $\delta_{I_k}^i$ is the equivalent error of the u_{ff} input node in the IDNN, which is computed by the BPA. Because u_{ff} is also the output of the FFNC, the equivalent error Eq. (30) can be used directly as the equivalent error for the network G in the BPA.

Training begins with small random values of weight parameters in the FFNC. This allows the feedforward control input to grow from a small random value and converge to the smallest solution of u_{ref}, which is preferred over all other possible solutions.

At the end of the training, the weight parameters in the FFNC are adjusted so that the output of the IDNN follows a given reference output. Training is finished when the average error between the IDNN output and the given reference output converges to a small value and FFNC has learned the steady-state inverse dynamics of the plant with the help of the IDNN:

$$
\begin{aligned}
y_{\text{ref}} &= f(y_{\text{ref}}, y_{\text{ref}}, \ldots, y_{\text{ref}}, u_{\text{ref}}, u_{\text{ref}}, \ldots, u_{\text{ref}}) \\
&\approx \hat{y}_{\text{ref}} = F(y_{\text{ref}}, y_{\text{ref}}, \ldots, y_{\text{ref}}, u_{\text{ff}}, u_{\text{ff}}, \ldots, u_{\text{ff}}, \mathbf{W}), \\
u_{\text{ref}} &\approx u_{\text{ff}} = G(y_{\text{ref}}, \mathbf{W}).
\end{aligned}
\tag{31}
$$

D. Feedback Neurocontroller

The role of the feedback neurocontroller (FBNC) is to stabilize tracking error dynamics when the plant output follows an arbitrarily given reference output, where it is assumed that the optimal control law exists and has a form characterized by a feedback action. This objective can be achieved by minimizing the quadratic performance index Eq. (2) discussed previously.

Noting that $u_{\text{ref}} \approx u_{\text{ff}}$ and $u_{(k)} = u_{\text{ff}} + u_{\text{fb}(k)}$, the performance index Eq. (2) can be modified as

$$
J = \sum_{k=1}^{N} J_k = \tfrac{1}{2} \sum_{k=1}^{N} \left(Q(y_{\text{ref}} - y_{(k+1)})^2 + R(u_{\text{fb}(k)})^2 \right),
\tag{32}
$$

where $u_{\text{fb}(k)}$ is the feedback control input.

From the NARMA model Eq. (1), the feedback control input can be viewed as an inverse mapping

$$
u_{\text{fb}(k)} = h(y_{\text{ref}}, y_{(k)}, y_{(k-1)}, \ldots, y_{(k-n+1)}, u_{(k-1)}, u_{(k-2)}, \ldots, u_{(k-m+1)}),
\tag{33}
$$

where y_{ref} indicates the nonlinear dependency of the input function on the set point. The corresponding FBNC can be represented as a nonlinear network H

$$
u_{\text{fb}(k)} = H(y_{\text{ref}}, y_{(k)}, y_{(k-1)}, \ldots, y_{(k-n+1)}, u_{(k-1)}, u_{(k-2)}, \ldots, u_{(k-m+1)}, \mathbf{W}).
\tag{34}
$$

Whereas the target value for the optimal feedback control $u_{\text{fb}(k)}$ is not available for training, the traditional BPA method is not applicable here. Therefore, the FBNC learns the control law by trial and error as it drives the IDNN to generate the equivalent error for backpropagation.

The learning process by trial and error consists of two parts. First, from the given initial state and an arbitrarily given reference set point, the combined

FFNC and FBNC drive the IDNN for N steps. Second, update the weight parameters in the FBNC using the equivalent error generated by the generalized backpropagation-through-time algorithm developed in the following section.

E. GENERALIZED BACKPROPAGATION-THROUGH-TIME ALGORITHM

GBTT, an extension of the BTT algorithm of Werbos [10], generates an equivalent error from a general quadratic cost function Eq. (32). Originally, the BTT was used for the cost function with output error only. Conversely, the GBTT is used for the general quadratic cost function Eq. (32) which includes not only output errors, but also input variables.

The GBTT is based upon output and input sensitivities of the cost function defined by

$$\delta_y^k \triangleq -\frac{\partial J}{\partial y_{(k)}}, \qquad k = 1, 2, 3, \ldots, N + 1. \tag{35}$$

$$\delta_u^k \triangleq -\frac{\partial J}{\partial u_{(k)}}, \qquad k = 0, 1, 2, \ldots, N. \tag{36}$$

Whereas for a fixed feedforward control,

$$\delta_{u_{\text{fb}}}^k = -\frac{\partial J}{\partial u_{\text{fb}(k)}} = -\frac{\partial J}{\partial u_{(k)}} \frac{\partial u_{(k)}}{\partial u_{\text{fb}(k)}} = -\frac{\partial J}{\partial u_{(k)}} \frac{\partial (u_{\text{ff}} + u_{\text{fb}(k)})}{\partial u_{\text{fb}(k)}} = -\frac{\partial J}{\partial u_{(k)}} = \delta_u^k, \tag{37}$$

the subscript fb will be dropped in the following development.

1. Output Sensitivity Equation

An output $y_{(k)}$ at an arbitrary time step k influences both the plant dynamics Eq. (1) and the inverse dynamics Eq. (31). Whereas the plant dynamics Eq. (1) is defined with n delayed output variables, an arbitrary output $y_{(k)}$ influences the plant dynamics for the next n steps, that is, $y_{(k+i)}$ is a function of $y_{(k)}$ for $i = 1, 2, \ldots, n$. Similarly, whereas the inverse dynamics Eq. (31) also has n delayed output variables, an output $y_{(k)}$ influences the input for the next n steps, that is, $u_{(k+i)}$ is a function of $y_{(k)}$ for $i = 0, 1, 2, \ldots, n - 1$.

Recall that the performance index Eq. (32) is defined on a finite interval, that is,

$$J = J(y_{(j+1)}, u_{(j)}; \ j = 1, 2, \ldots, N). \tag{38}$$

Thus, the gradient of J with respect to an output $y_{(k)}$ for some k is

$$
\frac{\partial J}{\partial y_{(k)}} = \sum_{\substack{i=1 \\ k+i \leq N+1}}^{n} \frac{\partial J}{\partial y_{(k+i)}} \frac{\partial y_{(k+i)}}{\partial y_{(k)}} + \sum_{\substack{i=0 \\ k+i \leq N}}^{n-1} \frac{\partial J}{\partial u_{(k+i)}} \frac{\partial u_{(k+i)}}{\partial y_{(k)}} - Q(y_{\text{ref}} - y_{(k)})
$$

$$
= \sum_{\substack{i=k+1 \\ i \leq N+1}}^{k+n} \frac{\partial J}{\partial y_{(i)}} \frac{\partial y_{(i)}}{\partial y_{(k)}} + \sum_{\substack{i=k \\ i \leq N}}^{k+n-1} \frac{\partial J}{\partial u_{(i)}} \frac{\partial u_{(i)}}{\partial y_{(k)}} - Q(y_{\text{ref}} - y_{(k)}). \tag{39}
$$

By using the definition of sensitivities, Eqs. (35) and (36),

$$
\delta_y^k = \sum_{\substack{i=k+1 \\ i \leq N+1}}^{k+n} \delta_y^i \frac{\partial y_{(i)}}{\partial y_{(k)}} + \sum_{\substack{i=k \\ i \leq N}}^{k+n-1} \delta_u^i \frac{\partial u_{(i)}}{\partial y_{(k)}} + Q(y_{\text{ref}} - y_{(k)}). \tag{40}
$$

Note that this output sensitivity equation (OSE) is dependent on the input sensitivities as well.

2. Input Sensitivity Equation

The input sensitivity equation (ISE) can be derived in a way similar to the OSE. Because the plant dynamics Eq. (1) is defined with m delayed input variables, whereas the inverse dynamics Eq. (31) is defined with $m - 1$ delayed input variables, $y_{(k+i)}$ is a function of $u_{(k)}$ for $i = 1, 2, \ldots, m$, and $u_{(k+i)}$ is a function of $u_{(k)}$ for $i = 1, 2, \ldots, m - 1$. Thus, the gradient of J with respect to an input $u_{\text{fb}(k)}$ for some k is

$$
\frac{\partial J}{\partial u_{(k)}} = \sum_{\substack{i=1 \\ k+i \leq N+1}}^{m} \frac{\partial J}{\partial y_{(k+i)}} \frac{\partial y_{(k+i)}}{\partial u_{(k)}} + \sum_{\substack{i=1 \\ k+i \leq N}}^{m-1} \frac{\partial J}{\partial u_{(k+i)}} \frac{\partial u_{(k+i)}}{\partial u_{(k)}} + R u_{\text{fb}(k)}
$$

$$
= \sum_{\substack{i=k+1 \\ i \leq N+1}}^{k+m} \frac{\partial J}{\partial y_{(i)}} \frac{\partial y_{(i)}}{\partial u_{(k)}} + \sum_{\substack{i=k+1 \\ i \leq N}}^{k+m-1} \frac{\partial J}{\partial u_{(i)}} \frac{\partial u_{(i)}}{\partial u_{(k)}} + R u_{\text{fb}(k)}. \tag{41}
$$

By using the definition of sensitivities, Eqs. (35) and (36),

$$
\delta_u^k = \sum_{\substack{i=k+1 \\ i \leq N+1}}^{k+m} \delta_y^i \frac{\partial y_{(i)}}{\partial u_{(k)}} + \sum_{\substack{i=k+1 \\ i \leq N}}^{k+m-1} \delta_u^i \frac{\partial u_{(i)}}{\partial u_{(k)}} - R u_{\text{fb}(k)}. \tag{42}
$$

This ISE is also dependent on the output sensitivities, and both are coupled to one another.

The plant dynamics Eq. (1) and the inverse dynamics Eq. (31) are not known; therefore, they are approximated by the corresponding networks, the IDNN F, and the feedback neurocontroller H, to yield

$$\delta_y^k = \sum_{\substack{i=k+1 \\ i \leq N+1}}^{k+n} \delta_y^i \frac{\partial F_i}{\partial \hat{y}_{(k)}} + \sum_{\substack{i=k \\ i \leq N}}^{k+n-1} \delta_u^i \frac{\partial H_i}{\partial \hat{y}_{(k)}} + Q(y_{\text{ref}} - \hat{y}_{(k)}), \tag{43}$$

$$\delta_u^k = \sum_{\substack{i=k+1 \\ i \leq N+1}}^{k+m} \delta_y^i \frac{\partial F_i}{\partial u_{(k)}} + \sum_{\substack{i=k+1 \\ i \leq N}}^{k+m-1} \delta_u^i \frac{\partial H_i}{\partial u_{(k)}} - R u_{\text{fb}(k)}. \tag{44}$$

Note that the last terms in OSE and ISE are, respectively, the error terms for the output and input variables, and the terms under summation operations are the error (or delta) terms backpropagated through the networks F and H. For example, $\delta_y^i(\partial F_i/\partial \hat{y}_{(k)})$ [or $\delta_u^i(\partial F_i/\partial u_{(k)})$] is the error δ_y^i (or δ_u^i) backpropagated through the network F to the input node $\hat{y}_{(k)}$ [or $u_{(k)}$].

The objective of the GBTT is to compute the sensitivity δ_u^k, which will be used as the equivalent error for training of FBNC. This can be achieved by solving the OSE Eq. (43) and ISE Eq. (44) backward starting from $j = N + 1$:

$j = N + 1$:

$$\delta_u^{N+1} = 0,$$
$$\delta_y^{N+1} = Q(y_{\text{ref}} - \hat{y}_{(N+1)});$$

$j = N$:

$$\delta_u^N = \delta_y^{N+1} \frac{\partial F_{N+1}}{\partial u_{(N)}} - R u_{\text{fb}(N)},$$
$$\delta_y^N = \delta_y^{N+1} \frac{\partial F_{N+1}}{\partial \hat{y}_{(N)}} + \delta_y^N \frac{\partial H_N}{\partial \hat{y}_{(N)}} + Q(y_{\text{ref}} - \hat{y}_{(N)});$$

$j = N - 1$:

$$\delta_u^{N-1} = \delta_y^{N+1} \frac{\partial F_{N+1}}{\partial u_{(N-1)}} + \delta_y^N \frac{\partial F_N}{\partial u_{(N-1)}} + \delta_u^N \frac{\partial H_N}{\partial u_{(N-1)}} - R u_{\text{fb}(N-1)},$$
$$\delta_y^{N-1} = \delta_y^{N+1} \frac{\partial F_{N+1}}{\partial \hat{y}_{(N-1)}} + \delta_y^N \frac{\partial F_N}{\partial \hat{y}_{(N-1)}} + \delta_u^N \frac{\partial H_N}{\partial \hat{y}_{(N-1)}} + \delta_u^{N-1} \frac{\partial H_{N-1}}{\partial \hat{y}_{(N-1)}}$$
$$+ Q(y_{\text{ref}} - \hat{y}_{(N-1)});$$

$j = N - 2$:

$$\delta_u^{N-2} = \delta_y^{N+1} \frac{\partial F_{N+1}}{\partial u_{(N-2)}} + \delta_y^N \frac{\partial F_N}{\partial u_{(N-2)}} + \delta_y^{N-1} \frac{\partial F_{N-1}}{\partial u_{(N-2)}} + \delta_u^N \frac{\partial H_N}{\partial u_{(N-2)}}$$
$$+ \delta_u^{N-1} \frac{\partial H_{N-1}}{\partial u_{(N-2)}} - R u_{\text{fb}(N-2)},$$

$$\delta_y^{N-2} = \delta_y^{N+1}\frac{\partial F_{N+1}}{\partial \hat{y}_{(N-2)}} + \delta_y^N \frac{\partial F_N}{\partial \hat{y}_{(N-2)}} + \delta_y^{N-1}\frac{\partial F_{N-1}}{\partial \hat{y}_{(N-2)}} + \delta_u^N \frac{\partial H_N}{\partial \hat{y}_{(N-2)}}$$

$$+ \delta_u^{N-1}\frac{\partial H_{N-1}}{\partial \hat{y}_{(N-2)}} + \delta_u^{N-2}\frac{\partial H_{N-2}}{\partial \hat{y}_{(N-2)}} + Q(y_{\text{ref}} - \hat{y}_{(N-2)});$$

$j = k:$ $\vdots$

$$\delta_u^k = \delta_y^{k+m}\frac{\partial F_{k+m}}{\partial u_{(k)}} + \cdots + \delta_y^{k+1}\frac{\partial F_{k+1}}{\partial u_{(k)}} + \delta_u^{k+m-1}\frac{\partial H_{k+m-1}}{\partial u_{(k)}} + \cdots$$

$$+ \delta_u^{k+1}\frac{\partial H_{k+1}}{\partial u_{(k)}} - Ru_{\text{fb}(k)},$$

$$\delta_y^k = \delta_y^{k+n}\frac{\partial F_{k+n}}{\partial \hat{y}_{(k)}} + \cdots + \delta_y^{k+1}\frac{\partial F_{k+1}}{\partial \hat{y}_{(k)}} + \delta_u^{k+n-1}\frac{\partial H_{k+n-1}}{\partial \hat{y}_{(k)}} + \cdots$$

$$+ \delta_u^{k+1}\frac{\partial H_{k+1}}{\partial \hat{y}_{(k)}} + \delta_u^k \frac{\partial H_k}{\partial \hat{y}_{(k)}} + Q(y_{\text{ref}} - \hat{y}_{(k)}).$$

3. Forward Simulator

Before solving the OSE and ISE, the error terms need to be generated. This can be done by driving the IDNN with the controllers FFNC and FBNC for N steps forward. This process is illustrated by the *forward simulator* shown in Fig. 6, where Δ is the tapped delay operator as defined in Fig. 1.

Starting from initial conditions, $y_{(1)}$, $u_{(0)}$, the forward simulator generates a sequence of inputs $u_{\text{fb}(1)}, u_{\text{fb}(2)}, \ldots, u_{\text{fb}(N)}$ and outputs $\hat{y}_{(2)}, \hat{y}_{(3)}, \ldots, \hat{y}_{(N+1)}$. These variables provide the error terms $(y_{\text{ref}} - \hat{y}_{(k)})$ and $u_{\text{fb}(k)}$ for any k.

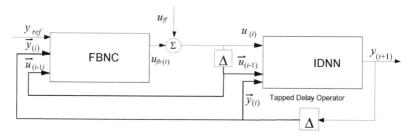

Figure 6   Forward simulator for GBTT.

4. Backward Simulator

The OSE and ISE are simulated backward to backpropagate the error terms. This can be performed by the *backward simulator* shown in Fig. 7, where *the summed advance operator* ∇ is defined as a "dual" of the tapped delay operator mapping from a sequence of vector input $\{\mathbf{x}_{(i)}\}$, where $\mathbf{x}_{(i)} = (x_{i,1}, x_{(i,2)}, \ldots, x_{(i,n)})$ to a scalar output defined as $x_{(i)} = \sum_{k=1}^{n} x_{(i+k,k)}$.

The backward simulator can be shown to be the dual of the forward simulator, which is then constructed by using the *duality principle*, that is, reversing the direction of arrows, interchanging the sums and nodes, and replacing the tapped delay operators with the summed advance operators. Using the errors generated by the forward simulator, the backward simulator runs backward starting from $i = N$. The backward simulator generates the sensitivities δ_y^{i+1} and δ_u^i. Noting that δ_u^i is in the output node of the FBNC, it is used as the equivalent error in computing the weight parameter adjustment in the FBNC, $\Delta \mathbf{W}^i$.

The process of the GBTT training algorithm is summarized as follows

1. Set the weight parameters of the FBNC with small random numbers.
2. Set the reference output and initial state with random numbers in the operation region of the plant.
3. Run the forward simulator for N steps forward from $i = 1$.
4. Using the operation result in step 3, run the backward simulator backward from $i = N$ to evaluate the equivalent error δ_u^i and the weight adjustment vector $\Delta \mathbf{W}^i$.
5. Update the weight parameters in the FBNC by using the average of the weight adjustment vectors found in step 4.
6. Go to step 2.

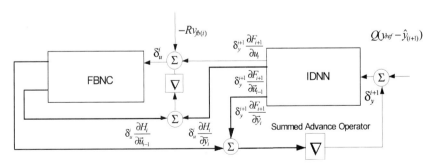

Figure 7   Backward simulator for GBTT.

176   *Young-Moon Park et al.*

Training the feedback controller is finished when the average decrease of the cost function converges to a small value for arbitrary reference outputs and initial conditions. Whereas the training algorithm is essentially a gradient descent method, the local minimum problem is possibile. However, this problem can be avoided by starting with different initial weight parameters or with a different number of nodes in the feedback controller [2, 18].

IV. CASE STUDIES

A. INVERTED PENDULUM CONTROL

1. Problem Description

A prototypical control problem that has been widely used for neural network application is the pole balancing, or the inverted pendulum, problem. The problem involves balancing a pole hinged on a cart as shown in Fig. 8. This problem is of interest because it describes an inherently unstable system and is representative of a wide class of problems with severe nonlinearity in a broad operation region. Additional details of the pole balancing experiments, including simulations, can be found in [14, 19–21]. In these works, the control problem was limited to the case of balancing the pole at the vertical position.

The control objective of the pole balancing problem in this chapter is to extend the results of previous efforts by keeping the pole at an arbitrary angle, not necessarily at the vertical position, while minimizing a general quadratic cost function. Whereas a steady acceleration is required to maintain the pole at an angle, we assume that the cart can travel indefinitely in either direction. The optimal track-

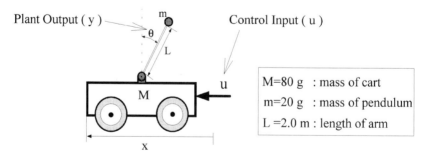

Figure 8   Inverted pendulum.

ing neurocontroller (OTNC) is constructed and trained by the proposed method to meet the following control objectives:

1. Set the pendulum to an arbitrarily given reference angle.
2. Minimize the quadratic cost function while tracking the reference angle.

The nonlinear differential equation of the plant dynamics is

$$(M + m)\ddot{x} + mL\cos(\theta)\ddot{\theta} - mL\sin(\theta)\dot{\theta}^2 = u, \\ m\ddot{x}\cos(\theta) + mL\ddot{\theta} = mg\sin(\theta), \tag{45}$$

where the variables and the parameters are defined as in Fig. 6. The dynamics has severe nonlinearity when the angle deviates from zero, in which case it is difficult to solve the control problem by any conventional method.

2. Design of the Optimal Tracking Neurocontroller

A paradigm for the IDNN is chosen by trial and error. It consists of two hidden layers with 40 nodes each, an input layer with 6 input nodes, and an output layer with 1 node. Three of the six input nodes are for output history, $y_{(k)}, y_{(k-1)}, y_{(k-2)}$, two are for input history, $u_{(k)}, u_{(k-1)}$, and one is for bias input, 1.0. Training patterns of the IDNN are generated from the mathematical model with random initial value and random input within the operation region of ± 1.1 rad. Discrete-time training patterns are obtained by applying the modified Euler method with a time step size of 0.13 s in simulation.

To avoid oscillation during the training stage, weight parameters are corrected from the average of corrections calculated for every 10 patterns. After training the IDNN for 1 h in a SUN-SPARC2 workstation, it is tested with arbitrary initial conditions and sinusoidal inputs of different amplitude, which are presented in Fig. 9. The IDNN approximates the plant very closely and is sufficient for training the neurocontrollers.

The feedforward neurocontroller (FFNC) has two hidden layers with 30 nodes each. The input layer has two nodes: one for reference output y_{ref} and one for the bias input. The output layer has one node for the control u_{ff}. The reference output is given randomly to be within the operation region of ± 0.9 rad to train the FFNC, which is coupled with the IDNN and the plant.

After training the FFNC for 1 h, it is tested with a reference output varying within the operation region. Although the plant tracks the time-varying reference output, the error remains small as shown in Fig. 10.

The feedback neurocontroller (FBNC) has two hidden layers with 30 nodes each. The input layer has five nodes: three for output history, $y_{(k)}, y_{(k-1)}, y_{(k-2)}$, one for previous input, $u_{(k-1)}$, and one for the bias input. The cost function for

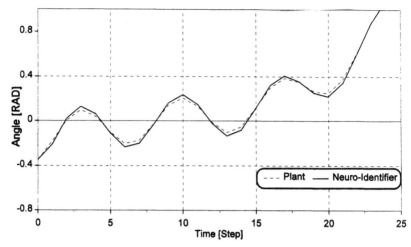

Figure 9   Training results of the IDNN: initial condition $\theta = -0.3$ and input $u = -0.9\cos(k)$.

the N-step ahead optimal controller is set as

$$J = \tfrac{1}{2} \sum_{k=1}^{N} \left(1.0(y_{\text{ref}} - y_{(k+1)})^2 + 0.3(u_{\text{fb}(k)})^2\right). \qquad (46)$$

The FBNC is trained once for an initial condition and a reference output, which are randomly selected while driving the plant for N steps. This training is repeated for other initial conditions and reference outputs. Each training is performed in two phases. First, the training is done with small $N(=3)$ because in the beginning the controller has little knowledge of control. This also prevents the pendulum from falling down. Then the step is increased gradually to $N = 15$. The second phase training is carried on with N fixed at 15.

3. Simulation and Discussion

After training the FBNC with the GBTT algorithm for 2 h, it is tested with several nonzero set points as presented in Fig. 11. It shows a larger overshoot for a larger set point. A larger overshoot corresponds to an operating condition with severe nonlinearity.

Figure 12 shows a case for a set point changing at each of 40 time steps. The trajectories of the plant output, reference output, and the control input are presented in the figure. Notice that the OTNC for a nonlinear system behaves in a way similar to the usual optimal tracking controller for a linear quadratic prob-

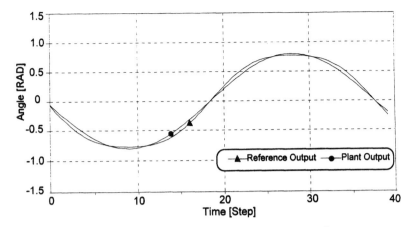

Figure 10   Training result of the feedforward neurocontroller.

lem [12]; the shapes of output trajectories are typical fast responses with reasonable overshoots. Figure 13 shows the corresponding feedforward and feedback control inputs for the changing set point. The FFNC generates the control input corresponding only to the reference output in steady state. On the other hand, the FBNC generates the control input corresponding to the regulating error between the reference and the plant outputs during transient.

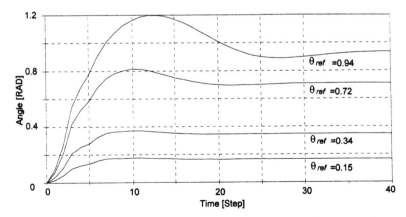

Figure 11   Angle trajectories of the inverted pendulum for different reference set points.

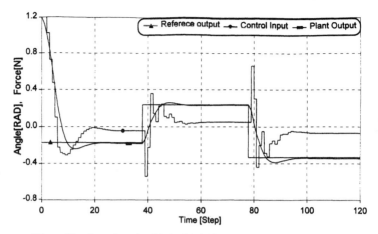

Figure 12   Control result with the OTNC for changing reference set point.

B. POWER SYSTEM CONTROL

1. Problem Description

The proposed OTNC is applied to the power plant control to enhance output performance in the electric frequency and the terminal voltage of a generator. The difficulty in the power plant control system design comes from handling the nonlinearity in the system model. However, OTNC avoids difficulties in solving the nonlinear control problems by using the neural networks' learning abilities.

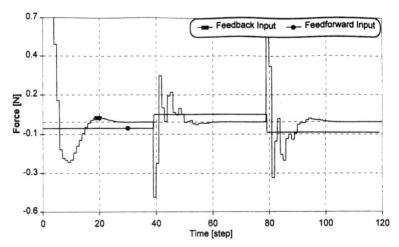

Figure 13   Feedforward and the feedback control inputs for a changing reference set point.

In normal steady state, the turbine power keeps balance with the electric air gap power, resulting in zero acceleration and a constant speed or frequency. Disturbances or load perturbations upset the balance. The power difference accelerates or decelerates the turbine generator. When there is a disturbance or load change, the electric frequency oscillates due to the lack of mechanical damping in the generators. Then the frequency finds a new steady state by using the combined effects of the frequency characteristics for loads and generators if the power system is stable. However, on occasion, a power system experiences low-frequency oscillations. This is one of the most important problems arising in a power system.

Noting the previous phenomenon, the control objective is summarized as: Minimize the low-frequency oscillations in the electric frequency of a power plant. To achieve the control objective, OTNC is applied to the voltage control loop of a power plant dynamic to minimize the well-known quadratic performance index as

$$J = \tfrac{1}{2} \sum_{k=0}^{Nh} \left\{ Q(\omega_{(k+1)} - \omega_{\text{ref}})^2 + R(u_{(k)})^2 \right\}, \tag{47}$$

where ω is the electric frequency of the plant, u is the output of a voltage regulator, ω_{ref} is normal synchronous speed, and u_0 is the corresponding steady-state control input. The weightings are set as $Q = 1.0$ and $R = 0.04$. This quadratic performance index not only keeps the output from oscillating, but also keeps the input from deviating from the prescribed value.

OTNC is applied to a simple power system network shown in Fig. 14. The power system consists of two power plants: one is a thermal unit and the other is a

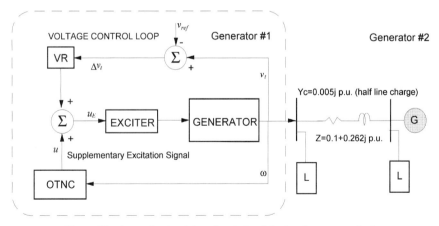

Figure 14 A neural network based control architecture for a power plant.

Table I

Bus Data for the Normal Loading Condition (0.8 p.u.)

Bus	v_t (p.u.)	θ (rad)	P_e (MVA)	Q_e (MVA)	P_l (MVA)	Q_l (MVA)
1	0.997	0.0475	120	−20	70	10
2	1.0	0.0	146	37.4	195	10

hydro unit. The line parameters of the network also are shown in the figure. From the point of view of the first generator, a normal loading operating condition is considered, and is given in Tables I and II.

The study power system has sustained typical low-frequency oscillations. The control objectives are to improve system damping to reduce the low-frequency oscillations in the outputs of the first power plant by applying proposed OTNC. The dynamic behavior of a power plant is determined by basic components (synchronous machine, exciter, power system networks, loads) that have significant effects on the low-frequency oscillation in an electric power system. The power system stabilizing control input is the supplementary excitation signal to the first power plant.

Figure 14 also shows proposed OTNC architecture where OTNC is added in the voltage control loop to produce a supplementary excitation signal.

For the study on the low-frequency oscillation problem, the state equations for the synchronous machine are written as third-order models in which the state variables are ω (rotor speed), δ (torque angle), and e'_q (voltage behind the transient reactance), where the change in flux linkage of the field winding is considered. The third-order model for a synchronous machine connected to a network at the jth bus is [26]

$$\frac{d\omega}{dt} = \frac{1}{M}(T_m - T_e + D(\omega_o - \omega)) \qquad \left(T_m = \frac{P_m}{\omega},\ T_e = \frac{P_e}{\omega}\right), \qquad (48)$$

$$\frac{d\delta}{dt} = \omega_b(\omega - \omega_o) \qquad (\omega_b = 2\pi f_0,\ \omega \approx 1), \qquad (49)$$

$$\frac{de'_q}{dt} = \frac{1}{T'_{do}}\left[E_{FD} - e'_q - \frac{(x_d - x'_d)}{x'_d}\cdot\left(e'_q - v_j\cos(\delta - \theta_j)\right)\right], \qquad (50)$$

where the variables are defined in [26].

Table II

Bus Data for the Heavy Loading Operating Condition (1.0 p.u.)

Bus	v_t (p.u.)	θ (rad)	P_e (MVA)	Q_e (MVA)	P_l (MVA)	Q_l (MVA)
1	0.993	0.0357	150	−20	100	10
2	1.0	0.0	146	37.4	195	10

In Eq. (50), the dynamics of e'_q is controlled by the field excitation voltage E_{FD} which is the output of a conventional exciter. The proposed OTNC is to be attached to control the exciter input.

The generating powers always satisfy the algebraic power balance constraint

$$P_{ei}(e'_q, v_i, \delta, \theta_i) + P_l(v_i, \theta_i) = P_{Ni}(v, \theta),$$
$$Q_{ei}(e'_q, v_i, \delta, \theta_i) + Q_l(v_i, \theta_i) = Q_{Ni}(v, \theta), \quad \text{for } i = 1, 2, \tag{51}$$

where P_{ei} and Q_{ei} are the real and reactive powers of the ith generator, P_{Ni} and Q_{Ni} are the net powers on the ith bus injected into the networks, and P_{li} and Q_{li} are the local loads on the ith bus. The loads can be modeled as a nonlinear function of system variables. The real and reactive powers are presented as

$$P_{ei}(e'_q, v_i, \delta, \theta_i) = \frac{e'_q v_i}{x'_d} \sin(\delta - \theta_i) + \frac{v_i^2(x'_d - x_q)}{2x'_d x_q} \sin(2(\delta - \theta_i)),$$

$$Q_{ei}(e'_q, v_i, \delta, \theta_i) = \frac{e'_q v_i \cos(\delta - \theta_i)}{x'_d} \tag{52}$$
$$- \frac{v_i^2(x_q \cos^2(\delta - \theta_i) + x'_q \sin^2(\delta - \theta_i))}{2x'_d x_q},$$

$$P_{li}(v_i, \theta_i) = P_{l0} \cdot \left(\frac{v_i}{v_{i0}}\right)^{\alpha_P} \cdot (1 + \beta_P \cdot \Delta f_i),$$
$$Q_{li}(v_i, \theta_i) = Q_{l0} \cdot \left(\frac{v_i}{v_{i0}}\right)^{\alpha_Q} \cdot (1 + \beta_Q \cdot \Delta f_i), \tag{53}$$

$$P_{Ni}(v, \theta) = \sum_{k=1}^{2} v_i v_k \big(g_{ik} \cos(\theta_i - \theta_k) + b_{ik} \sin(\theta_i - \theta_k)\big),$$
$$Q_{Ni}(v, \theta) = \sum_{k=1}^{2} v_i v_k \big(g_{ik} \sin(\theta_i - \theta_k) - b_{ik} \cos(\theta_i - \theta_k)\big), \tag{54}$$

Table III

Parameters of Generators

	T'_{do}	D	H	x_d	x'_d	x_q	x'_q	Self-base (MVA)
1st generator	4.0	1.0	4.46	1.25	0.6	0.9	0.6	150
2nd generator	6.3	1.0	5.5	0.94	0.4	0.65	0.4	150

where the parameters α, β and g_{ik}, b_{ik} are given to represent the load characteristics and the line admittance from ith bus to kth bus, respectively. The parameters for generators are given in Table III.

Typical IEEE governor and turbine models [26] are used: TGOV1 is used for the first generator and IEEEG2 is used for the second generator in the study power system. The IEEE exciter and voltage regulator model EXST1 is used for the generators.

2. Design of the Optimal Tracking Neurocontroller

OTNC consists of IDNN and FBNC, whereas the steady-state supplementary excitation signal is zero. A paradigm for the neural networks in OTNC is chosen by trial and error. IDNN and FBNC have one hidden layer with 40 nodes each. In IDNN, the input layer has 10 nodes: 5 for output history, $y_{(k)}, y_{(k-1)}, \ldots, y_{(k-4)}$, 4 for input history, $u_{(k)}, u_{(k-1)}, \ldots, u_{(k-3)}$, and 1 for the bias input. This network generates one output: $y_{(k+1)}$. The corresponding FBNC has 8 nodes in the input layer: 5 for output history, $y_{(k)}, y_{(k-1)}, \ldots, y_{(k-4)}$, 3 for input history, $u_{(k-1)}, \ldots, u_{(k-3)}$.

The proper choice of sampling period is very important [27]. The choice is dependent on the plant characteristics and the target oscillating mode to be damped by adding a digital controller. In this case, the frequency band of the plant output in the low-frequency oscillation is 1–2 Hz. That band is found by simulating the power system in Fig. 14 when there is a load change in the normal loading condition. Training patterns for IDNN are obtained with a time step size of 0.04 s by simulating the power system when there are load changes or line faults during a wide range operation. This sampling time allows at least 20 points sampling in a cycle of low-frequency oscillations whose frequency band is less than 1.25 Hz. It takes 20 min in an IBM-PC 486 to training the IDNN.

3. Simulation and Discussion

Two cases of disturbances are considered in a wide operation range. The types of disturbances in the simulations are typical in the real power system operation: one is a line fault and the other is an abrupt load level change. Two loading conditions are considered: normal loading condition for 0.8 (p.u.) in generating power and heavy loading condition for 1.0 (p.u.).

The power system simulations are done for three cases in the voltage control loop of the first generator dynamics. First, there is the conventional voltage regulator only; second, the conventional voltage regulator and the conventional PSS (STAB4 [26]); third, the conventional voltage regulator and the proposed OTNC.

a. Case 1: Tie-Line Impedance Change

This simulates a line fault disturbance. In this case, the tie line impedance between the two generators changes from 100 to 75% at $t = 1.2$ s, undergoing a transient state when 25% of the line impedance is grounded at half point and cleared at $t = 1.85$ s. This fault imitates one line outage when the tie line consists of four parallel line elements as in Fig. 14.

Figures 15 and 16 show trajectories of the speed deviation of the first generator when the power system is in the normal loading condition and heavy loading condition, respectively. They show that in the case of OTNC, the peaks are smaller and oscillations converge more quickly than the others.

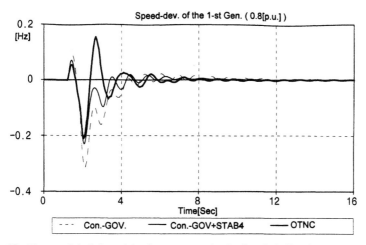

Figure 15 The speed deviation of the first generator for the line fault disturbance in the normal loading condition.

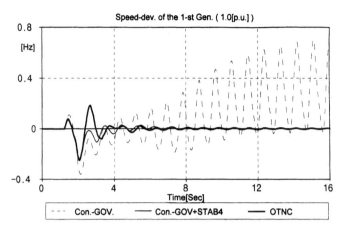

Figure 16   The speed deviation of the first generator for the line fault disturbance in the heavy loading condition.

b. Case 2: Electrical Load Level Change

This simulates a case of disturbance when there is a stepwise 5% increase in all loads in Fig. 14 at $t = 1.2$ s, 5% decrease at $t = 3.7$ s, 5% decrease at $t = 5$ s, and 5% increase at $t = 6.3$ s.

Figures 17 and 18 show the trajectories for the speed deviation of the first generator for load change disturbance when the power system is in the normal

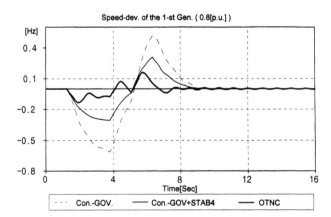

Figure 17   The speed deviation of the first generator for the load change disturbance in the normal loading condition.

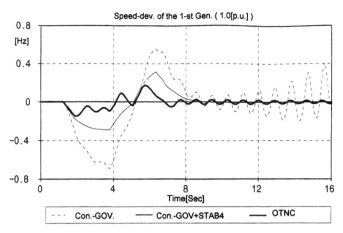

Figure 18 The speed deviation of the first generator for the load change disturbance in the heavy loading condition.

loading condition and in the heavy loading condition, respectively. They show that in the case of OTNC, the peaks of OTNC are smaller and oscillations converge more quickly than the others.

V. CONCLUSIONS

For an optimal tracking control problem for nonlinear dynamic plants, a new architecture—the optimal tracking neurocontroller (OTNC)—is developed using feedback and feedforward controls. First, the feedforward neurocontroller (FFNC) is introduced to solve the tracking problem with a nonzero set point. A novel training method for the FFNC is developed by using the concept of an inverse mapping to generate the feedforward control input corresponding to the output set point. Second, the feedback neurocontroller (FBNC) is designed to solve an optimal regulator problem with general quadratic cost function. A generalized backpropagation-through-time (GBTT) training algorithm is developed to train the FBNC. The proposed OTNC scheme is demonstrated in a typical nonlinear plant, in an inverted pendulum, and in a power system. Simulation results show good performance over a wide range of nonlinear operations and the possibility of using the OTNC for the optimal tracking control of other nonlinear systems.

ACKNOWLEDGMENTS

This work was supported in part by Korea Science and Engineering Foundation (KOSEF) and the National Science Foundation (NSF) under grants "Development of Power System Intelligent Co-ordinated Control" (INT-9605028) and "Free Model Based Intelligent Control of Power System" (ECS-9705105)

REFERENCES

[1] T. Poggio and F. Girosi. Networks for approximation and learning. *Proc. IEEE* 78:1481–1497, 1990.

[2] D. Nguyen and B. Widrow. The truck backer-upper: An example of self-learning in neural networks. *IEEE Control Systems Mag.* 10(3):18–23, 1990.

[3] F. C. Chen. Back-propagation neural networks for nonlinear self-tuning adaptive control. *IEEE Control Systems Mag.* 10(3):44–48, 1990.

[4] Y. Iiguni and H. Sakai. A nonlinear regulator design in the presence of system uncertainties using multilayered neural networks. *IEEE Trans. Neural Networks* 3:410–417, 1991.

[5] C. C. Ku and K. Y. Lee. Diagonal recurrent neural network for dynamic system control. *IEEE Trans. Neural Networks* 6:144–156, 1995.

[6] C. C. Ku, K. Y. Lee, and R. M. Edwards. Improved nuclear reactor temperature control using diagonal neural networks. *IEEE Trans. Nucl. Sci.* 39:2298–2308, 1992.

[7] C. C. Ku and K. Y. Lee. Diagonal recurrent neural network based control using adaptive learning rates. In *Proceedings of the IEEE Conference on Decision and Control*, December 1992, pp. 3485–3490.

[8] W. T. Miller, R. S. Sutton, and P. J. Werbos. *Neural Networks for Control*, pp. 28–65. The MIT Press, Cambridge, MA, 1990.

[9] T. Yamada and T. Yabuta. Neuro-controller using autotuning method for nonlinear functions. *IEEE Trans. Neural Networks* 3:595–601, 1992.

[10] P. J. Werbos. Backpropagation through time: What it does and how to do it. *Proc. IEEE* 78:1550–1560, 1990.

[11] K. S. Narendra and K. Parthasarathy. Identification and control of dynamic systems using neural networks. *IEEE Trans. Neural Networks* 1:4–28, 1990.

[12] H. Kwakernaak and R. Sivan. *Linear Optimal Control Systems*, pp. 201–319. Wiley, New York, 1972.

[13] U. diCaprio and P. P. Wang. A study of the output regulator problem for linear system with input vector. In *Proceedings of the Seventh Annual Allerton Conference on Circuits and System Theory*, 1969, pp. 186–188.

[14] J. E. Slotine. *Applied Nonlinear Control*. Prentice-Hall, Englewood Cliffs, NJ, 1991.

[15] A. Meystel and S. Uzzaman. The planning of tracking control via search. In *Proceedings of the International Symposium on Intelligent Control*, Chicago, 1993, pp. 554–559.

[16] M. I. Jordan and D. E. Rumelhart. Forward models: Supervised learning with a distal teacher. *Cognitive Sci.* 16:307–354, 1992.

[17] T. Troudet, S. Garg, D. Mattern, and W. Merrill. Towards practical control design using neural computation. In *Proceedings of the International Joint Conference on Neural Networks*, Seattle, 1991, pp. II-675–681.

[18] R. Hecht-Nilsen. *Neuro Computing*, pp. 115–119. Addison-Wesley, Reading, MA, 1989.

[19] A. P. Wieland. Evolving neural network controllers for unstable systems. In *Proceedings of the International Joint Conference on Neural Networks*, Singapore, 1992, pp. 214–219.

[20] V. Williams and K. Matsuoka. Learning to balance the inverted pendulum using neural networks. In *Proceedings of the International Joint Conference on Neural Networks*, Singapore, 1992, pp. 214–219.

[21] B. Friedland. *Control System Design: An Introduction to State-Space Methods*. McGraw-Hill, New York, 1986.

[22] Y. Yu. *Electric Power System Dynamics*, pp. 114–118. Academic, New York, 1983.

[23] F. P. deMello and C. A. Concordia. Concept of synchronous machine stability as affected by excitation control. *IEEE Trans. Power Apparatus and Systems*, PAS-103:316–319, 1969.

[24] O. I. Elgerd. *Electric Energy Systems Theory*. MaGraw-Hill, New York, 1971.

[25] Y. M. Park, M. S. Choi, and K. Y. Lee. A neural network-based power system stabilizer using power flow characteristics. *IEEE Trans. Energy Conversion* 11:435–441, 1996.

[26] T. E. Kostyniack. *PSS/E Program Operation Manual*, Power Technology, Inc., 1983.

[27] K. J. Astrom and B. Wittenmark. *Computer Controlled Systems: Theory and Design*, pp. 30–31. Prentice-Hall, Englewood Cliff, NJ, 1984.

On-Line Approximators for Nonlinear System Identification: A Unified Approach

Marios M. Polycarpou
Department of Electrical and Computer Engineering
University of Cincinnati
Cincinnati, Ohio 45221-0030

Due to their approximation capabilities and inherent adaptivity features, neural networks have been employed recently to model and control complex nonlinear dynamical systems. This chapter presents a unified approximation theory perspective to the design and analysis of nonlinear system identification schemes using neural network and other on-line approximation models. Depending on the location of the adjustable parameters, networks are classified into *linearly* and *nonlinearly* parametrized networks. Based on this classification, a unified procedure for modeling discrete-time and continuous-time dynamical systems using on-line approximators is developed and analyzed using Lyapunov stability theory. A projection algorithm is used to guarantee the stability of the overall system even in the presence of approximation errors.

I. INTRODUCTION

The construction of accurate mathematical models from observations of physical activity is an integral part of the general scientific methodology. The quest for suitable expansions to approximate the behavior of general nonlinear dynamical

Control and Dynamic Systems

systems has a long history. Classical models include, among others, the functional series of Volterra and Wiener and the Hammerstein model [1]. Although there are many theoretical results concerning these representations, practical difficulties hindered their wide application. The emergence of the neural network paradigm as a powerful tool for learning complex mappings from a set of examples has generated a great deal of excitement in using neural network models for identification and control of dynamical systems with unknown nonlinearities (see, for example, the edited books [2–4]. Due to their approximation capabilities as well as their inherent adaptivity features, artificial neural networks offer an appealing representation for modeling complex nonlinear dynamical systems.

Several training methods for adjusting the weights of neural networks in dynamic environments have been proposed in the literature. Most of these methods rely on the gradient minimization methodology and involve the computation of partial derivatives or sensitivity functions. In this respect, they are extensions of the backpropagation algorithm for feedforward neural networks [5]. Examples of such learning algorithms include the *recurrent backpropagation* [6], the *backpropagation-through-time* [7], the *real-time recurrent learning* [8], and the *dynamic backpropagation* [9]. Although such schemes may perform well in many cases, in general, there are no systematic analytical methods to ensure the stability, robustness, and performance properties of the overall system. In an attempt to overcome some of these problems, recent studies [10–16] have proposed design procedures based on Lyapunov's stability theory. The advantage of these training methods is that the learning algorithm is derived based on the Lyapunov synthesis method and therefore guarantees the stability of the system.

In the framework of approximation theory, neural networks represent just one class of function approximators. The recent widespread attention that neural networks have received is justified more in terms of practical considerations, like easy parallel implementation and fast adaptability, rather than theoretical reasons. In this chapter we present, in a tutorial fashion, a unified approximation theory perspective in the design and analysis of identification schemes using on-line approximation models. This allows the study under the same framework of not only a wide variety of neural networks, but also other type of approximators, such as polynomial expansions and spline functions, as well as adaptive fuzzy systems.

Depending on the location of the adjustable parameters, the network approximators are classified into *linearly* and *nonlinearly* parametrized approximators. This classification provides a common analytical framework for the study of the various architectures that belong to each class. Moreover, it illustrates that the design and analysis of linearly parametrized schemes is a special case of the nonlinearly parametrized methodology with a certain "higher-order term" being identically equal to zero. The proposed identification procedure guarantees stability of the overall system even in the presence of approximation errors. The projection algorithm is used to guarantee that the parameter estimates remain bounded. In the presence of approximation errors, upper bounds for the average output error are obtained in terms of the approximation error.

The chapter is organized as follows. Section II describes some types of network approximators that can be used as models of dynamical systems and considers the approximation of the input–output behavior of discrete-time dynamical systems. This is followed, in Section III, by the synthesis and analysis of learning algorithms for nonlinear system identification based on generic approximation models. Finally, Section IV develops a nonlinear system identification methodology for continuous-time systems using on-line approximators.

II. NETWORK APPROXIMATORS

Consider a single-input–single-output (SISO), discrete-time, time-invariant, nonlinear dynamical system of the form

$$y(k) = f\big(y(k-1), \ldots, y(k-n_y), u(k-1), \ldots, u(k-n_u)\big), \qquad (1)$$

where $k \in \mathbb{Z}^+$ is the discrete temporal variable, $u(k) \in \mathbb{R}^1$ and $y(k) \in \mathbb{R}^1$ are the input and output, respectively, at time k, and $f \colon \mathbb{R}^{n_y} \times \mathbb{R}^{n_u} \mapsto \mathbb{R}^1$ is an unknown smooth mapping defined on an open set of $\mathbb{R}^{n_y} \times \mathbb{R}^{n_u}$. The positive integers n_y and n_u denote the maximum lag in the system output and input, respectively. For notational convenience we let $n_z := n_y + n_u$ and

$$z(k-1) := \big[y(k-1), \ldots, y(k-n_y), u(k-1), \ldots, u(k-n_u)\big]^T \in \mathbb{R}^{n_z}.$$

Whereas f is unknown, the objective is to use some type of (smooth) network approximator $\mathcal{F}(z; \theta)$ to approximate $f(z)$. In the network formulation, $z \in \mathbb{R}^{n_z}$ is the input to the network and $\theta \in \mathbb{R}^q$ is a set of adjustable parameters in vector form. By changing the value of θ it is possible to change the input–output response of the network $\mathcal{F}$. Although the concept of an *approximator* in the form of a *network* is rather new, the underlying idea is the same as in the classical problem of approximation theory [17], which can be described as follows: Let $\mathcal{A}$ be a compact set in a metric space $\mathcal{B}$; then for $b \in \mathcal{B}$ find an element $a \in \mathcal{A}$ such that the distance between a and b is minimized.

In the modeling of dynamical systems, $\mathcal{B}$ is usually the space of continuous functions defined on some compact domain $\mathcal{Z}$ (denoted by $C[\mathcal{Z}]$), and $\mathcal{A}$ is the class of maps attained by $\mathcal{F}$ for different values of θ, that is,

$$\mathcal{A} = \big\{g(z)\colon g(z) = \mathcal{F}(z; \theta), \ \theta \in \mathbb{R}^q\big\}.$$

From an analytical viewpoint it is convenient to distinguish between linearly and nonlinearly parametrized approximation methods. In the case of *linearly parametrized approximators*, $\mathcal{F}$ is of the form

$$\mathcal{F}(z; \theta) = \Omega(z)^T \theta,$$

where θ is a q-dimensional vector and $\Omega \colon \mathbb{R}^{n_z} \mapsto \mathbb{R}^q$ is a smooth mapping. Approximators whose structure is such that the parameters appear in a nonlinear fashion are referred to as *nonlinearly parametrized approximators*. In the framework of approximation theory, linearly parametrized approximation corresponds to the special case where the set $\mathcal{A}$ is a *linear* subspace of $\mathcal{B}$.

A. UNIVERSAL APPROXIMATORS

Next we describe briefly some types of network approximators that can be used in the identification and control of dynamical systems. To simplify the notation, in our description we consider the case of approximating functions of a single variable. It is noted that the terms "parameter" and "weight" are used interchangeably.

1. Polynomials

Polynomial functions form the most extensively studied approximation method. The class of polynomial functions of (at most) degree n is given by

$$\mathcal{F}_n(z; \theta) := \left\{ \sum_{i=0}^{n} \theta_i z^i \colon \theta_i \in \mathbb{R}^1 \right\}.$$

Polynomials are linearly parametrized approximators and according to the well-known Weierstrass theorem [17], for any function $f \in C[\mathcal{Z}]$ and any $\varepsilon > 0$, there exists a polynomial $p \in \mathcal{F}_n$ (for n arbitrarily large) such that $\sup_{z \in \mathcal{Z}} |f(z) - p(z)| \leq \varepsilon$. It is worth noting that in the special case of $n = 1$, the polynomial expansion reduces to linear systems which constitute the best developed part of system theory.

2. Rational Functions

Another type of approximation method is rational functions or rational approximation. In this case,

$$\mathcal{F}_{n,m}(z; \theta, \vartheta) := \left\{ \frac{\sum_{i=0}^{n} \theta_i z^i}{\sum_{i=0}^{m} \vartheta_i z^i} \colon \theta_i, \vartheta_i \in \mathbb{R}^1 \right\},$$

with the restriction that the zeros of the denominator polynomial are outside the approximation region. In general, rational functions have greater approximation power than polynomial functions, in the sense that with the same number of parameters one is able to obtain better approximation accuracy [17]. Rational functions are nonlinearly parametrized approximators.

3. Spline Functions

Spline functions are examples of piecewise polynomial approximators. The main idea behind spline functions is that the approximation region is broken up into a finite number of subregions via the use of *knots*. In each subregion a polynomial of degree at most n is used, with the additional requirement that the overall function is $n - 1$ times differentiable. The most popular type of spline function is *cubic splines*, where $n = 3$, that is, cubic polynomial pieces that are joined so that the overall function is twice differentiable. Note that splines (with fixed knots) are linearly parametrized approximators. However, splines with variable knots are nonlinearly parametrized approximators. Spline functions possess several nice approximation properties, which is reflected in their wide application [18].

4. Multilayer Neural Networks

In recent years there has been a great deal of interest in approximation methods that are loosely based on models of biological signal activity. Although various neural network models have been proposed, by far the most popular is the class of multilayer networks with sigmoidal-type activation functions. In the case of a two-layer network (that is, one hidden layer),

$$\mathcal{F}_n(z; \theta, \vartheta, \varphi) := \left\{ \sum_{i=1}^n \theta_i \sigma(\vartheta_i z + \varphi_i) : \theta_i, \vartheta_i, \varphi_i \in \mathbb{R}^1 \right\},$$

where $\sigma \colon \mathbb{R} \mapsto \mathbb{R}$ is the sigmoidal activation function and n is the number of units (also called *nodes* or *neurons*) in the hidden layer. Theoretical works by several researchers (see, e.g., [19]) have shown that such networks can uniformly approximate any function $f \in C[\mathcal{Z}]$ to any degree of accuracy provided n is sufficiently large, or equivalently, provided the network has a sufficient number of neurons. Multilayer neural networks are nonlinearly parametrized approximators.

5. Radial-Basis-Function Networks

Another class of neural networks that has attracted considerable attention recently is the radial-basis-function (RBF) network model. The output of RBF networks is of the form

$$\mathcal{F}_n(z; \theta) := \left\{ \sum_{i=1}^{n} \theta_i g_i(z) : \theta_i \in \mathbb{R}^1 \right\},$$

where g_i is the output of the ith *basis function*. The Gaussian function $g_i(z) := \exp(-|z - c_i|^2/\sigma_i^2)$, where c_i and σ_i are the ith center and width, respectively, is usually chosen as the basis function; sometimes, *normalized* Gaussian functions are employed as basis functions. RBF networks are also capable of universal approximation [20]. In many aspects, the underlying structure of RBF networks is similar to that of spline functions. For example, if the centers and widths are kept fixed, then RBF networks are linearly parametrized approximators; if they are allowed to vary, then RBF networks become nonlinearly parametrized. Therefore, the centers of RBF networks correspond to the knots of the spline function formulation.

6. Adaptive Fuzzy Systems

The fuzzy logic paradigm provides yet another type of approximator. Fuzzy systems approximate functions by covering their graphs with fuzzy patches or fuzzy rules of the form "if antecedent conditions hold, then consequent conditions hold" [21]. The approximation increases in accuracy as the fuzzy patches increase in number and decrease in size. Fuzzy systems can also approximate any continuous function on a compact domain to any degree of accuracy. In adaptive fuzzy systems each fuzzy rule is weighted by adjustable parameters or weights. Fuzzy systems offer the possibility of using linguistic information, based for example on common sense or experts' knowledge, for control of systems where a mathematical model is hard to determine. Note that fuzzy systems are usually used in the approximation of the *inverse dynamics* rather than the system model: the fuzzy rules are usually of the form "if the system output satisfies conditions X, then the control input satisfies conditions U."

The preceding list of approximators, although not complete, includes many of the approximation methods used to model dynamical systems. The first three of the foregoing approximators are based on traditional approximation methods, whereas the rest were proposed in the context of "intelligent control" methods. The selection of the type and structure of approximation models is an important task that, ultimately, influences the performance of the modeling and control. Unfortunately, with the exception of some network growing and pruning approaches, no systematic procedure for selecting the network structure is available currently.

Although investigations comparing neural networks and other approximation models are still at a preliminary stage, from a systems engineering perspective, neural networks possess several properties that make them suitable for approximation of unknown nonlinearities in the context of system identification. Some of these properties are massive parallelism, fault tolerance, potential of analog hardware implementation, convenient adaptation capabilities, and good generalization features. Of course, these properties are not unique to neural networks. The key observation is that advances in computer technology have rendered possible the use of various types of approximation models (some dependent on biological modeling, some not) to construct and tune nonlinear dynamical models. The approach taken herein is to study on-line approximators in a unified framework. In the sequel, $\mathcal{F}$ denotes a generic, sufficiently smooth, network approximator.

B. UNIVERSAL APPROXIMATION
OF DYNAMICAL SYSTEMS

In this subsection we consider the problem of approximating the behavior of a dynamical system of the form given by (1), that is,

$$y(k) = f\big(z(k-1)\big). \tag{2}$$

In particular, we examine the following question: suppose that the network approximator $\mathcal{F}$ can approximate the function f to any degree of accuracy if provided with sufficiently large number of nodes or adjustable parameters. Do there exist parameter values θ^* such that the input–output response of the dynamical system

$$\hat{y}(k) = \mathcal{F}\big(\hat{z}\big((k-1)\big); \theta^*\big), \tag{3}$$
$$\hat{z}(k-1) := \big[\hat{y}(k-1), \ldots, \hat{y}(k-n_y), u(k-1), \ldots, u(k-n_u)\big]^T$$

approximates the input–output response of the system (2)?

The following lemma addresses this question. For well-posedness we assume that both functions $f(\cdot)$ and $\mathcal{F}(\cdot;\cdot)$ are smooth functions and also that $z(k) = \hat{z}(k) = 0$ for all $k < 0$.

LEMMA 1. *Suppose $y(0) = \hat{y}(0) = y^0$. Then given any $\varepsilon > 0$ and any finite integer $N > 0$, there exist a positive integer q^* and weight values $\theta^* \in \mathbb{R}^{q^*}$ such that for all $u(k) \in \mathcal{U}$, $k = 0, \ldots, N-1$, where $\mathcal{U}$ is a compact set, the outputs of the systems described by (2) and (3) satisfy*

$$\max_{0 \le k \le N} \big|y(k) - \hat{y}(k)\big| \le \varepsilon. \tag{4}$$

Proof. Let

$$Y(k-1) := \left[y(k-1), \ldots, y(k-n_y)\right]^T \in \mathbb{R}^{n_y},$$
$$U(k-1) := \left[u(k-1), \ldots, y(k-n_u)\right]^T \in \mathbb{R}^{n_u}.$$

Therefore $z(k-1) = [Y(k-1)^T, U(k-1)^T]^T$. Since f is smooth, there exists a constant M such that $|y(k) - y^0| \leq M$ for each $k = 1, \ldots, N$. Let $\mathcal{Z}$ be the compact set defined by

$$\mathcal{Z} := \{ z = (Y, U) \in \mathbb{R}^{n_z}: |Y_i - y^0| \leq M + \varepsilon,$$
$$U_j \in \mathcal{U}, \ i = 1, \ldots, n_y, \ j = 1, \ldots, n_u \}.$$

Now, since by assumption $\mathcal{F}$ is smooth, it satisfies a Lipschitz condition in the compact domain $\mathcal{Z}$; that is, there exists a constant λ such that for all $(Y^\alpha, U), (Y^\beta, U) \in \mathcal{Z}$,

$$\left| \mathcal{F}(Y^\alpha, U; \theta^*) - \mathcal{F}(Y^\beta, U; \theta^*) \right| \leq \lambda \sum_{i=1}^{n_y} \left| y^\alpha(k-i) - y^\beta(k-i) \right|. \quad (5)$$

Let $e(k) := y(k) - \hat{y}(k)$. Then from (2) and (3), we have

$$\begin{aligned}
e(k) &= f\big(z(k-1)\big) - \mathcal{F}\big(\hat{z}(k-1); \theta^*\big) \\
&= \left[f\big(z(k-1)\big) - \mathcal{F}\big(z(k-1); \theta^*\big) \right] \\
&\quad + \left[\mathcal{F}\big(z(k-1); \theta^*\big) - \mathcal{F}\big(\hat{z}(k-1); \theta^*\big) \right].
\end{aligned} \quad (6)$$

Whereas $\mathcal{F}$ is a universal approximator, there exist an integer q^* and a parameter vector $\theta^* \in \mathbb{R}^{q^*}$ such that

$$\max_{z \in \mathcal{Z}} \left| f(z) - \mathcal{F}(z; \theta^*) \right| \leq \delta^*, \quad (7)$$

where $\delta^* > 0$ can be made arbitrarily small. Therefore, in view of (5), (6), and (7),

$$\begin{aligned}
|e(k)| &\leq \delta^* + \lambda \sum_{i=1}^{n_y} \left| e(k-i) \right| \\
&= \delta^* + \lambda \sum_{i=k-n_y}^{k-1} \left| e(i) \right| \\
&\leq \delta^* + \lambda \sum_{i=1}^{k-1} \left| e(i) \right| \\
&\leq \delta^* (1 + \lambda)^{k-1},
\end{aligned}$$

where the last inequality is obtained by the discrete-time version of the Bellman–Gronwall lemma [22]. Now, if δ^* is chosen as

$$\delta^* = \frac{1}{(1+\lambda)^{N-1}}\varepsilon, \tag{8}$$

then for each $k = 1, \ldots, N$,

$$\left|e(k)\right| \le \frac{1}{(1+\lambda)^{N-k}}\varepsilon \le \varepsilon.$$

Going backward, it can be verified easily that if (8) is satisfied, then $\hat{z}(k-1) \in \mathcal{Z}$ for each $k = 1, \ldots, N$. This concludes the proof. ■

Note that Lemma 1 is strictly an existence result; it states that if a universal (static) network approximator is allowed to have an arbitrarily large number of nodes, then there exists an optimal set of parameter values, denoted by θ^*, such that the network model approximates a general dynamical system arbitrarily closely. In practice, it may not be possible (or desirable) to construct such a large network. Furthermore, the optimal parameter set θ^* will be unknown. Therefore, we are interested in parameter estimation methods that work "reliably" even in the presence of approximation errors. In Section III, we develop discrete-time parameter adaptive laws for recursive adjustment of the parameters $\theta(k)$, under the presence of approximation errors. Section IV develops a similar learning methodology for identification of continuous-time nonlinear systems.

C. PROBLEM FORMULATION

The system described by (1) can be rewritten in the form

$$y(k) = \mathcal{F}\big(z(k-1); \theta^*\big) + \nu(k-1), \tag{9}$$

where $\nu(k)$ denotes the *approximation error* (also referred to as *network reconstruction error*), which is defined as

$$\nu(k-1) := f\big(z(k-1)\big) - \mathcal{F}\big(z(k-1); \theta^*\big). \tag{10}$$

The optimal weight or parameter vector θ^* is an "artificial" quantity required only for analytical purposes. We choose θ^* as the value of θ that minimizes the distance between f and $\mathcal{F}$ for all z in some compact learning domain $\mathcal{Z}$, subject to the additional restriction that θ^* belongs to the hypersphere $\mathcal{B}(M_\theta) := \{\theta \in \mathbb{R}^q : |\theta| \le M_\theta\}$, where M_θ is a large design constant; that is,

$$\theta^* := \arg\min_{|\theta|\le M_\theta}\left\{\sup_{z\in\mathcal{Z}}\left|f(z) - \mathcal{F}(z;\theta)\right|\right\}. \tag{11}$$

In the development of parameter adjustment laws for the estimates of θ^*, we restrict the estimates (using a projection algorithm) within this hypersphere; by doing so we avoid any numerical problems that may otherwise arise due to very large parameter values. Furthermore, the projection algorithm prevents the parameters from drifting to infinity, which is a phenomenon that may occur with standard adaptive laws in the presence of modeling or approximation errors [23, 24].

The network approximation error, denoted by $v(k)$, is a critical quantity that represents the minimum possible deviation between the unknown nonlinear function f and the input–output function of the approximator $\mathcal{F}$. In general, increasing the number of adjustable weights (denoted by q) reduces the approximation error. Universal approximation results for neural networks indicate that if q is sufficiently large, then v can be made arbitrarily small on a compact region. In practical systems v will be nonzero. Its value depends on many design variables, such as the type of network-approximation model, the number of adjustable parameters, the number of layers (in the case of multilayer networks), and also the "size" of the compact sets $\mathcal{Z}$ and $\mathcal{B}(M_\theta)$. For example, the constraint that θ^* belongs to the hypersphere $\mathcal{B}(M_\theta)$ may yield a larger value of v than that obtained with no constraints on θ^*. However, if the design constant M_θ is (sufficiently) large, then it is clear that any such increase will be small. This indicates a design trade-off on how large M_θ is chosen: if M_θ is very large, then we may have numerical problems due to large parameter values; if M_θ is not large enough, then we may have unacceptably large approximation errors.

In general, if the network approximator $\mathcal{F}$ is constructed appropriately, then v will be a small quantity. Unfortunately, at present, the factors that influence how well networks are constructed, such as the number of adjustable parameters and number of layers, are chosen for the most part by trial and error or other ad hoc techniques. Therefore, an attractive feature of the synthesis and analysis procedure developed here is that no knowledge regarding the value or an upper bound for v is required. However, it is clear that the smaller v is, the better approximation accuracy can be achieved, and consequently, a better mathematical model can be constructed.

III. LEARNING ALGORITHM

In this section we study the construction of discrete-time identification schemes based on the system model described by

$$y(k) = \mathcal{F}\big(z(k-1); \theta^*\big) + v(k-1). \tag{12}$$

As is common in identification procedures, we will assume that the sequences $y(k)$ and $u(k)$ are uniformly bounded. This implies that there exists a compact set $\mathcal{Z} \subset \mathbb{R}^{n_z}$ such that $z(k) \in \mathcal{Z}$ for all $k \geq 0$.

Based on (12) we consider the identification model

$$\hat{y}(k) = \mathcal{F}\big(z(k-1); \theta(k-1)\big), \tag{13}$$

where $\theta(k) \in \mathbb{R}^q$ is the estimate of θ^* at the discrete-time k, and $\hat{y}(k)$ is the output of the identifier at time k. The overall identification configuration is shown in block diagram representation in Fig. 1. The delay circuit computes previous values of the output y and input u; these tapped delay values are the inputs to the network approximator $\mathcal{F}$. The output error $e(k) := \hat{y}(k) - y(k)$ is used to tune the adjustable parameters θ.

From (12) and (13), the output error, or as is sometimes called, *prediction error*, satisfies

$$e(k) = \mathcal{F}\big(z(k-1); \theta(k-1)\big) - \mathcal{F}\big(z(k-1); \theta^*\big) - v(k-1). \tag{14}$$

By considering the Taylor series expansion of $\mathcal{F}(z(k-1); \theta(k-1))$ around $(z(k-1), \theta^*)$, (14) can be expressed as

$$e(k) = \xi(k-1)^T \phi(k-1) + \mathcal{F}_0\big(z(k-1); \theta(k-1)\big) - v(k-1), \tag{15}$$

where $\phi(k) := \theta(k) - \theta^*$ is the *parameter estimation error* and $\xi \in \mathbb{R}^q$ is the *sensitivity function* between the output of the network and the adjustable parameters or weights:

$$\xi(k) := \frac{\partial^T \mathcal{F}(z(k); \theta(k))}{\partial \theta}. \tag{16}$$

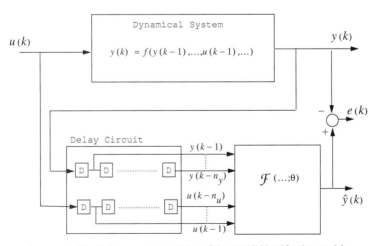

Figure 1   Block diagram representation of the overall identification model.

The notation $(\partial^T \mathcal{F}(p))/\partial\theta$ denotes the transpose of the partial derivative $\partial\mathcal{F}/\partial\theta := [\partial\mathcal{F}/\partial\theta_1 \cdots \partial\mathcal{F}/\partial\theta_q]$ evaluated at the point p.

The function $\mathcal{F}_0(z; \theta)$ in (15) satisfies

$$\mathcal{F}_0(z; \theta) := \mathcal{F}(z; \theta) - \mathcal{F}(z; \theta^*) - \frac{\partial^T \mathcal{F}(z; \theta)}{\partial\theta}(\theta - \theta^*). \tag{17}$$

Therefore, $\mathcal{F}_0$ represents the higher-order terms (with respect to θ) of the Taylor series expansion; hence, from (17) we obtain that $\mathcal{F}_0$ satisfies [25]

$$\left|\mathcal{F}_0(z; \theta)\right| \le \delta(z, \theta)|\phi|, \qquad \forall z \in \mathcal{Z}, \ \forall \theta \in \mathcal{B}(M_\theta), \tag{18}$$

where[1]

$$\delta(z, \theta) := \sup_{\bar{\theta} \in [\theta, \theta^*]} \left\{ \left| \frac{\partial^T \mathcal{F}(z; \bar{\theta})}{\partial\theta} - \frac{\partial^T \mathcal{F}(z; \theta)}{\partial\theta} \right| \right\}. \tag{19}$$

According to the definition of δ as given by (19), it is clear that for each $z \in \mathcal{Z}$ we have $\lim_{\theta \to \theta^*} \delta(z, \theta) = 0$.

It is worth noting that the higher-order term $\mathcal{F}_0$ encapsulates the nonlinear parametrization structure of the network, or from an approximation theory perspective, it reflects the fact that the set of approximants $\mathcal{F}$ is a nonlinear subspace. In the case of linearly parametrized networks, the higher-order term $\mathcal{F}_0$ is identically equal to zero.

A. WEIGHT ADAPTATION

The approach taken in this paper is to design parameter adaptive laws for $\theta(k)$ based on the linearized part e_L of the error equation (15), that is,

$$e_L(k - 1) = \xi(k - 1)^T \phi(k - 1), \tag{20}$$

and analyze the stability and convergence properties based on the overall error equation

$$e(k) = e_L(k - 1) + q(k - 1), \tag{21}$$

$$q(k - 1) := \mathcal{F}_0\big(z(k - 1); \theta(k - 1)\big) + \nu(k - 1). \tag{22}$$

Neglecting the higher-order term $\mathcal{F}_0$ in the derivation of the parameter adaptive law can be interpreted in two ways: from a stability theory viewpoint, it is equivalent to linearization techniques, whose main principle is that for ϕ sufficiently small, the linearized term $\xi^T \phi$ dominates the higher-order term $\mathcal{F}_0$. On the other hand, from an optimization theory perspective, the linearized part of the

---

[1]In (19) we use the notation $[\theta, \theta^*] := \{\vartheta \in \mathbb{R}^q : \vartheta = \lambda\theta + (1 - \lambda)\theta^*, \ 0 \le \lambda \le 1\}$.

error equation is obtained by using the gradient method in the minimization of a quadratic cost function. Based on these observations we consider the weight adaptive law

$$\theta(k) = \begin{cases} \mu(k), & \text{if } |\mu(k)| \leq M_\theta, \\ \mathcal{P}[\mu(k)], & \text{if } |\mu(k)| > M_\theta, \end{cases} \qquad (23)$$

$$\mu(k) := \theta(k-1) - \frac{\gamma_0 e(k)}{\beta_0 + |\xi(k-1)|^2} \xi(k-1),$$
$$\beta_0 > 0, \ 0 < \gamma_0 < 2, \qquad (24)$$

$$\mathcal{P}[\mu(k)] := \frac{M_\theta}{|\mu(k)|} \mu(k). \qquad (25)$$

The constant γ_0 is known as *learning rate* or *adaptive gain*; both β_0 and γ_0 are design constants. The projection operation $\mathcal{P}[\cdot]$ [26], as described by (25), modifies the standard adaptive law to prevent the adjustable parameters from drifting to infinity [27]. According to (23), if the new parameter estimates are inside the hypersphere $\mathcal{B}(M_\theta)$, then the standard adaptive law $\theta(k) = \mu(k)$ is used; if the new parameter estimates are directed toward the outside of the hypersphere, then they are projected back onto the boundary by using (25). Therefore, the adaptive law (23)–(25) guarantees that $|\theta(k)| \leq M_\theta$ for all $k \geq 0$, provided of course that $|\theta(0)| \leq M_\theta$. Based on the adaptive law (23)–(25), the following stability properties can be obtained.

THEOREM 2.   *The parameter adaptive law (23)–(25) guarantees that $e(k)$ and $\theta(k)$ are uniformly bounded and there exist constants λ_1 and λ_2 such that for any finite integer N,*

$$\sum_{k=1}^{N} |e(k)|^2 \leq \lambda_1 + \lambda_2 \sum_{k=1}^{N} |q(k-1)|^2, \qquad (26)$$

where q is given by (22).

Proof.   The parameter adaptive law (23)–(25) is first rewritten as

$$\theta(k) = \mu(k) - I_\mu \left(\frac{|\mu(k)| - M_\theta}{|\mu(k)|} \right) \mu(k), \qquad (27)$$

where I_μ denotes the indicator function defined as

$$I_\mu = \begin{cases} 1, & \text{if } |\mu(k)| > M_\theta, \\ 0, & \text{if } |\mu(k)| \leq M_\theta. \end{cases}$$

Consider the positive definite function

$$V(\phi(k)) := |\phi(k)|^2 = |\theta(k) - \theta^*|^2. \qquad (28)$$

Using (27) we have

$$
\begin{aligned}
V\big(\phi(k)\big) &= \big|\mu(k) - \theta^*\big|^2 - I_\mu\left(\frac{|\mu(k)| - M_\theta}{|\mu(k)|}\right) \\
&\quad \times \big[|\mu(k)|^2 - 2\mu(k)^T\theta^* + |\mu(k)|M_\theta\big] \\
&= \big|\mu(k) - \theta^*\big|^2 - I_\mu\left(\frac{|\mu(k)| - M_\theta}{|\mu(k)|}\right) \\
&\quad \times \big[|\mu(k) - \theta^*|^2 + |\mu(k)|M_\theta - |\theta^*|^2\big] \\
&= \big|\mu(k) - \theta^*\big|^2 - I_\mu\rho(k),
\end{aligned}
\tag{29}
$$

where

$$
\rho(k) := \left(\frac{|\mu(k)| - M_\theta}{|\mu(k)|}\right)\big[|\mu(k) - \theta^*|^2 + |\mu(k)|M_\theta - |\theta^*|^2\big].
\tag{30}
$$

Next we proceed to show that $I_\mu\rho(k) \geq 0$ for all $k > 0$. Suppose that $I_\mu = 1$; otherwise the inequality holds trivially. In this case $|\mu(k)| > M_\theta$. Whereas, by definition, $|\theta^*| \leq M_\theta$ we have that

$$
|\mu(k)|M_\theta - |\theta^*|^2 > M_\theta{}^2 - |\theta^*|^2 \geq 0.
$$

This shows that $\rho(k) > 0$; hence $I_\mu\rho(k) \geq 0$.

Next we compute $|\mu(k) - \theta^*|^2$. From (24) we have

$$
\begin{aligned}
|\mu(k) - \theta^*|^2 &= \left|\phi(k-1) - \frac{\gamma_0}{\beta_0 + |\xi(k-1)|^2}e(k)\xi(k-1)\right|^2 \\
&= V\big(\phi(k-1)\big) - \frac{\gamma_0}{\beta_0 + |\xi(k-1)|^2} \\
&\quad \times \left(2e(k)\phi(k-1)^T\xi(k-1) - \frac{\gamma_0 e^2(k)|\xi(k-1)|^2}{\beta_0 + |\xi(k-1)|^2}\right) \\
&= V\big(\phi(k-1)\big) - \frac{\gamma_0}{\beta_0 + |\xi(k-1)|^2} \\
&\quad \times \left(e^2(k)\left[2 - \frac{\gamma_0|\xi(k-1)|^2}{\beta_0 + |\xi(k-1)|^2}\right] - 2e(k)q(k-1)\right),
\end{aligned}
$$

where the last equality is obtained by using (15) and (21). Now define

$$
\alpha(k-1) := \frac{\gamma_0}{\beta_0 + |\xi(k-1)|^2}\left[2 - \frac{\gamma_0|\xi(k-1)|^2}{\beta_0 + |\xi(k-1)|^2}\right].
$$

Note that since $0 < \gamma_0 < 2$, we have $\alpha(k-1) > 0$ for all $k > 0$. Therefore

$$
\begin{aligned}
\left| \mu(k) - \theta^* \right|^2 = V\big(\phi(k-1)\big) &- \frac{\alpha(k-1)}{2} e^2(k) - \frac{\alpha(k-1)}{2} \\
&\times \left[e^2(k) - \frac{4}{\alpha(k-1)} e(k) q(k-1) \right].
\end{aligned}
$$

By completing the square we obtain

$$
\begin{aligned}
\left| \mu(k) - \theta^* \right|^2 = V\big(\phi(k-1)\big) &- \frac{\alpha(k-1)}{2} e^2(k) - \frac{\alpha(k-1)}{2} \\
&\times \left(e(k) - \frac{2}{\alpha(k-1)} q(k-1) \right)^2 + \frac{2}{\alpha(k-1)} q^2(k-1). \quad (31)
\end{aligned}
$$

Now, if we let $\Delta_V(k) := V(\phi(k)) - V(\phi(k-1))$ and substitute (31) in (29), we obtain

$$
\begin{aligned}
\Delta_V(k) = &-\frac{\alpha(k-1)}{2} e^2(k) - \frac{\alpha(k-1)}{2} \left(e(k) - \frac{2}{\alpha(k-1)} q(k-1) \right)^2 \\
&+ \frac{2}{\alpha(k-1)} q^2(k-1) - I_\mu \rho(k) \\
\leq &-\frac{\alpha(k-1)}{2} e^2(k) + \frac{2}{\alpha(k-1)} q^2(k-1). \quad (32)
\end{aligned}
$$

Let $\alpha_0 := \inf_k \alpha(k-1)$. From (32) we obtain

$$
e^2(k) \leq -\frac{2}{\alpha_0} \Delta_V(k) + \frac{4}{\alpha_0^2} q^2(k-1).
$$

By summing both sides from $k = 1$ to $k = N$, where N is a finite integer, we have

$$
\sum_{k=1}^{N} e^2(k) \leq \frac{2}{\alpha_0} \big(|\phi(0)|^2 - |\phi(N)|^2 \big) + \frac{4}{\alpha_0^2} \sum_{k=1}^{N} q^2(k-1)
$$

$$
\leq \lambda_1 + \lambda_2 \sum_{k=1}^{N} q^2(k-1).
$$

This concludes the proof. ■

According to Theorem 2, in any discrete-time interval $[0, N]$, the "energy" of the prediction error e is (at most) of the same order as the "energy" of q; that is, the sum of the network approximation error and the magnitude of the higher-order term $\mathcal{F}_0$. This points out the relationship between the output error and both the approximation error and higher-order terms. Specifically, it indicates that the "larger" the magnitude of the higher-order term $\mathcal{F}_0$, the "larger" the prediction

error will be; this is to be expected of course, because in the derivation of the
adaptive law, the higher-order term $\mathcal{F}_0$ was neglected.

B. Linearly Parametrized Approximators

Now consider the special case that the network employed is a linearly
parametrized approximator; that is,

$$\mathcal{F}\big(z(k); \theta(k)\big) = \Omega\big(z(k)\big)^T \theta(k) = \zeta(k)^T \theta(k),$$

where $\zeta(k) := \Omega(z(k))$ and $\Omega: \mathbb{R}^{n_z} \mapsto \mathbb{R}^q$ is a smooth function. For example, in
the case of RBF networks with fixed centers and widths, $\zeta(k)$ is the output of the
basis functions at time k. A crucial difference between linearly and nonlinearly
parametrized approximators is that for linearly parametrized models, the higher-
order term $\mathcal{F}_0$ is equal to zero. Therefore the error equation (15) becomes

$$e(k) = \phi(k-1)^T \zeta(k-1) - \nu(k-1). \tag{33}$$

The corresponding adaptive law for adjusting the parameters of the network is the
same as (23)–(25) with $\xi(k-1)$ being replaced by $\zeta(k-1)$. The following corollary to Theorem 2 describes the stability properties of the identification scheme
in the special case that linearly parametrized networks are employed.

Corollary 3.   *In the special case of linearly parametrized networks the
parameter adaptive law (23)–(25) guarantees that:*

   (i) *$e(k)$ and $\theta(k)$ are uniformly bounded.*
   (ii) *There exist constants λ_1 and λ_2 such that for any finite integer N,*

$$\sum_{k=1}^{N} |e(k)|^2 \leq \lambda_1 + \lambda_2 \sum_{k=1}^{N} |\nu(k-1)|^2. \tag{34}$$

   (iii) *If $\nu \in l_2$, that is, $\sum_{k=0}^{\infty} |\nu(k)|^2 < \infty$, then $e \in l_2$ and $\lim_{k \to \infty} e(k) = 0$.*

Proof.   The proof of parts (i) and (ii) follows directly from Theorem 2 by
letting $\mathcal{F}_0 = \delta = 0$. To prove part (iii) note that if $\nu \in l_2$, then from (34) it follows
that $e \in l_2$; this, in turn, implies that $\lim_{k \to \infty} e(k) = 0$.   ∎

Remark 4.   A comparison of Theorem 2 and Corollary 3 may, at first sight,
suggest that a linearly parametrized approximator, in general, results in better performance than nonlinearly parametrized approximation models due to the higher-
order term $\mathcal{F}_0$ being equal to zero. However, this is not necessarily a valid observation because the approximation error $\nu(k)$ is different for the two cases. In fact,
it is known that some nonlinearly parametrized approximators have in general
greater approximation power than linearly parametrized approximators. For example, recent work by Barron [28] indicates that for certain classes of functions,

sigmoidal neural network models (with one hidden layer) can achieve a specified approximation accuracy with a number of nodes which is linearly dependent
on the dimension of the input vector. This result is very important because it is
well known that linearly parametrized approximation methods suffer from the so-
called *curse of dimensionality* problem; this expression (originally used by Bellman [29]) describes the fact that an exponentially increasing number of nodes
are needed to approximate mappings of increasing input dimension. Other than
the approximation power, a comparison study of different network approximators involves the investigation of issues such as adaptivity features, mathematical
tractability, smoothness of the approximation model, and its implementation properties. With the present rate of improvement in parallel computing, the issues of
hardware implementation and fast weight adjustment become increasingly more
significant.

Remark 5. The projection algorithm is just one of many possible methods for
dealing with the parameter drift problem. In the adaptive control literature there
are several other modifications that can be used to achieve a similar objective [24],
as for example, the σ modification, the dead zone, and ε modification. However,
the projection algorithm is simple to implement, requires no upper bounds on the
modeling error, and has the intuitive interpretation of restricting the parameters
to some prescribed limits. Moreover, in hardware implementations of network architectures, each weight will (by construction) have some lower and upper bound
and therefore the implementation of the projection algorithm also has a convenient
physical interpretation. It is worth noting that instead of restricting the parameter
estimates within a hypersphere, a similar projection methodology can be applied
to restrict the parameter estimates within a *hypercube* of the form

$$\mathcal{K}(M_\theta) := \left\{ \theta \in \mathbb{R}^q \colon |\theta_i| \leq M_\theta, i = 1, \ldots, q \right\}$$

or, more generally, a hyperrectangle

$$\mathcal{R}(M_{\theta_i}) := \left\{ \theta \in \mathbb{R}^q \colon |\theta_i| \leq M_{\theta_i}, i = 1, \ldots, q \right\}.$$

Remark 6. As shown in Corollary 3, in the ideal case of no approximation
error (or, more generally, if $v \in l_2$), the linearly parametrized approximation
scheme with the adaptive law described by (23)–(25) guarantees that the prediction error $e(k)$ converges to zero. This does *not* imply that the parameter estimates
$\theta(k)$ converge to θ^*. To guarantee that the parameter estimation error $\phi(k)$ converges to zero, $\zeta(k)$ in (33) needs to satisfy a *persistency of excitation* condition.
A discrete-time signal $\zeta(k) \in \mathbb{R}^q$ is said to be persistently exciting in $\mathbb{R}^n$ if there
exist positive constants α_0, α_1, and N such that for all $k \geq 0$,

$$\alpha_0 I \leq \sum_{\kappa=k}^{k+N} \zeta(\kappa)\zeta(\kappa)^T \leq \alpha_1 I,$$

where I denotes the $q \times q$ identity matrix [30]. In contrast to linear systems where the persistency of excitation condition has been transformed into a condition on the input signal, in nonlinear systems this condition cannot be verified *a priori*. Note that for nonlinear systems, even if $u(k)$ is persistently exciting, it not necessarily true that $y(k)$ will be persistently exciting; furthermore, even if $z(k)$ is persistently exciting, it is not necessarily true that $\zeta(k) = \Omega(z(k))$ will be persistently exciting.

Remark 7. For notational simplicity the learning rate γ_0 in the adaptive law (23)–(25) is a scalar. It can be easily verified that the analysis is still valid if γ_0 in (24) is replaced by a positive definite matrix $\Gamma_0 \in \mathbb{R}^{q \times q}$ provided the largest eigenvalue $\lambda_{\max}(\Gamma_0)$ satisfies $\lambda_{\max}(\Gamma_0) < 2$.

C. MULTIVARIABLE SYSTEMS

The design and analysis procedure of the previous subsections can be expanded from the the single-input–single-output (SISO) case to systems that are multi-input–multi-output (MIMO). Consider the system

$$Y(k) = F\big(Y(k-1), \ldots, Y(k-n_Y), U(k-1), \ldots, U(k-n_U)\big), \qquad (35)$$

where $U(k) \in \mathbb{R}^m$ and $Y(k) \in \mathbb{R}^n$ are the inputs and outputs, respectively, and $F \colon \mathbb{R}^p \mapsto \mathbb{R}^n$ is a smooth vector field with $p := n \cdot n_Y + m \cdot n_U$. The integers n_Y and n_U are the maximum time lags of the output and input, respectively, that affect the current output. By using the notation

$$\omega(k-1)^T := \big[Y(k-1)^T \cdots Y(k-n_Y)^T U(k-1)^T \cdots U(k-n_U)^T\big] \in \mathbb{R}^p,$$

Eq. (35) is expressed in the compact form

$$Y(k) = F\big(\omega(k-1)\big). \qquad (36)$$

As before, the objective is to approximate the unknown mapping $F(\omega)$ in (36) by some type of network approximator with output $\mathcal{F}(\omega; \theta)$, where θ denotes an adjustable parameter vector. Therefore, (36) is rewritten as

$$Y(k) = \mathcal{F}\big(\omega(k-1); \theta^*\big) + N(k-1), \qquad (37)$$

$$N(k-1) := F\big(\omega(k-1)\big) - \mathcal{F}\big(\omega(k-1); \theta^*\big), \qquad (38)$$

$$\theta^* := \arg \min_{|\theta| \le M_\theta} \left\{ \sup_{\omega \in \Omega} \big|F(\omega) - \mathcal{F}(\omega; \theta)\big| \right\}, \qquad (39)$$

where M_θ is a large design constant and $\Omega \subset \mathbb{R}^p$ is a compact set that contains $\omega(k)$ for every $k \ge 0$. The existence of such a compact set is assured by the assumption that the system is bounded-input–bounded-output stable.

Based on (37) we consider the identification model described by

$$\widehat{Y}(k) = \mathcal{F}\big(\omega(k-1); \theta(k-1)\big), \tag{40}$$

which results in a prediction error $E(k) := \widehat{Y}(k) - Y(k)$ that satisfies

$$\begin{aligned}
E(k) &= \Xi(k-1)^T \phi(k-1) + Q(k-1), \\
Q(k-1) &:= \mathcal{F}_0\big(\omega(k-1); \theta(k-1)\big) - N(k-1), \\
\Xi(k) &:= \frac{\partial^T \mathcal{F}(\omega(k); \theta(k))}{\partial \theta} \in \mathbb{R}^{q \times n}.
\end{aligned} \tag{41}$$

As in the previous section, $\mathcal{F}_0$ denotes the sum of the higher-order terms, which satisfy a relationship of the form (18); the variables $Q(k-1)$ and $\Xi(k-1)$ correspond to $q(k-1)$ and $\xi(k-1)$, respectively, of the SISO case.

Next we proceed to develop a corresponding parameter adaptive law for the MIMO case. In the sequel, we use the Frobenius matrix norm [31]

$$\|A\| := \big(\text{trace}(AA^T)\big)^{1/2} = \left(\sum_i \sum_j |a_{ij}|^2 \right)^{1/2}.$$

Based on the same procedure as in Section III.A, we consider the adaptive law

$$\theta(k) = \begin{cases} \mu(k), & \text{if } |\mu(k)| \le M_\theta, \\ \mathcal{P}[\mu(k)], & \text{if } |\mu(k)| > M_\theta, \end{cases} \tag{42}$$

$$\mu(k) := \theta(k-1) - \frac{\gamma_0}{\beta_0 + \|\Xi(k-1)\|^2} \Xi(k-1)E(k),$$

$$\beta_0 > 0, \ 0 < \gamma_0 < 2, \tag{43}$$

$$\mathcal{P}[\mu(k)] := \frac{M_\theta}{|\mu(k)|} \mu(k). \tag{44}$$

The multivariable version of Theorem 2 for the algorithm given by (42)–(44) is given next.

THEOREM 8.   *The parameter adaptive law (42)–(44) guarantees that $E(k)$ and $\theta(k)$ are uniformly bounded and there exist constants λ_1 and λ_2 such that for any finite integer N,*

$$\sum_{k=1}^{N} |e(k)|^2 \le \lambda_1 + \lambda_2 \sum_{k=1}^{N} |Q(k-1)|^2. \tag{45}$$

The proof is a straightforward extension of the proof of Theorem 2 and is therefore omitted.

Remark 9.   In the special case of linearly parametrized network approximators, the higher-order term $\mathcal{F}_0$ is identically equal to zero.

Remark 10. The "curse of dimensionality" problem (see Remark 4) is partic-ularly important in the approximation of multivariable systems. For example, for a dynamical system with three inputs and three outputs and maximum time lags $n_Y = n_U = 3$, the network approximator has an input vector of dimension 18! If linearly parametrized networks are employed, then the computational demands, both in memory and computational time, are tremendous. Therefore, especially in cases such as this, multilayer neural networks with sigmoidal activation func-tions (or, more generally, nonlinearly parametrized approximators) are well worth investigating.

IV. CONTINUOUS-TIME IDENTIFICATION

In this section we consider the identification of continuous-time nonlinear sys-tems of the form

$$\dot{x} = f(x) + g(x)u, \tag{46}$$

where $u \in \mathbb{R}$ is the input, $x \in \mathbb{R}^n$ is the state, which is assumed to be available for measurement, and f and g are unknown, smooth vector fields defined on an open set of $\mathbb{R}^n$. The preceding class of continuous-time nonlinear systems is called *affine systems* because in Eq. (46) the control input u appears linearly with respect to g. There are several reasons to consider the class of affine systems. First, most of the systems encountered in engineering are, by nature or design, affine systems. Second, most of the nonlinear control techniques, including feedback lineariza-tion, are developed for affine systems. Finally, we note that nonaffine systems can be converted to affine systems by passing the input through integrators [32]. This procedure is known as *dynamic extension*.

The problem of identification consists of choosing an appropriate identifica-tion model and adjusting the parameters of the model according to some adaptive law such that the response $\hat{x}(t)$ of the model to an input signal $u(t)$ (or a class of input signals) approximates the response $x(t)$ of the real system to the same input. Whereas a mathematical characterization of a system is often a prerequisite to analysis and controller design, system identification is important not only to understand and predict the behavior of the system, but also to obtain an effective control law. In this section we consider continuous-time identification schemes that are based on the setting shown in Fig. 2, which is known as the *series-parallel configuration* [23].

As is common in identification procedures, we assume that the state $x(t)$ is bounded for all admissible bounded inputs $u(t)$. Note that even though the real system is bounded-input–bounded-state (BIBS) stable, there is no *a priori* guar-antee that the output $\hat{x}$ of the identification model or that the adjustable parameters

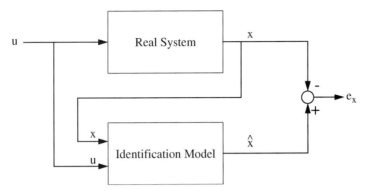

Figure 2 A general configuration for identification of nonlinear dynamical systems based on the series-parallel model.

in the model remain bounded. Stability of the overall scheme depends on the particular identification model that is used as well as on the parameter adjustment rules that are chosen. This section is concerned with the development of identification models, based on sigmoidal and RBF neural networks, and the derivation of adaptive laws that guarantee stability of the overall identification structure.

The unknown nonlinearities $f(x)$ and $g(x)$ are parametrized by static neural networks with outputs $\hat{f}(x, \theta_f)$, and $\hat{g}(x, \theta_g)$, respectively, where $\theta_f \in \mathbb{R}^{n_f}$ and $\theta_g \in \mathbb{R}^{n_g}$ are the adjustable weights and n_f and n_g denote the number of weights in the respective neural network approximation of f and g. By adding and subtracting the terms $\hat{f}$ and $\hat{g}u$, the nonlinear system described by (46) is rewritten as

$$\dot{x} = \hat{f}(x, \theta_f^*) + \hat{g}(x, \theta_g^*)u + \left[f(x) - \hat{f}(x, \theta_f^*)\right] + \left[g(x) - \hat{g}(x, \theta_g^*)\right]u, \quad (47)$$

where θ_f^* and θ_g^* denote the optimal weight values (in the L_∞-norm sense) in the approximation of $f(x)$ and $g(x)$, respectively, for x belonging to a compact set $\mathcal{X} \subset \mathbb{R}^n$. For a given class of bounded input signals u, the set $\mathcal{X}$ is such that it contains all possible trajectories $x(t)$. The set $\mathcal{X}$ is, in general, not known *a priori* and is used only for analysis purposes.

Although the "optimal" weights θ_f^* and θ_g^* in (47) could take arbitrarily large values, from a practical perspective we are interested only in weights that belong to a (large) compact set. Therefore we consider "optimal" weights θ_f^* and θ_g^* that belong to the convex compact sets $\mathcal{B}(M_f)$ and $\mathcal{B}(M_g)$, respectively, where M_f and M_g are design constants and $\mathcal{B}(M) := \{\theta: |\theta| \leq M\}$ denotes a ball of radius M. In the adaptive law, the estimates of θ_f^* and θ_g^*, which are the adjustable weights in the approximation networks, are also restricted to $\mathcal{B}(M_f)$ and $\mathcal{B}(M_g)$, respectively, through the use of a projection algorithm. By doing so, numerical problems that may arise due to having weight values that are too large are avoided; furthermore, the projection algorithm prevents the weights from drifting to infinity, which is a phenomenon that may occur with standard adaptive laws.

To summarize, the optimal weight vector θ_f^* is defined as the element in $\mathcal{B}(M_f)$ that minimizes $|f(x) - \hat{f}(x, \theta_f)|$ for $x \in \mathcal{X} \subset \mathbb{R}^n$; that is,

$$\theta_f^* := \arg \min_{\theta_f \in \mathcal{B}(M_f)} \left\{ \sup_{x \in \mathcal{X}} |f(x) - \hat{f}(x, \theta_f)| \right\}. \tag{48}$$

Similarly, θ_g^* is defined as

$$\theta_g^* := \arg \min_{\theta_g \in \mathcal{B}(M_g)} \left\{ \sup_{x \in \mathcal{X}} |g(x) - \hat{g}(x, \theta_g)| \right\}. \tag{49}$$

Finally, note that if the optimal weights are not unique, then θ_f^* (and correspondingly θ_g^*) denotes an arbitrary (but fixed) element of the set of optimal weights.

Equation (47) is now expressed in compact form as

$$\dot{x} = \hat{f}(x, \theta_f^*) + \hat{g}(x, \theta_g^*)u + v(t), \tag{50}$$

where $v(t)$ denotes the *approximation error*, defined as

$$v(t) := \left[f(x(t)) - \hat{f}(x(t), \theta_f^*) \right] + \left[g(x(t)) - \hat{g}(x(t), \theta_g^*) \right] u(t).$$

Whereas by assumption, $u(t)$ and $x(t)$ are bounded, the modeling error $v(t)$ is bounded by some constant v_0, where

$$v_0 := \sup_{t \geq 0} \left| \left(f(x(t)) - \hat{f}(x(t), \theta_f^*) \right) + \left(g(x(t)) - \hat{g}(x(t), \theta_g^*) \right) u(t) \right|.$$

The value of v_0 depends on many factors, such as the type of neural network that is used and the number of weights and layers, as well as the "size" of the compact sets $\mathcal{X}$, $\mathcal{B}(M_f)$, and $\mathcal{B}(M_g)$. For example, the constraint that the optimal weights θ_f^* and θ_g^* belong to the sets $\mathcal{B}(M_f)$ and $\mathcal{B}(M_g)$, respectively, may increase the value of v_0. However, if the constants M_f and M_g are large, then any increase will be small.

Remark 11. In general, it is impossible to know what values of M_f and M_g are appropriate for the optimal weights θ_f^* and θ_g^* to be included in the sets $\mathcal{B}(M_f)$ and $\mathcal{B}(M_g)$, respectively. Subsequently, it is important to note that the preceding design procedure for the problem formulation does not require this information. In the case that the unconstrained optimal weights do not belong to the chosen hypersphere $\mathcal{B}(M_f)$, then a new set of "constrained" optimal weights are chosen, as described by (48) and (49), that belong to the chosen hypersphere. Of course, by constraining the optimal weights within $\mathcal{B}(M_f)$ the approximation error may be increased. However, by choosing M_f sufficiently large, then any increase in approximation error will be minimized. The preceding formulation describes mathematically the intuitive trade-off on the size of allowed parameter space: if the parameter size is allowed to be very large, then we may have numerical problems due to large, parameter values; if the size of the parameter space is constrained too much, then we may have unacceptably large approximation errors.

By replacing the unknown nonlinearities with feedforward neural network models, we have essentially rewritten the system (46) in the form (50), where the parameters θ_f^* and θ_g^* and the approximation error $\nu(t)$ are unknown, but the underlying structure of $\hat{f}$ and $\hat{g}$ is known. Based on (50), we next develop and analyze various types of continuous-time identification schemes. For illustration purposes we consider Gaussian RBF networks and multilayer network models with sigmoidal nonlinearities.

A. RADIAL-BASIS-FUNCTION NETWORK MODELS

We first consider the case where the network architectures employed for modeling f and g are RBF networks with fixed centers and widths. In this case, we have a linearly parametrized network; therefore, the functions $\hat{f}$ and $\hat{g}$ in (50) take the form

$$\hat{f} = W_1^* \xi(x), \qquad \hat{g} = W_2^* \zeta(x), \tag{51}$$

where W_1^* and W_2^* are $n \times n_1$ and $n \times n_2$ matrices, respectively, representing in the spirit of (48) and (49), the optimal weight values, subject to the constraints $\|W_1^*\|_F \leq M_1$ and $\|W_2^*\|_F \leq M_2$. The norm $\| \cdot \|_F$ denotes the Frobenius matrix norm. In the preceding representation the network weights are formulated in matrix form for notational convenience. The constants n_1 and n_2 are the number of kernel units in each approximation and the vector fields $\xi(x) \in \mathbb{R}^{n_1}$ and $\zeta(x) \in \mathbb{R}^{n_2}$, which we refer to as *regressors*, are Gaussian-type functions, defined elementwise as

$$\xi_i(x) = \exp\left(-|x - c_{1i}|^2/\sigma_{1i}^2\right), \qquad i = 1, 2, \ldots, n_1,$$
$$\zeta_j(x) = \exp\left(-|x - c_{2j}|^2/\sigma_{2j}^2\right), \qquad j = 1, 2, \ldots, n_2.$$

For analysis, it is crucial that the centers c_{1i} and c_{2j} and widths σ_{1i} and σ_{2j}, $i = 1, \ldots, n_1$, $j = 1, \ldots, n_2$, are chosen *a priori*. By doing so, the only adjustable weights are W_1 and W_2, which appear linearly with respect to the nonlinearities ξ and ζ, respectively. Based on "local tuning" training techniques several researchers have suggested methods for appropriately choosing the centers and widths of the radial basis functions [33, 34]. In this section, we simply assume that c_{1i}, c_{2j}, σ_{1i}, and σ_{2j} are chosen *a priori* and kept fixed during adaptation of W_1 and W_2.

Generally, in the problem of identification of nonlinear dynamical systems one is usually interested in obtaining an accurate model in a (possibly large) neighborhood $\mathcal{N}(x^0)$ of an equilibrium point $x = x^0$ and therefore it is intuitively evident that the centers should be clustered around x^0 in this neighborhood. Clearly, the number of kernel units and the position of the centers and widths will affect the approximation capability of the model and consequently the value of the approximation error $v(t)$. Hence, current and future research dealing with effectively choosing these quantities is also relevent to the topics discussed in this work.

By substituting (51) in (50) we obtain

$$\dot{x} = W_1^* \xi(x) + W_2^* \zeta(x)u + v. \tag{52}$$

Based on the RBF network model described by (52), we next develop parameter update laws for stable identification using various techniques derived from the Lyapunov synthesis approach and also basic optimization methods.

1. Lyapunov Synthesis Method

The RBF network model (52) is rewritten in the form

$$\dot{x} = -\alpha x + \alpha x + W_1^* \xi(x) + W_2^* \zeta(x)u + v, \tag{53}$$

where $\alpha > 0$ is a scalar (design) constant. Based on (53) we consider the identification model

$$\dot{\hat{x}} = -\alpha \hat{x} + \alpha x + W_1 \xi(x) + W_2 \zeta(x)u, \tag{54}$$

where W_1 and W_2 are the estimates of W_1^* and W_2^*, respectively, whereas $\hat{x}$ is the output of the identification model. The identification model (54), which we refer to as *RBF error filtering model*, is similar to estimation schemes developed in [35]. The RBF error filtering model is depicted in Fig. 3. As can be seen from the figure, this identification model consists of two RBF network architectures in parallel and n first-order stable filters $h(s) = 1/(s + \alpha)$.

If we define $e_x := \hat{x} - x$, the state error, and $\Phi_1 := W_1 - W_1^*$, $\Phi_2 := W_2 - W_2^*$, the weight estimation errors, then from (53) and (54) we obtain the error equation

$$\dot{e}_x = -\alpha e_x + \Phi_1 \xi(x) + \Phi_2 \zeta(x)u - v. \tag{55}$$

The Lyapunov synthesis method consists of choosing an appropriate Lyapunov function candidate V and selecting weight adaptive laws so that the time derivative $\dot{V}$ satisfies $\dot{V} \leq 0$. The Lyapunov method as a technique for deriving stable adaptive laws can be traced as far back as the 1960s in the early literature of adaptive control theory for linear systems [36, 37]. In our case, an adaptive law for generating the parameter estimates $W_1(t)$ and $W_2(t)$ is developed by considering the Lyapunov function candidate

$$\begin{aligned} V(e_x, \Phi_1, \Phi_2) &= \frac{1}{2}|e_x|^2 + \frac{1}{2\gamma_1}\|\Phi_1\|_F^2 + \frac{1}{2\gamma_2}\|\Phi_2\|_F^2 \\ &= \frac{1}{2}e_x^T e_x + \frac{1}{2\gamma_1}\text{tr}\{\Phi_1\Phi_1^T\} + \frac{1}{2\gamma_2}\text{tr}\{\Phi_2\Phi_2^T\}, \end{aligned} \tag{56}$$

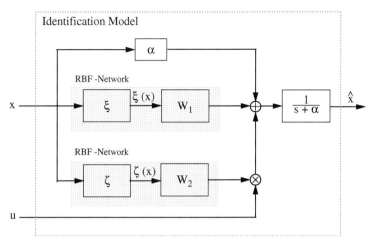

Figure 3 A block diagram representation of the error filtering identification model developed using RBF networks.

where γ_1 and γ_2 are positive constants. These constants appear in the adaptive laws and are referred to as *learning rates* or *adaptive gains*.

Using (55), the time derivative of V in (56) is expressed as

$$\dot{V} = -\alpha e_x^T e_x + \xi^T \Phi_1^T e_x + \frac{1}{\gamma_1} \mathrm{tr}\{\dot{\Phi}_1 \Phi_1^T\} + \zeta^T \Phi_2^T e_x u + \frac{1}{\gamma_2} \mathrm{tr}\{\dot{\Phi}_2 \Phi_2^T\} - e_x^T v.$$

Using properties of the trace, such as

$$\xi^T \Phi_1^T e_x = \mathrm{tr}\{\xi^T \Phi_1^T e_x\} = \mathrm{tr}\{e_x \xi^T \Phi_1^T\},$$

we obtain

$$\dot{V} = -\alpha |e_x|^2 + \mathrm{tr}\left\{ e_x \xi^T \Phi_1^T + \frac{1}{\gamma_1} \dot{\Phi}_1 \Phi_1^T \right\} + \mathrm{tr}\left\{ e_x u \zeta^T \Phi_2^T + \frac{1}{\gamma_2} \dot{\Phi}_2 \Phi_2^T \right\} - e_x^T v$$

$$= -\alpha |e_x|^2 + \frac{1}{\gamma_1} \mathrm{tr}\left\{ (\gamma_1 e_x \xi^T + \dot{\Phi}_1) \Phi_1^T \right\}$$

$$+ \frac{1}{\gamma_2} \mathrm{tr}\left\{ (\gamma_2 e_x u \zeta^T + \dot{\Phi}_2) \Phi_2^T \right\} - e_x^T v. \tag{57}$$

Whereas W_1^* and W_2^* are constant, we have that $\dot{W}_1 = \dot{\Phi}_1$ and $\dot{W}_2 = \dot{\Phi}_2$. Therefore it is clear from (57) that if the parameter estimates W_1 and W_2 are generated according to the adaptive laws

$$\dot{W}_1 = -\gamma_1 e_x \xi^T \quad \text{and} \quad \dot{W}_2 = -\gamma_2 e_x u \zeta^T, \tag{58}$$

then (57) becomes

$$\dot{V} = -\alpha|e_x|^2 - e_x^T v \le -\alpha|e_x|^2 + v_0|e_x|. \qquad (59)$$

If there is no approximation error (i.e., $v_0 = 0$), then from (59), $\dot{V}$ is negative semidefinite; hence stability of the overall identification scheme is guaranteed. However, in the presence of approximation error, if $|e_x| < v_0/\alpha$, then it is possible that $\dot{V} > 0$, which implies that the weights $W_1(t)$ and $W_2(t)$ may drift to infinity with time. This problem, which is referred to as *parameter drift*, is well known in the adaptive control literature [38]. Parameter drift also has been encountered in empirical studies of neural network learning, where it is usually referred to as *weight saturation*.

To avoid parameter drift, $W_1(t)$ and $W_2(t)$ are confined to the sets $\{W_1: \|W_1\|_F \le M_1\}$ and $\{W_2: \|W_2\|_F \le M_2\}$, respectively, through the use of a projection algorithm. In particular, the standard adaptive laws described by (58) are modified to

$$\dot{W}_1 = \begin{cases} -\gamma_1 e_x \xi^T, & \text{if } \{\|W_1\|_F < M_1\} \text{ or } \{\|W_1\|_F = M_1 \\ & \text{and } e_x^T W_1 \xi \ge 0\}, \\ \mathcal{P}\{-\gamma_1 e_x \xi^T\}, & \text{if } \{\|W_1\|_F = M_1 \text{ and } e_x^T W_1 \xi < 0\}, \end{cases} \qquad (60)$$

$$\dot{W}_2 = \begin{cases} -\gamma_2 e_x u \zeta^T, & \text{if } \{\|W_2\|_F < M_2\} \text{ or } \{\|W_2\|_F = M_2 \\ & \text{and } e_x^T W_2 \zeta u \ge 0\}, \\ \mathcal{P}\{-\gamma_2 e_x u \zeta^T\}, & \text{if } \{\|W_2\|_F = M_2 \text{ and } e_x^T W_2 \zeta u < 0\}, \end{cases} \qquad (61)$$

where $\mathcal{P}\{\cdot\}$ denotes the projection onto the supporting hyperplane, defined as

$$\mathcal{P}\{-\gamma_1 e_x \xi^T\} := -\gamma_1 e_x \xi^T + \gamma_1 \frac{e_x^T W_1 \xi}{\|W_1\|_F^2} W_1, \qquad (62)$$

$$\mathcal{P}\{-\gamma_2 e_x u \zeta^T\} := -\gamma_2 e_x u \zeta^T + \gamma_2 \frac{e_x^T W_2 \zeta u}{\|W_2\|_F^2} W_2. \qquad (63)$$

Therefore, if the initial weights are chosen such that $\|W_1(0)\|_F \le M_1$ and $\|W_2(0)\|_F \le M_2$, then we have $\|W_1(t)\|_F \le M_1$ and $\|W_2(t)\|_F \le M_2$ for all $t \ge 0$. This can be readily established by noting that whenever $\|W_1\|_F = M_1$ (and correspondingly for $\|W_2\|_F = M_2$), then

$$\frac{d}{dt}\{\|W_1(t)\|_F^2 - M_1^2\} \le 0,$$

which implies that the parameter estimate is directed toward the inside or the surface of the ball $\{W_1: \|W_1\|_F \le M_1\}$. It is worth noting that the projection modification causes the adaptive law to be discontinuous. However, the trajectory behavior on the discontinuity hypersurface is "smooth" and hence existence of a solution, in the sense of Carathèodory [39], is assured. The issue of existence and uniqueness of solutions in adaptive systems is treated in detail in [40].

With the adaptive laws (60) and (61), Eq. (57) becomes

$$\dot{V} = -\alpha|e_x|^2 - e_x^T v + I_1^* \, \mathrm{tr}\left\{ \frac{e_x^T W_1 \xi}{\|W_1\|_F^2} W_1 \Phi_1^T \right\}$$

$$+ I_2^* \, \mathrm{tr}\left\{ \frac{e_x^T W_2 \zeta u}{\|W_2\|_F^2} W_2 \Phi_2^T \right\}$$

$$\leq -\alpha|e_x|^2 - e_x^T v + I_1^* \frac{e_x^T W_1 \xi}{\|W_1\|_F^2} \, \mathrm{tr}\left\{ W_1 \Phi_1^T \right\}$$

$$+ I_2^* \frac{e_x^T W_2 \zeta u}{\|W_2\|_F^2} \, \mathrm{tr}\left\{ W_2 \Phi_2^T \right\}, \tag{64}$$

where I_1^* and I_2^* are indicator functions defined as $I_1^* = 1$ if the conditions $\|W_1\|_F = M_1$ and $e_x^T W_1 \xi < 0$ are satisfied and $I_1^* = 0$ otherwise (and correspondingly for I_2^*). The following lemma establishes that the additional terms introduced by the projection can only make $\dot{V}$ more negative, which implies that the projection modification guarantees boundedness of the weights without affecting the rest of the stability properties established in the absence of projection.

LEMMA 12. *Based on the adaptive laws (60) and (61) the following inequalities hold:*

(i) $I_1^*(e_x^T W_1 \xi / \|W_1\|_F^2) \, tr\{W_1 \Phi_1^T\} \leq 0$.
(ii) $I_2^*(e_x^T W_2 \zeta u / \|W_2\|_F^2) \, tr\{W_2 \Phi_2^T\} \leq 0$.

Proof. We prove here only part (i) because the proof of part (ii) follows by the same reasoning. Suppose $\|W_1\|_F = M_1$ and $e_x^T W_1 \xi < 0$. If this is not the case, then $I_1^* = 0$ and the inequality holds trivially. The term $\mathrm{tr}\{W_1 \Phi_1^T\}$ can be expressed as

$$\mathrm{tr}\{W_1 \Phi_1^T\} = \mathrm{tr}\{(\Phi_1 + W_1^*)\Phi^T\}$$

$$= \mathrm{tr}\{\tfrac{1}{2}\Phi_1 \Phi_1^T + (\tfrac{1}{2}\Phi_1 \Phi_1^T + W_1^* \Phi^T)\}$$

$$= \tfrac{1}{2}\|\Phi_1\|_F^2 + \tfrac{1}{2}\|W_1\|_F^2 - \tfrac{1}{2}\|W_1^*\|_F^2.$$

Therefore

$$I_1^* \frac{e_x^T W_1 \xi}{\|W_1\|_F^2} \, \mathrm{tr}\{W_1 \Phi_1^T\} = \frac{1}{2M_1^2}(e_x^T W_1 \xi)(\|\Phi_1\|_F^2 + M_1^2 - \|W_1^*\|_F^2). \tag{65}$$

Whereas the optimal weights W_1^* satisfy $\|W_1^*\|_F \leq M_1$, the last term in (65) is positive and therefore

$$I_1^* \frac{e_x^T W_1 \xi}{\|W_1\|_F^2} \, \text{tr}\{W_1 \Phi_1^T\} \leq 0. \qquad \blacksquare$$

Now, using Lemma 12, (64) becomes

$$\dot{V} \leq -\alpha |e_x|^2 - e_x^T v \leq -\alpha |e_x|^2 + v_0 |e_x|. \qquad (66)$$

Based on (66), we next summarize the properties of the weight adaptive laws (60) and (61). The proof of the following theorem employs well-known techniques from the adaptive control literature. In the sequel, the notation $z \in L_2$ means $\int_0^\infty |z(t)|^2 \, dt < \infty$, whereas $z \in L_\infty$ implies $\sup_{t \geq 0} |z(t)| < \infty$.

THEOREM 13.   *Consider the error filtering identification scheme* (54). *The weight adaptive laws given by* (60) *and* (61) *guarantee the following properties*:

(a) *For $v_0 = 0$ (no approximation error) we have*

- $e_x, \hat{x}, \Phi_1, \Phi_2 \in L_\infty$, $e_x \in L_2$;
- $\lim_{t \to \infty} e_x(t) = 0$, $\lim_{t \to \infty} \dot{\Phi}_1(t) = 0$, $\lim_{t \to \infty} \dot{\Phi}_2(t) = 0$.

(b) *For $\sup_{t \geq 0} |v(t)| \leq v_0$ we have*

- $e_x, \hat{x}, \Phi_1, \Phi_2 \in L_\infty$;
- *there exist constants k_1 and k_2 such that for any finite time t,*

$$\int_0^t |e_x(\tau)|^2 d\tau \leq k_1 + k_2 \int_0^t |v(\tau)|^2 d\tau.$$

Proof.   (a) With $v_0 = 0$, Eq. (66) becomes

$$\dot{V} \leq -\alpha |e_x|^2 \leq 0. \qquad (67)$$

Hence $V \in L_\infty$, which from (56) implies $e_x, \Phi_1, \Phi_2 \in L_\infty$. Furthermore, $\hat{x} = e_x + x$ is also bounded. Whereas V is a nonincreasing function of time and bounded from below, the $\lim_{t \to \infty} V(t) = V_\infty$ exists. Therefore, integrating (67) from 0 to ∞ yields

$$\int_0^\infty |e_x(\tau)|^2 d\tau \leq \frac{1}{\alpha}[V(0) - V_\infty] < \infty,$$

which implies that $e_x \in L_2$. By the definition of the Gaussian radial basis function, the regressor vectors $\xi(x)$ and $\zeta(x)$ are bounded for all x and by assumption u is also bounded. Hence from (55) we have that $\dot{e}_x \in L_\infty$. Whereas $e_x \in L_2 \cap L_\infty$ and $\dot{e}_x \in L_\infty$, using Barbalat's lemma [23, 38] we conclude that $\lim_{t \to \infty} e_x(t) = 0$. Now, using the boundedness of $\xi(t)$ and the convergence of

$e_x(t)$ to zero, we have that $\dot{\Phi}_1 = \dot{W}_1$ also converges to zero. Similarly, $\dot{\Phi}_2 \to 0$ as $t \to \infty$.

(b) The projection algorithm guarantees that $\|W_1\|_F \leq M_1$ and $\|W_2\|_F \leq M_2$. Therefore the weight estimation errors are also bounded; that is, $\Phi_1, \Phi_2 \in L_\infty$. From (66) it is clear that if $|e_x| > v_0/\alpha$, then $\dot{V} < 0$ which implies that $e_x \in L_\infty$ and consequently $\hat{x} \in L_\infty$. To prove the second part, we proceed to complete the square in (66):

$$\dot{V} \leq -\frac{\alpha}{2}|e_x|^2 - \frac{\alpha}{2}\left[|e_x|^2 + \frac{2}{\alpha}e_x^T v\right]$$

$$\leq -\frac{\alpha}{2}|e_x|^2 + \frac{1}{2\alpha}|v|^2.$$

Therefore, by integrating both sides and using the fact that $V \in L_\infty$ we obtain

$$\int_0^t \left|e_x(\tau)\right|^2 d\tau \leq \frac{2}{\alpha}\left[V(0) - V(t)\right] + \frac{1}{\alpha^2}\int_0^t \left|v(\tau)\right|^2 d\tau$$

$$\leq k_1 + k_2 \int_0^t \left|v(\tau)\right|^2 d\tau,$$

where $k_1 := (2/\alpha)\sup_{t \geq 0}\{V(0) - V(t)\}$ and $k_2 := 1/\alpha^2$. ∎

Remark 14. For notational simplicity the preceding identification scheme was developed with the filter pole α and the learning rates γ_1 and γ_2 being scalars. It can be verified readily that the analysis is still valid if $-\alpha$ in (54) is replaced by a Hurwitz matrix $A \in \mathbb{R}^{n \times n}$, and γ_1 and γ_2 in the parameter update laws are replaced by positive definite learning rate matrices $\Gamma_1 \in \mathbb{R}^{n_1 \times n_1}$ and $\Gamma_2 \in \mathbb{R}^{n_2 \times n_2}$, respectively.

Remark 15. Similar to the discrete-time case, nothing can be concluded about the convergence of the weights to their optimal values. To guarantee convergence, $\xi(x)$ and $\zeta(x)u$ need to satisfy a *persistency of excitation* condition. In the case of continuous-time systems a signal $z(t) \in \mathbb{R}^n$ is said to be persistently exciting in $\mathbb{R}^n$ if there exist positive constants α_0, α_1, and T such that

$$\alpha_0 I \leq \int_t^{t+T} z(\tau)z^T(\tau)\,d\tau \leq \alpha_1 I, \qquad \forall t \geq 0.$$

2. Optimization Methods

To derive stable optimization-based weight adjustment rules we need to develop an identification model in which the output error e_x is related to the weight estimation errors Φ_1 and Φ_2 in a simple algebraic fashion. To achieve this objective we consider filtered forms of the regressors ξ and ζ and the state x. We start by rewriting (53) in the filter form

$$x = \frac{\alpha}{s+\alpha}[x] + W_1^*\frac{1}{s+\alpha}[\xi(x)] + W_2^*\frac{1}{s+\alpha}[\zeta(x)u] + \frac{1}{s+\alpha}[v], \quad (68)$$

where s denotes the differential (Laplace) operator.[2] The notation $h(s)[z]$ is to be interpreted as the output of the filter $h(s)$ with z as the input. Equation (68) is now expressed in the compact form

$$x = x_f + W_1^*\xi_f + W_2^*\zeta_f + v_f, \quad (69)$$

where x_f, ξ_f, and ζ_f are generated by filtering x, ξ, and ζu, respectively:

$$\dot{x}_f = -\alpha x_f + \alpha x, \qquad x_f(0) = 0,$$
$$\dot{\xi}_f = -\alpha\xi_f + \xi, \qquad \xi_f(0) = 0,$$
$$\dot{\zeta}_f = -\alpha\zeta_f + \zeta u, \qquad \zeta_f(0) = 0.$$

Whereas $v(t)$ is bounded by v_0, the filtered approximation error $v_f := 1/(s+\alpha)[v]$ is another bounded disturbance signal; that is, $|v_f(t)| \le v_0/\alpha$.

Based on (69), we consider the identification model

$$\hat{x} = x_f + W_1\xi_f + W_2\zeta_f. \quad (70)$$

This model, which will be referred to as the *regressor filtering* identification model, is shown in Fig. 4 in block diagram representation. The regressor filtering scheme requires $n + n_1 + n_2$ first-order filters, which is considerably more than the n filters required in the error filtering scheme. As can be seen from Fig. 4, the filters appear inside the RBF networks, forming a *dynamic RBF network*. This neural network architecture can be applied directly to model dynamical systems in the same way that RBF networks are used as models of static mappings.

Using the regressor filtering scheme, the output error $e_x := \hat{x} - x$ satisfies

$$e_x = W_1\xi_f + W_2\zeta_f + x_f - x = \Phi_1\xi_f + \Phi_2\zeta_f - v_f, \quad (71)$$

where $\Phi_1 = W_1 - W_1^*$ and $\Phi_2 = W_2 - W_2^*$ are the weight estimation errors. Hence by filtering the regressors, an algebraic relationship between the output error e_x and the weight estimation errors Φ_1 and Φ_2 is obtained. In the framework of the optimization approach [41], adaptive laws for W_1 and W_2 are obtained by minimizing an appropriate cost functional with respect to each element of W_1 and W_2. Here we consider an *instantaneous* cost functional with a constraint

[2]In deriving (68) we assumed (without loss of generality) zero initial condition for the state, that is, $x^0 = 0$. Note that if $x^0 \neq 0$, then the initial condition will appear in the identification model so that it gets cancelled in the error equation; this is possible because x is available for measurement.

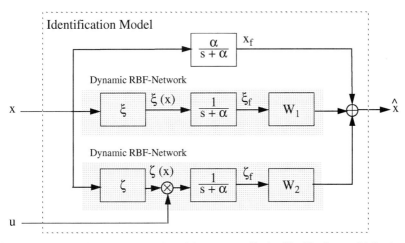

Figure 4 A block diagram representation of the regressor filtering identification model developed using RBF networks.

on the possible values that W_1 and W_2 can take. This leads to the constrained minimization problem:

$$\begin{aligned} \text{minimize} \quad & \mathcal{J}(W_1, W_2) = \tfrac{1}{2}e_x^T e_x \\ \text{subject to} \quad & \|W_1\|_F \le M_1, \\ & \|W_2\|_F \le M_2. \end{aligned} \tag{72}$$

Using the *gradient projection method* [41], the adaptive laws for continuous adjustment of the weights W_1 and W_2 are obtained as

$$\dot{W}_1 = \begin{cases} -\gamma_1 e_x \xi_f^T, & \text{if } \{\|W_1\|_F < M_1\} \text{ or } \{\|W_1\|_F = M_1 \\ & \text{and } e_x^T W_1 \xi_f \ge 0\}, \\ \mathcal{P}\{-\gamma_1 e_x \xi_f^T\}, & \text{if}\{\|W_1\|_F = M_1 \text{ and } e_x^T W_1 \xi_f < 0\}, \end{cases} \tag{73}$$

$$\dot{W}_2 = \begin{cases} -\gamma_2 e_x \zeta_f^T, & \text{if } \{\|W_2\|_F < M_2\} \text{ or } \{\|W_2\|_F = M_2 \\ & \text{and } e_x^T W_2 \zeta_f \ge 0\}, \\ \mathcal{P}\{-\gamma_2 e_x \zeta_f^T\}, & \text{if } \{\|W_2\|_F = M_2 \text{ and } e_x^T W_2 \zeta_f < 0\}. \end{cases} \tag{74}$$

where $\mathcal{P}\{\cdot\}$ denotes the projection operation. The weight adaptive laws (73) and (74) have the same form as (60) and (61), which were derived by the Lyapunov method, with the exception that the output error e_x and the regressor vectors ξ and ζ are defined differently.

The counterpart of Theorem 13 concerning the stability properties of the regressor filtering identification scheme with the adaptive laws (73) and (74), obtained using the gradient projection method, is described by the following result.

THEOREM 16. *Consider the regressor filtering identification scheme described by (70). The weight adaptive laws given by (73) and (74) guarantee the following properties:*

(a) *For $v(t) = 0$ (no approximation error) we have*

- $e_x, \hat{x}, \Phi_1, \Phi_2 \in L_\infty, \ e_x \in L_2$;
- $\lim_{t \to \infty} e_x(t) = 0, \ \lim_{t \to \infty} \dot{\Phi}_1(t) = 0, \ \lim_{t \to \infty} \dot{\Phi}_2(t) = 0.$

(b) *For $\sup_{t \geq 0} |v(t)| \leq v_0$ we have*

- $e_x, \hat{x}, \Phi_1, \Phi_2 \in L_\infty$;
- *there exist constants k_1 and k_2 such that for any finite t,*

$$\int_0^t |e_x(\tau)|^2 \, d\tau \leq k_1 + k_2 \int_0^t |v(\tau)|^2 \, d\tau.$$

Proof. Consider the Lyapunov function candidate

$$V(\Phi_1, \Phi_2) = \frac{1}{2\gamma_1} \|\Phi_1\|_F^2 + \frac{1}{2\gamma_2} \|\Phi_2\|_F^2 = \mathrm{tr}\left\{ \frac{1}{2\gamma_1} \Phi_1 \Phi_1^T + \frac{1}{2\gamma_2} \Phi_2 \Phi_2^T \right\}. \quad (75)$$

Using (71), Lemma 12, and the fact that $\dot{\Phi}_1 = \dot{W}_1$ and $\dot{\Phi}_2 = \dot{W}_2$, the time derivative of V along (73) and (74) can be expressed as

$$\dot{V} = \mathrm{tr}\left\{ -e_x \xi_f^T \Phi_1^T - e_x \zeta_f^T \Phi_2^T \right\} + I_1^* \frac{e_x^T W_1 \xi_f}{\|W_1\|_F^2} \mathrm{tr}\left\{ W_1 \Phi_1^T \right\}$$

$$+ I_2^* \frac{e_x^T W_2 \zeta_f}{\|W_2\|_F^2} \mathrm{tr}\left\{ W_2 \Phi_2^T \right\}$$

$$\leq -\mathrm{tr}\left\{ e_x (\xi_f^T \Phi_1^T + \zeta_f^T \Phi_2^T) \right\}$$

$$= -\mathrm{tr}\left\{ e_x (e_x^T + v_f^T) \right\} = -|e_x|^2 - v_f^T e_x \leq -|e_x|^2 + \frac{v_0}{\alpha} |e_x|. \quad (76)$$

(a) If $v(t) = 0$, then

$$\dot{V} \leq -|e_x|^2 \leq 0. \quad (77)$$

Therefore $V \in L_\infty$, which from (75) implies that $\Phi_1, \Phi_2 \in L_\infty$. Using this, together with the boundedness of ξ_f and ζ_f in (71) gives $e_x, \hat{x} \in L_\infty$. Furthermore, by integrating both sides of (77) from 0 to ∞ it can be shown that $e_x \in L_2$. Now by taking the time derivative of e_x in (71) we obtain

$$\dot{e}_x = \dot{\Phi}_1 \xi_f + \Phi_1 \dot{\xi}_f + \dot{\Phi}_2 \zeta_f + \Phi_2 \dot{\zeta}_f. \quad (78)$$

Whereas $\dot{\Phi}_1, \xi_f, \Phi_1, \dot{\xi}_f, \dot{\Phi}_2, \zeta_f, \Phi_2, \dot{\zeta}_f \in L_\infty$, (78) implies that $\dot{e}_x \in L_\infty$ and thus using Barbalat's lemma we conclude that $\lim_{t\to\infty} e_x(t) = 0$. Using the boundedness of ξ_f and ζ_f, it can be readily verified that $\dot{\Phi}_1$ and $\dot{\Phi}_2$ also converge to zero.

(b) Suppose $\sup_{t\geq 0} |v(t)| \leq v_0$. The projection algorithm guarantees that W_1 and W_2 are bounded, which implies $\Phi_1, \Phi_2 \in L_\infty$. Whereas ξ_f and ζ_f are also bounded, from (71) we obtain $e_x \in L_\infty$ and also $\hat{x} \in L_\infty$. The proof of the second part follows directly along the same lines as its counterpart in Theorem 13. ∎

Remark 17. As in the Lyapunov method, the scalar learning rates γ_1 and γ_2 in the parameter update laws (73) and (74) can be replaced by positive definite matrices Γ_1 and Γ_2.

Remark 18. Minimization of an appropriate *integral cost* using Newton's method yields the recursive least-squares algorithm. In the least-squares algorithm the learning rate matrices Γ_1 and Γ_2 are time varying and are adjusted concurrently with the weights. Unfortunately, the least-squares algorithm is computationally very expensive and, especially in neural network modeling, where the number of units n_1 and n_2 are usually large, updating the matrices Γ_1 and Γ_2, which consist of n_1^2 and n_2^2 entries, respectively, makes this algorithm impractical.

B. MULTILAYER NETWORK MODELS

In this section we consider the case where the network structures employed in the approximation of $f(x)$ and $g(x)$ are multilayer neural networks with sigmoidal-type activation functions. In this case the on-line approximator is nonlinearly parametrized. Although this is the most commonly used class of neural network models in empirical studies, there are few analytical results concerning the stability properties of such networks in problems dealing with learning in dynamic environments. The main difficulty in analyzing the behavior of recurrent network architectures with feedforward multilayer neural networks as subsystems arises due to the fact that the adjustable weights appear nonaffinely with respect to the nonlinearities of the network structure.

The approach followed in this paper relies on developing and analyzing an error identification scheme based on the Lyapunov synthesis method described in Section IV.A. The analysis proceeds through the use of a Taylor series expansion around the optimal weights. The adaptive law is designed based on the first-order (linear) approximation of the Taylor series expansion. In this framework, the RBF error filtering scheme presented in Section IV.A constitutes a special case of the analysis developed in this section, with the higher-order terms not present and the regressor being independent of the weight values. As a consequence of the presence of higher-order terms, the results obtained here are weaker, in the sense that, even if there is no approximation error, it cannot be guaranteed that the output error, will converge to zero.

The significance of this analysis is based on (1) proving that all the signals in the proposed identification scheme remain bounded, (2) obtaining an upper bound on the output error in terms of the approximation error and the high-order terms, and (3) developing a unified approach to synthesizing and analyzing stable dynamic learning configurations using different types of neural network architectures.

Consider the system (50), where θ_f^* and θ_g^* are the optimal weights in the minimization of the approximation errors $|f(x) - \hat{f}(x, \theta_f^*)|$ and $|g(x) - \hat{g}(x, \theta_g^*)|$, respectively, for $x \in \mathcal{X}$ and subject to the constraints $|\theta_f^*| \leq M_f$, $|\theta_g^*| \leq M_g$, where M_f and M_g are (large) design constants. The functions $\hat{f}$ and $\hat{g}$ are the outputs of multilayer neural networks with sigmoidal nonlinearities between layers. We start by adding and subtracting αx in (50), where α is a positive design constant. This gives

$$\dot{x} = -\alpha x + \alpha x + \hat{f}(x, \theta_f^*) + \hat{g}(x, \theta_g^*)u + v. \tag{79}$$

Based on (79) we consider the error filtering identification model

$$\dot{\hat{x}} = -\alpha \hat{x} + \alpha x + \hat{f}(x, \theta_f) + \hat{g}(x, \theta_g)u, \tag{80}$$

where $\theta_f \in \mathbb{R}^{n_f}$ and $\theta_g \in \mathbb{R}^{n_g}$ are the estimates of the optimal weights θ_f^* and θ_g^*, respectively. The constants n_f and n_g are the number of weights in each multilayer neural network model. The error filtering identification scheme described by (80) is shown in block diagram representation in Fig. 5. From (79) and (80), the output error $e_x = \hat{x} - x$ satisfies the differential equation

$$\dot{e}_x = -\alpha e_x + \left[\hat{f}(x, \theta_f) - \hat{f}(x, \theta_f^*)\right] + \left[\hat{g}(x, \theta_g) - \hat{g}(x, \theta_g^*)\right]u - v. \tag{81}$$

To obtain an adaptive law for the weights θ_f it is convenient to consider the first-order approximation of the difference $\hat{f}(x, \theta_f) - \hat{f}(x, \theta_f^*)$. Using the Taylor series expansion,[3] the difference $\hat{f}(x, \theta_f) - \hat{f}(x, \theta_f^*)$ can be expressed as

$$\hat{f}(x, \theta_f) - \hat{f}(x, \theta_f^*) = \frac{\partial \hat{f}}{\partial \theta_f}(x, \theta_f) \cdot \left(\theta_f - \theta_f^*\right) + \hat{f}_0(x, \theta_f), \tag{82}$$

where $\hat{f}_0(x, \theta_f)$ represents the higher-order terms (with respect to θ_f) of the expansion. If we define the weight estimation error as $\phi_f := \theta_f - \theta_f^*$, then from

---

[3]Throughout the analysis we require that the network outputs $\hat{f}(x, \theta_f)$ and $\hat{g}(x, \theta_g)$ are smooth functions of their arguments. This can be achieved easily if the sigmoid used is a smooth function. The logistic function and the hyperbolic tangent are examples of popular sigmoids that also satisfy this smoothness condition.

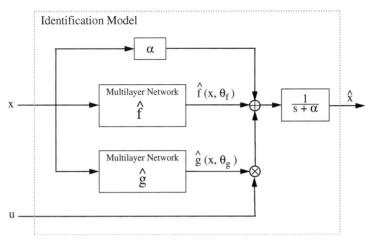

Figure 5 A block diagram representation of the error filtering identification model developed using multilayer sigmoidal neural networks.

(82) we have that

$$\hat{f}_0(x, \theta_f) = \hat{f}(x, \theta_f) - \hat{f}(x, \theta_f^*) - \frac{\partial \hat{f}}{\partial \theta_f}(x, \theta_f) \cdot \phi_f.$$

By the mean value theorem for vector valued functions, it can be shown [25] that for each $x \in \mathcal{X}$ and $\theta_f \in \mathcal{B}(M_f)$, we have

$$\left| \hat{f}_0(x, \theta_f) \right| \leq \tilde{\delta}_f(x, \theta_f) |\phi_f|,$$

where

$$\tilde{\delta}_f(x, \theta_f) := \sup_{\bar{\theta}_f \in [\theta_f, \theta_f^*]} \left\| \frac{\partial \hat{f}}{\partial \theta_f}(x, \bar{\theta}_f) - \frac{\partial \hat{f}}{\partial \theta_f}(x, \theta_f) \right\|. \tag{83}$$

Note that the set $[\theta_f, \theta_f^*]$ denotes $[\theta_f, \theta_f^*] := \{\vartheta \in \mathbb{R}^{n_f} : \vartheta = \lambda\theta_f + (1 - \lambda)\theta_f^*, \ 0 \leq \lambda \leq 1\}$ and $\mathcal{B}(M_f)$ denotes a ball of radius M_f. Now, let

$$\delta_f := \sup_{\substack{x \in \mathcal{X} \\ \theta_f \in \mathcal{B}(M_f)}} \tilde{\delta}_f(x, \theta_f). \tag{84}$$

Hence

$$\left| \hat{f}_0(x, \theta_f) \right| \leq \delta_f |\phi_f|, \qquad \forall x \in \mathcal{X}, \ \forall \theta_f \in \mathcal{B}(M_f). \tag{85}$$

Using the same procedure we have

$$\hat{g}(x, \theta_g) - \hat{g}(x, \theta_g^*) = \frac{\partial \hat{g}}{\partial \theta_g}(x, \theta_g) \cdot \phi_g + \hat{g}_0(x, \theta_g), \tag{86}$$

where $\phi_g := \theta_g - \theta_g^*$ and $\hat{g}_0(x, \theta_g)$ satisfies

$$\left| \hat{g}_0(x, \theta_g) \right| \le \delta_g |\phi_g|, \qquad \forall x \in \mathcal{X}, \ \forall \theta_g \in \mathcal{B}(M_g), \tag{87}$$

and the constant δ_g is defined as

$$\delta_g := \sup_{\substack{x \in \mathcal{X} \\ \theta_g \in \mathcal{B}(M_g)}} \left\{ \sup_{\bar{\theta}_g \in [\theta_g, \theta_g^*]} \left\| \frac{\partial \hat{g}}{\partial \theta_g}(x, \bar{\theta}_g) - \frac{\partial \hat{g}}{\partial \theta_g}(x, \theta_g) \right\| \right\}. \tag{88}$$

From now on, for notational simplicity we define the *regressors*

$$\xi(x, \theta_f) := \frac{\partial \hat{f}}{\partial \theta_f}(x, \theta_f), \qquad \zeta(x, \theta_g) := \frac{\partial \hat{g}}{\partial \theta_g}(x, \theta_g),$$

where ξ and ζ are $n \times n_f$ and $n \times n_g$ matrices, respectively, representing the sensitivity functions between the output of the network and the adjustable weights. It is worth noting that these sensitivity functions are exactly the functions that are calculated by the standard (static) backpropagation algorithm.

Now, by replacing (82) and (86) in (81) we obtain

$$\dot{e}_x = -\alpha e_x + \xi(x, \theta_f)\phi_f + \zeta(x, \theta_g)\phi_g u + \hat{f}_0(x, \theta_f) + \hat{g}_0(x, \theta_g)u - v. \tag{89}$$

Once expressed in the form (89), it is clear that this error equation is a generalization of the corresponding error equation for RBF networks. In particular, if the higher-order terms $\hat{f}_0$ and $\hat{g}_0$ are identically zero and the regressors ξ and ζ are independent of θ_f and θ_g, respectively, then (89) becomes of the same form as the error equation for RBF networks. In deriving adaptive laws for adjusting the weights θ_f and θ_g, we neglect the higher-order terms $\hat{f}_0$ and $\hat{g}_0$. This is consistent with linearization techniques, whose main principle is that for ϕ_f and ϕ_g sufficiently small the linearized terms $\xi\phi_f$ and $\zeta\phi_g$ dominate the higher-order terms $\hat{f}_0$ and $\hat{g}_0$. Based on these observations we consider and then analyze the following adaptive laws for adjusting the weights θ_f and θ_g:

$$\dot{\theta}_f = \begin{cases} -\gamma_f \xi^T e_x, & \text{if } \{|\theta_f| < M_f\} \text{ or } \{|\theta_f| = M_f \\ & \text{and } e_x^T \xi \theta_f \ge 0\}, \\ \mathcal{P}\{-\gamma_f \xi^T e_x\}, & \text{if } \{|\theta_f| = M_f \text{ and } e_x^T \xi \theta_f < 0\}, \end{cases} \tag{90}$$

$$\dot{\theta}_g = \begin{cases} -\gamma_g \zeta^T e_x u, & \text{if } \{|\theta_g| < M_g\} \text{ or } \{|\theta_g| = M_g \\ & \text{and } e_x^T \zeta \theta_g u \geq 0\}, \\ \mathcal{P}\{ -\gamma_g \zeta^T e_x u \}, & \text{if } \{|\theta_g| = M_g \text{ and } e_x^T \zeta \theta_g u < 0\}, \end{cases} \tag{91}$$

where $\mathcal{P}\{\cdot\}$ again denotes the projection operation, which is computed as

$$\mathcal{P}\{ -\gamma_f \xi^T e_x \} := -\gamma_f \xi^T e_x + \gamma_f \frac{e_x^T \xi \theta_f}{|\theta_f|^2} \theta_f,$$

$$\mathcal{P}\{ -\gamma_g \zeta^T e_x u \} := -\gamma_g \zeta^T e_x u + \gamma_g \frac{e_x^T \zeta \theta_g u}{|\theta_g|^2} \theta_g.$$

The weight adjustment laws (90) and (91) are the usual adaptive laws obtained by the Lyapunov synthesis method, with the projection algorithm modification for preventing the weights from drifting to infinity.

Remark 19. An intuitive interpretation of the preceding adaptive laws, based on more familiar optimization techniques, can be obtained as follows: by setting $\dot{e}_x = 0$ in the error equation (81) and solving for the quasi-steady-state response e_{ss} of e_x we have

$$e_{ss} = \frac{1}{\alpha} [(\hat{f}(x, \theta_f) - \hat{f}(x, \theta_f^*)) + (\hat{g}(x, \theta_g) - \hat{g}(x, \theta_g^*)) u - v]. \tag{92}$$

Based on (92), if we minimize the quadratic cost functional

$$\mathcal{J}_{ss}(\theta_f, \theta_g) = \frac{\alpha^2}{2} |e_{ss}|^2$$

subject to $|\theta_f| \leq M_f$ and $|\theta_g| \leq M_g$, by using the gradient projection method, we obtain weight adjustment laws of the form (90) and (91), with the exception that e_x is replaced by its quasi-steady-state response e_{ss}. This indicates the close relationship between dynamic backpropagation-type algorithms and the adaptive laws described in this work.

The next theorem establishes the stability of the proposed identification scheme.

THEOREM 20. *Consider the error filtering identification model* (80). *The adaptive laws* (90) *and* (91) *guarantee that*

- $e_x, \hat{x}, \phi_f, \phi_g \in L_\infty$;
- *there exist constants k_1 and k_2 such that*

$$\int_0^t |e_x(\tau)|^2 d\tau \leq k_1 + k_2 \int_0^t (\delta_f + \delta_g + |v(\tau)|)^2 d\tau.$$

Proof.   Consider the Lyapunov function candidate

$$V(e_x, \phi_f, \phi_g) = \frac{1}{2}e_x^T e_x + \frac{1}{2\gamma_f}\phi_f^T \phi_f + \frac{1}{2\gamma_g}\phi_g^T \phi_g. \tag{93}$$

Using (89), (90), and (91), the time derivative of V in (93) can be expressed as

$$\dot{V} = -\alpha|e_x|^2 + e_x^T f_0 + e_x^T g_0 u - e_x^T v$$
$$+ I_1^* \frac{\theta_f^T \xi^T e_x}{|\theta_f|^2}\theta_f^T \phi_f + I_2^* \frac{\theta_g^T \zeta^T e_x u}{|\theta_g|^2}\theta_g^T \phi_g, \tag{94}$$

where the last two terms in (94) are due to the projection modification. Using the same procedure as in Lemma 12, it can be shown that these terms are nonpositive. Therefore (94) becomes

$$\dot{V} \leq -\alpha|e_x|^2 + |e_x||f_0| + u_0|e_x||g_0| + v_0|e_x|$$
$$\leq -\alpha|e_x|^2 + \delta_f|e_x||\phi_f| + \delta_g u_0|e_x||\phi_g| + v_0|e_x|, \tag{95}$$

where u_0 and v_0 are the bounds for $u(t)$ and $v(t)$, respectively, and δ_f and δ_g are as defined in Eqs. (83) and (88), respectively. Whereas $\theta_f \in \mathcal{B}(M_f)$ and $\theta_g \in \mathcal{B}(M_g)$, it is clear that $|\phi_f| \leq 2M_f$ and $|\phi_g| \leq 2M_g$; hence, (95) can be written in the form

$$\dot{V} \leq -\alpha|e_x|^2 + \beta|e_x|, \tag{96}$$

where $\beta := 2\delta_f M_f + 2u_0\delta_g M_g + v_0$. Therefore, for $e_x > \beta/\alpha$, we have $\dot{V} < 0$, which implies that $e_x \in L_\infty$ and consequently $\hat{x} \in L_\infty$. The proof of the second part follows along the same lines as its counterpart in Theorem 13.   ∎

Remark 21.   To retain the generality of the result, in the foregoing analysis we have not assumed any special structure for $\hat{f}(x, \theta_f)$ and $\hat{g}(x, \theta_g)$ except that these functions are sufficiently smooth. Given a specific number of layers and type of sigmoidal nonlinearity, estimates of δ_f and δ_g can be computed.

Remark 22.   As a consequence of the weights appearing nonaffinely, the regressor filtering and optimization techniques descibed in Section IV.A.2 for RBF networks cannot be applied, at least directly, in the case that the identification model is based on multilayer networks.

V. CONCLUSIONS

The principal contribution of this paper is the synthesis and stability analysis of neural network based schemes for identifying discrete-time and continuous-time dynamical systems with unknown nonlinearities. Two wide classes of neural network architectures have been considered: (1) multilayer neural networks with sigmoidal-type activation functions, as an example of nonlinearly parametrized approximators, and (2) RBF networks, as an example of linearly parametrized approximators. Although these two classes of networks are evidently constructed differently, we have examined them in a common framework as approximators of the unknown nonlinearities of the system. These results are intended to complement the numerous empirical results concerning learning and control that have appeared in recent neural network literature. Furthermore, these results unify the learning techniques used by connectionists and adaptive control theorists. Bridging the gap between the two areas would certainly be beneficial to both fields.

REFERENCES

[1] S. A. Billings. Identification of nonlinear systems—a survey. *IEE Proc.* 127:272–285, 1980.

[2] T. W. Miller, S. T. Sutton, III, and P. J. Werbos, Eds. *Neural Networks for Control*. MIT Press, Cambridge, MA, 1990.

[3] D. A. White and D. A. Sofge, Eds. *Handbook of Intelligent Control: Neural, Fuzzy, and Adaptive Approaches*. Van Nostrand–Reinhold, New York, 1993.

[4] M. M. Gupta and D. H. Rao, Eds. *Neuro-Control Systems: Theory and Applications*. IEEE Press, New York, 1994.

[5] D. Rumelhart, D. Hinton, and G. Williams. Learning internal representations by error propagation. In *Parallel Distributed Processing* (D. Rumelhart and F. McClelland, Eds.), Vol. 1. MIT Press, Cambridge, MA, 1986.

[6] F. J. Pinenda. Generalization of back propagation to recurrent networks. *Phys. Rev. Lett.* 59:2229–2232, 1988.

[7] P. J. Werbos. Backpropagation through time: what it does and how to do it. *Proc. IEEE* 78:1550–1560, 1990.

[8] R. J. Williams and D. Zipser. A learning algorithm for continually running fully recurrent neural networks. *Neural Comput.* 1:270–280, 1989.

[9] K. S. Narendra and K. Parthasarathy. Gradient methods for the optimization of dynamical systems containing neural networks. *IEEE Trans. Neural Networks* 2:252–262, 1991.

[10] M. Polycarpou and P. Ioannou. Modeling, identification and stable adaptive control of continuous-time nonlinear dynamical systems using neural networks. In *Proceedings of the 1992 American Control Conference*, June 1992, pp. 36–40.

[11] R. M. Sanner and J.-J. E. Slotine. Gaussian networks for direct adaptive control. *IEEE Trans. Neural Networks* 3:837–863, 1992.

[12] E. Kosmatopoulos, M. Polycarpou, M. Christodoulou, and P. Ioannou. High-order neural network structures for identification of dynamical systems. *IEEE Trans. Neural Networks* 6:422–431, 1995.

[13] F. L. Lewis, K. Liu, and A. Yesildirek. Multilayer neural net robot controller with guaranteed tracking performance. In *Proceedings of the Conference on Decision and Control*, December 1993, pp. 2785–2791.

[14] F.-C. Chen and H. K. Khalil. Adaptive control of a class of nonlinear discrete-time systems using neural networks. *IEEE Trans. Automat. Control* 40:791–801, 1995.

[15] M. M. Polycarpou. Stable adaptive neural control scheme for nonlinear systems. *IEEE Trans. Automat. Control* 41:447–451, 1996.

[16] E. Tzirkel-Hancock and F. Fallside. Stable control of nonlinear systems using neural networks. *Internat. J. Robust Nonlinear Control* 2:67–81, 1992.

[17] M. J. D. Powell. *Approximation Theory and Methods*. Cambridge Univ. Press, Cambridge, 1981.

[18] L. L. Schumaker. *Spline Functions: Basic Theory*. Wiley, New York, 1981.

[19] K. Hornik, M. Stinchcombe, and H. White. Multilayer feedforward networks are universal approximators. *Neural Networks* 2:359–366, 1989.

[20] E. J. Hartman, J. D. Keeler, and J. M. Kowalski. Layered neural networks with Gaussian hidden units as universal approximations. *Neural Computation* 2:210–215, 1990.

[21] B. Kosko. *Neural Networks and Fuzzy Systems: A Dynamical Systems Approach to Machine Intelligence*. Prentice-Hall, Englewood Cliffs, NJ, 1991.

[22] C. A. Desoer and M. Vidyasagar. *Feedback Systems: Input–Output Properties*. Academic Press, New York, 1975.

[23] K. S. Narendra and A. M. Annaswamy. *Stable Adaptive Systems*. Prentice-Hall, Englewood Cliffs, NJ, 1989.

[24] P. A. Ioannou and A. Datta. Robust adaptive control: a unified approach. *Proc. IEEE* 79:1736–1768, 1991.

[25] J. D. Depree and C. W. Swartz. *Introduction to Real Analysis*. Wiley, New York, 1988.

[26] D. G. Luenberger. *Optimization by Vector Space Methods*. Wiley, New York, 1969.

[27] G. C. Goodwin and D. Q. Mayne. A parameter estimation perspective of continuous time model reference adaptive control. *Automatica* 23:57–70, 1987.

[28] A. R. Barron. Universal approximation bounds for superpositions of a sigmoidal function. *IEEE Trans. Inform. Theory* 39:930–945, 1993.

[29] R. E. Bellman. *Adaptive Control Processes*. Princeton Univ. Press, Princeton, NJ, 1961.

[30] G. C. Goodwin and K. S. Sin. *Adaptive Filtering, Prediction and Control*. Prentice-Hall, Englewood Cliffs, NJ, 1984.

[31] G. H. Golub and C. F. Van Loan. *Matrix Computations*, 2nd ed. The John Hopkins Univ. Press, Baltimore, 1989.

[32] H. Nijmeijer and A. J. van der Schaft. *Nonlinear Dynamical Control Systems*. Springer-Verlag, New York, 1990.

[33] J. Moody and C. J. Darken. Fast learning in networks of locally-tuned processing units. *Neural Comput.* 1:281–294, 1989.

[34] T. Holcomb and M. Morari. Local training of radial basis function networks: towards solving the hidden unit problem. In *Proceedings of the American Control Conference*, 1991, pp. 2331–2336.

[35] L. Praly, G. Bastin, J.-B. Pomet, and Z. P. Jiang. Adaptive stabilization of nonlinear systems. In *Foundations of Adaptive Control* (P. V. Kokotovic, Ed.), pp. 347–433. Springer-Verlag, Berlin, 1991.

[36] P. C. Parks. Lyapunov redesign of model reference adaptive control systems. *IEEE Trans. Automat. Control* AC-11:362–367, 1966.

[37] P. Kudva and K. S. Narendra. Synthesis of an adaptive observer using Lyapunov's direct method. *Internat. J. Control* 18:1201–1210, 1973.

[38] P. A. Ioannou and J. Sun. *Robust Adaptive Control*. Prentice-Hall, Englewood Cliffs, NJ, 1995.

[39] J. K. Hale. *Ordinary Differential Equations*. Wiley-Interscience, New York, 1969.

[40] M. M. Polycarpou and P. A. Ioannou. On the existence and uniqueness of solutions in adaptive control systems. *IEEE Trans. Automat. Control* 38:474–479, 1993.

[41] D. G. Luenberger. *Linear and Nonlinear programming*. Addison-Wesley, Reading, MA, 1984.

The Determination of Multivariable Nonlinear Models for Dynamic Systems

S. A. Billings

Department of Automatic Control
and Systems Engineering
University of Sheffield
Sheffield S1 3JD, England

S. Chen

Department of Electrical and
Electronic Engineering
University of Portsmouth
Portsmouth PO1 3DJ, England

I. INTRODUCTION

Modeling and identification constitute one of the major areas of control engineering, and the theory and practice of linear system identification are now well established [1, 2]. During the past decade, efforts have been focused on developing coherent and concise methods of nonlinear system modeling and identification [3–7], and more recently artificial neural networks have been applied to complex nonlinear dynamic systems [8–12]. There are two basic components in any system identification problem: determining the model structure and estimating or fitting the model parameters. These two tasks are critically influenced by the kind of model employed. Parameter estimation is relatively straightforward if the model structure is known a priori, but this information is rarely available in practice and has to be learned.

A general principle of system modeling is that the model should be no more complex than is required to capture the underlying system dynamics. This concept, known as the *parsimonious principle*, is particularly relevant in nonlinear model building because the size of a nonlinear model can easily become explosively large. An overcomplicated model may simply fit the noise in the training

Control and Dynamic Systems

data, resulting in overfitting. An overfitted model does not capture the underlying system structure well and will perform badly on new data. In neural network terminology, the model is said to have a poor generalization property. The NARMAX (nonlinear autoregressive moving average model with exogenous inputs) methodology, developed by the authors and others, is based on the philosophy that determining the model structure is a critical part of the identification process. Many of the NARMAX techniques can readily be applied to neural network modeling.

During the resurgence in artificial neural network research in the 1980s, there was a tendency in neural network modeling to simply fit a large network model to data and to disregard structure information. Often models consisting of hundreds of parameters were used. It was thought that because neural networks had excellent representation capabilities and coded information into neuron connection weights, there was less need for physical insight into the underlying functional relationship and dependencies, and model structure determination became less critical. Soon people realized that a huge network model not only requires a very long training time and has poor generalization performance, but also has little value in system analysis and design. The parsimonious principle is therefore just as important in neural network modeling.

Many techniques have been developed to improve the generalization properties of neural network models. These include pruning to reduce the size of a complex network model [13–15], automatic construction algorithms to build parsimonious models for certain types of neural networks [16–19], regularization [20–22], and combining regularization with construction algorithms [18, 23–26]. Selection of the best and smallest nonlinear model from the set of all possible candidates is a hard optimization problem that is generally intractable or computationally too expensive. Rather than insisting on the true optimal solution, a pragmatic approach is usually adopted in practice to find a suboptimal solution. In fact, all of the successful construction algorithms attempt to find a sufficiently good suboptimal nonlinear model structure in this sense.

In a nonlinear model, the relationship between model outputs and model inputs is nonlinear by definition. But the relationship between the model outputs and the free adjustable model parameters can be either nonlinear or linear. Identification schemes can therefore be classified into two categories, nonlinear-in-the-parameters or linear-in-the-parameters. Neural networks are generally nonlinear-in-the-parameters. A typical example is the multilayer perceptron (MLP) model. Some basis function neural networks such as the radial basis function (RBF) network, B-spline networks, and neural-fuzzy models can be configured in a linear-in-the-parameters structure.

The nonlinear-in-the-parameters neural network models are generally more compact and require fewer parameters compared with the linear-in-the-parameters models. In this sense the latter are often said to suffer from the "curse of dimensionality." But this problem arises largely because no attempt has been made to determine a parsimonious representation from the data. Learning for a nonlinear-in-the-parameters neural network is generally much more difficult than that for a linear-in-the-parameters neural network. More importantly, structure determination or the selection of a parsimonious model is extremely complex when the model is nonlinear-in-the-parameters and few results are available. Several powerful construction algorithms that automatically seek parsimonious models have been reported for basis function neural networks.

The remainder of this chapter is organized as follows. Section II introduces the NARMAX system representation, which provides a framework for nonlinear system modeling and identification. Conventional or nonneural NARMAX modeling approaches are summarized in Section III, where the similarity between the conventional approach and neural network modeling is emphasized. Section IV discusses a few popular neural network models employed in nonlinear system identification and classifies these models into the nonlinear-in-the-parameters and linear-in-the-parameters categories. Section V presents a gradient-based identification technique for neural network models that are nonlinear-in-the-parameters, and Section VI discusses construction algorithms for building parsimonious neural network models that are linear-in-the-parameters. Issues of identifiability and the implications for nonlinear system modeling, together with a scheme for constructing local models, is treated in Section VII. Some concluding remarks and further reserach topics are given in Section VIII. Throughout this chapter, simulated and real data are used to illustrate the concepts and techniques discussed.

II. THE NONLINEAR SYSTEM REPRESENTATION

For the class of discrete-time multivariable nonlinear dynamic systems, depicted in Fig. 1, the general input–output relationship can be written as

$$y(k) = f_s(y(k-1), \ldots, y(k-n_y), u(k-1), \ldots, u(k-n_u), e(k-1), \ldots,$$
$$e(k-n_e)) + e(k), \tag{1}$$

where

$$y(k) = \begin{bmatrix} y_1(k) \\ \vdots \\ y_m(k) \end{bmatrix}, \quad u(k) = \begin{bmatrix} u_1(k) \\ \vdots \\ u_r(k) \end{bmatrix}, \quad e(k) = \begin{bmatrix} e_1(k) \\ \vdots \\ e_m(k) \end{bmatrix} \tag{2}$$

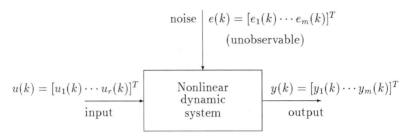

Figure 1 Block diagram of multivariable nonlinear dynamic system.

are the system m-dimensional output, r-dimensional input, and m-dimensional noise vectors, respectively; n_y, n_u, and n_e are lags in the output, input, and noise, respectively; and $f_s(\cdot)$ is some m-dimensional vector-valued nonlinear function. The noise $e(k)$ is a zero mean independent sequence. This nonlinear system representation is known as the NARMAX model [3, 6]. Nonlinear time series can be viewed as a special case of this general nonlinear system and can be represented by the NARMA model [27]:

$$y(k) = f_s\big(y(k-1), \ldots, y(k-n_y), e(k-1), \ldots, e(k-n_e)\big) + e(k). \qquad (3)$$

The NARMAX model provides a general basis for the development of nonlinear system identification techniques.

The functional form $f_s(\cdot)$ for a real-world system is generally very complex and unknown. Any practical modeling must be based on a chosen model set of known functions. Obviously, this model set should be capable of approximating the underlying process to within an acceptable accuracy, and neural networks are just a class of such "functional approximators" [28–31]. In addition, an efficient identification procedure must be developed for the selection of a parsimonious model structure because the dimension of a nonlinear model can easily become extremely large. Without efficient subset selection, the resulting model often has little practical value.

In general, recurrent or feedback neural networks are required to model the general NARMAX system. In some approximations, it may be possible to reduce the NARMAX model to a simplified form:

$$y(k) = f_s\big(y(k-1), \ldots, y(k-n_y), u(k-1), \ldots, u(k-n_u)\big) + e(k). \qquad (4)$$

For autonomous systems or time series, the model (4) can be further reduced to

$$y(k) = f_s\big(y(k-1), \ldots, y(k-n_y)\big) + e(k). \qquad (5)$$

The model in (4) is obviously less general than the model in (1). Identification schemes and system analysis are simpler, however, when based on the model (4), and many feedforward neural networks are sufficient to identify this simplified NARMAX model.

A main assumption for the NARMAX system is that the "system state space" has a finite dimension. This agrees with a basic result of dynamical systems theory [32], which states that if the attractor of a dynamical system is contained within a finite-dimensional manifold, then an embedding of the manifold can be constructed from time series observations of the dynamics on the manifold. The dynamical system induced by the embedding is differentiably equivalent to the one being observed. In fact, the lag n_y in (5) corresponds to the embedding vector dimension. Dynamical systems theory provides a firm basis for nonlinear time series modeling and many nonlinear signal-processing applications, and it has strong connections with nonlinear system modeling and identification.

The choice of identification scheme is critically influenced by the chosen model form. Consider a general nonlinear model:

$$\hat{y}(k) = f_m\big(\mathbf{x}(k); \Theta\big), \qquad (6)$$

where Θ is the vector of adjustable model parameters. If the model is used to identify the NARMAX system (1), the model input vector $\mathbf{x}(k)$ is

$$\mathbf{x}(k) = \big[y^T(k-1)\cdots y^T(k-n_y)\, u^T(k-1)\cdots u^T(k-n_u)\epsilon^T(k-1)\cdots$$
$$\epsilon^T(k-n_e)\big]^T, \qquad (7)$$

where

$$\epsilon(k) = y(k) - \hat{y}(k) \qquad (8)$$

is the prediction error. If the model is used to identify the simplified NARMAX system (4),

$$\mathbf{x}(k) = \big[y^T(k-1)\cdots y^T(k-n_y)\, u^T(k-1)\cdots u^T(k-n_u)\big]^T. \qquad (9)$$

According to the relationship between the model output and adjustable model parameters, various models can be classified into two categories, namely, nonlinear-in-the-parameters and linear-in-the-parameters models.

III. THE CONVENTIONAL NARMAX METHODOLOGY

The NARMAX modeling methodology, developed by the authors and others, is centered around the belief that the determination of model structure is a crucial part of the identification process. The NARMAX methodology attempts to break a system identification problem down into the following aspects:

> Nonlinearity detection—Determine whether a nonlinear model is needed.
> Structure determination—Select which terms to include in the model.
> Parameter estimation—Estimate the values of the model parameters.
> Model validation—Test whether the model is adequate.
> Model application—Use the model for prediction and system analysis.

These components form an identification toolkit that allows the user to build a concise mathematical description of the system from data. A crucial component of this iterative identification procedure is the determination of the model structure. This ensures that a parsimonious model is obtained.

Several conventional or nonneural nonlinear models have been developed for the NARMAX modeling approach. These typically include the polynomial model [7], the rational model [33], and the extended model set [34]. The polynomial model is derived by a polynomial expansion of $\mathbf{x}(k)$:

$$\hat{y}_i(k) = p_i\big(\mathbf{x}(k); \Theta_i\big), \qquad 1 \leq i \leq m, \tag{10}$$

where $p_i(\cdot)$ is a polynomial function of $\mathbf{x}(k)$, and Θ_i denotes the parameter vector consisting of the coefficients to be identified. The rational model is given as the quotient of two polynomial expansions of $\mathbf{x}(k)$:

$$\hat{y}_i(k) = \frac{p_{N_i}(\mathbf{x}(k); \Theta_{N_i})}{p_{D_i}(\mathbf{x}(k); \Theta_{D_i})}, \qquad 1 \leq i \leq m. \tag{11}$$

The extended model set refers to a general modeling approach that first forms some fixed nonlinear expansions to obtain various nonlinear model terms and then linearly combines the resulting terms. The same idea was given a neural network interpretation in [35], and the resulting model was called the functional link network (FLN).

A. STRUCTURE DETERMINATION AND PARAMETER ESTIMATION

When identifying nonlinear systems with unknown structure, it is important to avoid losing significant terms that must be included in the final model. Consequently, the experimenter is forced to start with a large model set. The key to the NARMAX modeling approach is an efficient procedure for subset model selection that aims to select only the significant model terms to form a parsimonious subset model. The subset selection scheme used will depend on the particular model form employed.

1. Nonlinear-in-the-Parameters Models

A class of parameter estimation algorithms widely used for nonlinear models is the prediction error algorithms [5, 33]. This is a class of gradient-based algorithms that minimize the cost function

$$J_1(\Theta) = \log\big(\det\big(Q(\Theta)\big)\big), \tag{12}$$

where $\det(\cdot)$ denotes the determinant,

$$Q(\Theta) = \frac{1}{N} \sum_{k=1}^{N} \epsilon(k; \Theta)\epsilon^T(k; \Theta), \tag{13}$$

and N is the number of data samples. Here we have made the dependency of the prediction error on the model parameters explicit by using the notation $\epsilon(k; \Theta)$. Let the dimension of Θ be n_Θ, that is, $\Theta = [\theta_1 \cdots \theta_{n_\Theta}]^T$. The parameter estimate that minimizes the performance criterion $J_1(\Theta)$ can be obtained by using the Gauss–Newton algorithm based on the gradient of $J_1(\Theta)$,

$$\nabla J_1(\Theta) = \left[\frac{\partial J_1}{\partial \Theta}\right]^T = \left[\frac{\partial J_1}{\partial \theta_1} \cdots \frac{\partial J_1}{\partial \theta_{n_\Theta}}\right]^T, \tag{14}$$

and the Hessian,

$$H_1(\Theta) \approx \frac{\partial^2 J_1}{\partial \Theta^2} + \rho I = \frac{\partial}{\partial \Theta}\left[\frac{\partial J_1}{\partial \Theta}\right]^T + \rho I, \tag{15}$$

where ρ is a small positive constant and I is the identity matrix of appropriate dimension. The addition of ρI to the Hessian matrix avoids the possibility of ill-conditioning. The gradient and the Hessian can be computed as follows [1]:

$$\frac{\partial J_1}{\partial \theta_i} = \frac{2}{N} \sum_{k=1}^{N} \epsilon^T(k; \Theta) Q^{-1}(\Theta) \frac{\partial \epsilon(k; \Theta)}{\partial \theta_i}, \qquad 1 \le i \le n_\Theta, \tag{16}$$

$$\frac{\partial^2 J_1}{\partial \theta_i \partial \theta_j} \approx \frac{2}{N} \sum_{k=1}^{N} \frac{\partial \epsilon^T(k; \Theta)}{\partial \theta_i} Q^{-1}(\Theta) \frac{\partial \epsilon(k; \Theta)}{\partial \theta_j}, \qquad 1 \le i, j \le n_\Theta. \tag{17}$$

The Hessian $H_1(\Theta)$ also plays an important role in subset model selection. In fact, $H_1(\Theta)$ contains sufficient information regarding the significance of each parameter. Deleting a parameter corresponds to removing a row and a column in the Hessian matrix. This is because subset model selection can be formulated as an optimization problem with the performance criterion [4]

$$C = N \log\big(\det\big(Q(\overline{\Theta})\big)\big) + n_{\bar{\Theta}} \chi_\alpha(1), \tag{18}$$

where $\overline{\Theta}$ is the parameter vector of the particular model and $n_{\overline{\Theta}}$ its dimension, and $\chi_\alpha(1)$ is the critical value of the χ^2 distribution with one degree of freedom and a given significance level α. The best model selected from all of the competing models is the one that minimizes the C-criterion (18). The first term in (18) indicates the model performance, and the second term is a complexity measure that penalizes large models. An appropriate value for $\chi_\alpha(1)$ can be shown to be 4.0 [4]. The Hessian of the C-criterion (18), which is the same as $H_1(\Theta)$ except for a scalar N, is what is actually needed to efficiently evaluate the C-criterion value of a candidate model [4, 36].

For a full model size of n_Θ, the number of competing models is 2^{n_Θ}. The true optimal model that minimizes (18) must be selected from all of these 2^{n_Θ} models. It is obvious that this is computationally impossible to do, even for a moderate n_Θ. A practical method for selecting parsimonious models is based on backward elimination, which eliminates some parameters of a large model according to information provided by $H_1(\Theta)$. This method is similar to the stepwise backward elimination scheme in the statistical literature [37], and can generally find a suboptimal subset model. The details of a numerical algorithm, which implements this backward elimination algorithm using $H_1(\Theta)$ to compute the values of the C-criterion, can be found in [36]. We use the following simulated system taken from [33] to illustrate the combined approach of structure determination and parameter estimation.

EXAMPLE 1.   The data were generated from the simulated single-input single-output system,

$$y(k) = \frac{y(k-1) + u(k-1) + y(k-1)u(k-1) + y(k-1)e(k-1)}{1 + y^2(k-1) + u(k-1)e(k-1)}$$
$$+ e(k). \tag{19}$$

The input $u(k)$ was an independent and uniformly distributed sequence with mean zero and variance 1.0, and the noise $e(k)$ was a Gaussian white sequence with mean zero and variance 0.01. Notice that if the system is nonlinear, the possibility of nonlinear noise terms must be accommodated in the model. Six hundred points of the input–output data were used in the identification. A rational model with the lags $n_y = n_u = n_e = 1$ and a quadratic numerator and denominator were used. The full model had 20 terms (shown in Table I).

The prediction error algorithm was first used to obtain the parameter estimate $\widehat{\Theta}$ for the full model. The full model Hessian, evaluated at $\widehat{\Theta}$, was then used in the backward elimination procedure to eliminate redundant parameters in the model. The elimination process is shown in Table II, where it is seen that the process stopped after stage 13 because the C-criterion reached a minimum value, giving rise to a subset model of seven parameters. The prediction error algorithm was

Table I
Full Rational Model Set for Example 1

Numerator p_N	Denominator p_D
constant	constant
$y(k-1)$	$y(k-1)$
$u(k-1)$	$u(k-1)$
$\epsilon(k-1)$	$\epsilon(k-1)$
$y^2(k-1)$	$y^2(k-1)$
$y(k-1)u(k-1)$	$y(k-1)u(k-1)$
$y(k-1)\epsilon(k-1)$	$y(k-1)\epsilon(k-1)$
$u^2(k-1)$	$u^2(k-1)$
$u(k-1)\epsilon(k-1)$	$u(k-1)\epsilon(k-1)$
$\epsilon^2(k-1)$	$\epsilon^2(k-1)$

again used to fine-tune the subset model parameters. The final subset model, given in Table III, is obviously a good estimate.

The combined approach of structure determination and parameter estimation based on the prediction error estimation method, discussed above, is very general

Table II
Model Reduction Using Backward Elimination for Example 1

Elimination step	Eliminated parameter	C-criterion value
	Full model	-2678.4
1	$u(k-1)\epsilon(k-1)$	-2682.4
2	$y^2(k-1)$	-2686.3
3	$u^2(k-1)$	-2690.0
4	Constant	-2693.5
5	$u(k-1)^+$	-2697.3
6	$\epsilon^2(k-1)$	-2700.6
7	$\epsilon(k-1)^+$	-2703.3
8	$y(k-1)^+$	-2705.7
9	$y(k-1)\epsilon(k-1)^+$	-2708.2
10	$y(k-1)u(k-1)^+$	-2710.3
11	$u^2(k-1)^+$	-2712.0
12	$\epsilon(k-1)$	-2714.2
<u>13</u>	$\epsilon^2(k-1)^+$	<u>-2715.4</u>
14	$y(k-1)\epsilon(k-1)$	-2695.5

The underline indicates where the procedure stops and $+$ denotes a denominator parameter.

Table III

Final Subset Rational Model for Example 1

	Model terms	Parameter estimates
Numerator p_N	$y(k-1)$	0.61074
	$u(k-1)$	0.61245
	$y(k-1)u(k-1)$	0.60968
	$y(k-1)\epsilon(k-1)$	0.58427
Denominator p_D	Constant	0.60820
	$y^2(k-1)$	0.61379
	$u(k-1)\epsilon(k-1)$	0.55981

and can be applied to any nonlinear model. An alternative cost function for the prediction error estimation method is

$$J_2(\Theta) = \text{trace}\big(Q(\Theta)\big) = \frac{1}{N}\sum_{k=1}^{N}\epsilon^T(k;\Theta)\epsilon(k;\Theta). \tag{20}$$

A similar algorithm that combines structure determination and parameter estimation can be derived based on this cost function. In the case of single-output systems, that is, $m = 1$, the two cost functions (12) and (20) become identical. A drawback of the approach discussed here is that, when the full model size n_Θ is very large, the computation of the full model Hessian can be expensive.

A multilayered neural network model is usually nonlinear-in-the-parameters. Therefore it is not surprising that the prediction error estimation method can readily be applied to constructing neural network models [9, 11, 38]. In fact, the famous back-propagation learning method [39] for neural networks can be viewed as a special case of the prediction error algorithm [38]. The importance of the parsimonious principle is now widely recognized in the neural network community. Weight elimination has been suggested as a method for reducing the size of large network models, and the process is known as pruning. Approaches adopted in pruning often have their roots in more traditional methods of subset model selection. For example, the so-called optimal brain damage method [13] uses the diagonal elements of the cost function Hessian in weight elimination.

2. Linear-in-the-Parameters Models

The polynomial model and the extended model set are examples of linear-in-the-parameters models. A linear-in-the-parameters model can generally be obtained by performing some fixed nonlinear functional transforms or expansions of the inputs before combining the resulting terms linearly. Specifically, a set of given functional expansions maps the input space onto a new space of increased dimension n_L,

$$\mathbf{x}(k) \to \left[\phi_1\big(\mathbf{x}(k)\big) \cdots \phi_{n_L}\big(\mathbf{x}(k)\big)\right]^T. \tag{21}$$

The model outputs are obtained as linear combinations of the new bases $\phi_i(\mathbf{x}(k))$, $1 \leqslant i \leqslant n_L$,

$$\hat{y}_i(k) = \sum_{j=1}^{n_L} \theta_{j,i} \phi_j\big(\mathbf{x}(k)\big), \qquad 1 \leq i \leq m. \tag{22}$$

To qualify for a linear-in-the-parameters model, the value of each given basis function must depend only on the input $\mathbf{x}(k)$, so that $\phi_i(\mathbf{x}(k))$ contains no other adjustable parameters.

An advantage of the model (22) is that the standard least-squares identification method can readily be applied to estimate the parameters $\theta_{j,i}$. In practice, the model dimension n_L can become excessively large. Consider, for example, the polynomial model, which uses the set of monomials of $\mathbf{x}(k)$ as bases. If the dimension of $\mathbf{x}(k)$ is 8, a degree-5 polynomial expansion will produce a model basis set of dimension $n_L = 1286$. Other choices of model bases [34] can also induce the problem of excessive model dimension. Subset selection is therefore essential, and an efficient subset selection procedure has been derived based on the orthogonal least-squares (OLS) method [7]. Given the full set of n_L candidate bases, the OLS algorithm selects significant model bases one by one in a forward regression manner until an adequate subset model is constructed. The selection procedure is made simple and efficient by exploiting an orthogonal property.

Specifically, the cost function (20) is adopted for a combined subset model selection and parameter estimation. The system outputs, the model outputs, and the prediction errors for $1 \leq k \leq N$ can be collected together in the matrix form

$$\mathbf{Y} = \mathbf{\Phi}\mathbf{\Theta} + \mathbf{E}, \tag{23}$$

where

$$\mathbf{Y} = [\mathbf{y}_1 \cdots \mathbf{y}_m] = \begin{bmatrix} y_1(1) & \cdots & y_m(1) \\ \vdots & \vdots & \vdots \\ y_1(N) & \cdots & y_m(N) \end{bmatrix} \tag{24}$$

$$\mathbf{E} = [\mathbf{e}_1 \cdots \mathbf{e}_m] = \begin{bmatrix} \epsilon_1(1) & \cdots & \epsilon_m(1) \\ \vdots & \vdots & \vdots \\ \epsilon_1(N) & \cdots & \epsilon_m(N) \end{bmatrix} \tag{25}$$

$$\mathbf{\Phi} = [\mathbf{\Phi}_1 \cdots \mathbf{\Phi}_{n_L}] = \begin{bmatrix} \phi_1(\mathbf{x}(1)) & \cdots & \phi_{n_L}(\mathbf{x}(1)) \\ \vdots & \vdots & \vdots \\ \phi_1(\mathbf{x}(N)) & \cdots & \phi_{n_L}(\mathbf{x}(N)) \end{bmatrix} \tag{26}$$

$$\mathbf{\Theta} = \begin{bmatrix} \theta_{1,1} & \cdots & \theta_{1,m} \\ \vdots & \vdots & \vdots \\ \theta_{n_L,1} & \cdots & \theta_{n_L,m} \end{bmatrix}. \tag{27}$$

Let an orthogonal decomposition of the regression matrix $\mathbf{\Phi}$ be $\mathbf{\Phi} = \mathbf{WA}$. The system (23) can be rewritten as

$$\mathbf{Y} = \mathbf{WG} + \mathbf{E}, \tag{28}$$

where

$$\mathbf{G} = \begin{bmatrix} g_{1,1} & \cdots & g_{1,m} \\ \vdots & \vdots & \vdots \\ g_{n_L,1} & \cdots & g_{n_L,m} \end{bmatrix} = \mathbf{A\Theta}. \tag{29}$$

It can be shown that

$$N J_2(\mathbf{\Theta}) = \mathrm{trace}(\mathbf{E}^T \mathbf{E}) = \mathrm{trace}(\mathbf{Y}^T \mathbf{Y}) - \sum_{j=1}^{n_L} \left(\sum_{i=1}^{m} g_{j,i}^2 \right) \mathbf{w}_j^T \mathbf{w}_j, \tag{30}$$

where the new bases $\mathbf{w}_j$ are columns of $\mathbf{W}$. Define the error reduction ratio due to $\mathbf{w}_l$ as

$$[err]_l = \left(\sum_{i=1}^{m} g_{l,i}^2 \right) \mathbf{w}_l^T \mathbf{w}_l / \mathrm{trace}(\mathbf{Y}^T \mathbf{Y}). \tag{31}$$

Based on this ratio, significant model terms can be selected in a forward-selection procedure. At the lth stage, a model term is selected among the $n_L - l + 1$ candidates if it produces the largest value of $[err]_l$ to add to the previously selected $(l - 1)$-term model. The selection procedure can be terminated at the n_S stage when

$$1 - \sum_{l=1}^{n_S} [err]_l < \eta, \tag{32}$$

where $0 < \eta < 1$ is a preset desired tolerance, giving rise to an n_S-term subset model. Alternatively, the procedure can be terminated when the criterion

$$N \log(J_2(\mathbf{\Theta})) + n_S \chi_\alpha(1) \tag{33}$$

reaches a minimum, where $\chi_\alpha(1)$ is as defined previously at (18).

The details of the OLS algorithm for forward subset selection can be found in [7]. It should be emphasized that the OLS algorithm is an efficient way of implementing forward subset selection and, like any forward subset selection method, it does not guarantee that the best n_S-term model will be found from the n_L-term full model. This is not a serious deficiency, however, because the subset model found is usually very good. Furthermore, the optimal solution will certainly be too expensive to compute, because n_L is often very large.

Notice that the aim here is to select a subset model consisting of a set of the original model bases, say, $\{\Phi_{j_l}, 1 \leq l \leq n_S\}$. The set of the selected orthogonal bases $\{w_l, 1 \leq l \leq n_S\}$ corresponds precisely to a subset of the original model bases $\{\Phi_{j_l}, 1 \leq l \leq n_S\}$, that is, the subspace spanned by $\{w_l, 1 \leq l \leq n_S\}$ is the same space spanned by $\{\Phi_{j_l}, 1 \leq l \leq n_S\}$. This is reflected in the fact that A is an upper triangular matrix with unit diagonal elements. In many signal-processing applications, the objective is to transform the original signal space Φ onto a new space by some orthogonal transformation and to tackle the problem on a transformed subspace. This can be achieved, for example, by using singular value decomposition (SVD) [40]. A subset of the orthonormal bases $\{v_l, 1 \leq l \leq n_S\}$, which correspond to the first n_S largest eigenvalues, is selected to form the required subspace. Because each v_l is a linear combination of all of the original bases $\Phi_j, 1 \leq j \leq n_L$, it is not known which subset $\{\Phi_{j_l}, 1 \leq l \leq n_S\}$ exactly represents the subspace spanned by $\{v_l, 1 \leq l \leq n_S\}$. When a subset model consisting of a subset of the original model bases is required, the OLS algorithm has clear advantages.

If the stopping criterion (32) is employed, the chosen tolerance η will influence both the modeling accuracy and the complexity of the final subset model. It is obvious that ideally η should be larger than but very close to the ratio trace$(E^T E)/$trace$(Y^T Y)$. Because trace$(E^T E)$ is not known a priori, an appropriate value of η usually must be found, and this can be achieved by the following iterative learning procedure. An initial guess is assigned to η. Once a model is selected, an estimate for trace$(E^T E)$ can be computed, and because trace$(Y^T Y)$ is known from the data, an improved η can be chosen. We use the following simulated system taken from [34] to illustrate this learning strategy and to demonstrate OLS subset selection.

EXAMPLE 2.   The data were generated from the simulated single-input single-output system:

$$
\begin{aligned}
y(k) = {} & 0.5y(k-1) + u(k-2) + 0.1u^2(k-1) \\
& + 0.5e(k-1) + 0.2u(k-1)e(k-2) + e(k),
\end{aligned} \tag{34}
$$

where the system noise $e(k)$ was a Gaussian white noise with mean zero and variance $\sigma_e^2 = 0.04$, and the system input $u(k)$ was a uniformly distributed independent sequence with mean zero and variance 1.0. A data set of 500 input–output

Table IV

Iterative OLS Procedure of Subset Model Selection for Example 2

	Model terms selected	Parameter estimates	$[err]_l$
Initial iteration	$u(k-2)$	0.10110E+1	0.67104E+0
$\eta = 0.032$	$y(k-1)$	0.63448E+0	0.28703E+0
	$u^2(k-1)$	0.86768E−1	0.85520E−2
	$y(k-3)u(k-2)$	−0.20542E−1	0.63062E−3
	$u^3(k-3)$	−0.64772E−1	0.55407E−3
	$y(k-2)$	−0.68662E−1	0.15841E−2
	σ_ϵ^2	0.50169E−1	
	$\sigma_\epsilon^2/\sigma_y^2$	0.30603E−1	
1st iteration	$u(k-2)$	0.10073E+1	0.67104E+0
$\eta = 0.03$	$y(k-1)$	0.50464E+0	0.28703E+0
	$u^2(k-1)$	0.92469E−1	0.85520E−2
	$\epsilon(k-1)$	0.40841E+0	0.49355E−2
	σ_ϵ^2	0.44924E−1	
	$\sigma_\epsilon^2/\sigma_y^2$	0.27404E−1	
2nd iteration	$u(k-2)$	0.10052E+1	0.67103E+0
$\eta = 0.027$	$y(k-1)$	0.50226E+0	0.28703E+0
	$u^2(k-1)$	0.90645E−1	0.85520E−2
	$\epsilon(k-1)$	0.49454E+0	0.61062E−2
	$u(k-1)\epsilon(k-2)$	0.22055E+0	0.14793E−2
	σ_ϵ^2	0.40719E−1	
	$\sigma_\epsilon^2/\sigma_y^2$	0.24839E−1	
3rd iteration	$u(k-2)$	0.10045E+1	0.67104E+0
$\eta = 0.025$	$y(k-1)$	0.50171E+0	0.28703E+0
	$u^2(k-1)$	0.90395E−1	0.85520E−2
	$\epsilon(k-1)$	0.54986E+0	0.68936E−2
	$u(k-1)\epsilon(k-2)$	0.25074E+0	0.17351E−2
	σ_ϵ^2	0.40230E−1	
	$\sigma_\epsilon^2/\sigma_y^2$	0.24540E−1	
4th iteration	$u(k-2)$	0.10033E+1	0.67047E+0
$\eta = 0.025$	$y(k-1)$	0.50289E+0	0.28735E+0
	$u^2(k-1)$	0.90966E−1	0.855640E−2
	$\epsilon(k-1)$	0.54796E+0	0.69244E−2
	$u(k-1)\epsilon(k-2)$	0.23848E+0	0.16174E−2
	σ_ϵ^2	0.41020E−1	
	$\sigma_\epsilon^2/\sigma_y^2$	0.25073E−1	
5th iteration	$u(k-2)$	0.10032E+1	0.67047E+0
$\eta = 0.0251$	$y(k-1)$	0.50281E+0	0.28735E+0
	$u^2(k-1)$	0.91097E−1	0.85640E−2
	$\epsilon(k-1)$	0.54843E+0	0.69317E−2
	$u(k-1)\epsilon(k-2)$	0.23785E+0	0.16098E−2
	σ_ϵ^2	0.41019E−1	
	$\sigma_\epsilon^2/\sigma_y^2$	0.25072E−1	

pairs was used in identification. A polynomial model with input

$$\mathbf{x}(k) = \begin{bmatrix} y(k-1) & y(k-2) & y(k-3) & u(k-1) & u(k-2) \\ u(k-3) & \epsilon(k-1) & \epsilon(k-2) & \epsilon(k-3) \end{bmatrix}^T \quad (35)$$

and degree-3 polynomial expansion was used to fit the data. The full model set contained $n_L = 220$ terms. The iterative procedure for the subset model selection using the OLS algorithm is summarized in Table IV.

Initially, we did not have the residual sequence and could not use the residual variance σ_ϵ^2 as an estimate of the noise variance σ_e^2. We also could not form any model terms containing $\epsilon(k-j)$. Thus, at the initial iteration, the model input was assumed to be

$$\mathbf{x}(k) = \begin{bmatrix} y(k-1) & y(k-2) & y(k-3) & u(k-1) & u(k-2) & u(k-3) \end{bmatrix}^T. \quad (36)$$

The desired tolerance was initially guessed to be $\eta = 0.032$, and the OLS algorithm selected a subset model with an estimated σ_ϵ^2. This initial iteration generated a residual sequence and allowed the assignment of a new tolerance, $\eta = 0.03$. The iterative procedure was completed after the fifth iteration, because the subset models selected at the fourth and fifth iterations were identical, and the variances of the resulting residual sequences were hardly changing.

For the class of linear-in-the-parameters models, automatic construction algorithms that are capable of building parsimonious models from data are crucial for overcoming the curse of dimensionality. The OLS algorithm described here is one such algorithm. On-line versions of these procedures are now available to provide adaptive model selection and estimation for nonlinear systems [41, 42]. Certain neural network learning problems can also be formulated within the linear-in-the-parameters framework. An example is the RBF network. When the centers and widths of a RBF network are fixed, learning becomes a linear-in-the-parameter problem [43]. The OLS algorithm can readily be applied to constructing parsimonious RBF networks [16].

B. MODEL VALIDATION

Model validation is an important step in any identification or modeling procedure. For linear system identification, if the model structure and parameter values are correct, the residual $\epsilon(k)$ will be uncorrelated with past inputs and outputs. In the case of single-input single-output ($r = m = 1$) systems, therefore, an identified linear model is regarded as adequate if the autocorrelation function of $\epsilon(k)$

and the cross-correlation function of $\epsilon(k)$ and $u(k)$ satisfy

$$\left.\begin{array}{ll} \mathcal{R}_{\epsilon\epsilon}(\tau) = 0, & \tau \neq 0 \\ \mathcal{R}_{\epsilon u}(\tau) = 0, & \text{for all } \tau \end{array}\right\}. \tag{37}$$

For validating single-input single-output nonlinear models, (37) is clearly insufficient. The known principle of nonlinear model validation is to test if the residuals are unpredictable from all of the linear and nonlinear combinations of past inputs and outputs. Three higher order correlation tests have been suggested in addition to the tests in (37) to validate whether a nonlinear model is correct [33, 44].

These correlation-based tests can also be applied to validation of multi-input multi-output nonlinear models, but a large number of correlation plots will be required. A more compact validation procedure has been developed for multivariable nonlinear models using the following tests [45]:

$$\left.\begin{array}{ll} \mathcal{R}_{\beta\beta}(\tau) = 0, & \tau \neq 0 \\ \mathcal{R}_{v\beta}(\tau) = 0, & \text{for all } \tau \\ \mathcal{R}_{\gamma\omega}(\tau) = 0, & \tau \neq 0 \\ \mathcal{R}_{\zeta\omega}(\tau) = 0, & \text{for all } \tau \end{array}\right\}, \tag{38}$$

where

$$\left.\begin{array}{l} \beta(k) = \epsilon_1(k) + \epsilon_2(k) + \cdots + \epsilon_m(k) \\ v(k) = u_1(k) + u_2(k) + \cdots + u_r(k) \\ \gamma(k) = \epsilon_1^2(k) + \epsilon_2^2(k) + \cdots + \epsilon_m^2(k) \\ \omega(k) = y_1(k)\epsilon_1(k) + y_2(k)\epsilon_2(k) + \cdots + y_m(k)\epsilon_m(k) \\ \zeta(k) = u_1^2(k) + u_2^2(k) + \cdots + u_r^2(k) \end{array}\right\}. \tag{39}$$

Alternatively, model validation can be performed based on χ^2 tests [4, 46], but the correlation tests discussed here are much easier to implement.

In practice, the correlation tests are computed in normalized form, and confidence limits are used to determined whether the tests have been violated. Experience has shown that if these tests are used in conjunction with a combined approach of structure determination and parameter estimation, they often give the experimenter a great deal of information regarding deficiencies in the fitted model and help to indicate which terms should be included in the model to improve the fit. Some examples of real-data identification coupled with the model validity tests using various models including neural networks can be found in [9, 33, 47, 48].

IV. NEURAL NETWORK MODELS

Generally, artificial neural networks refer to a computational paradigm in which a large number of simple computational units known as "neurons" (or simply "nodes"), interconnected to form a network, perform complex computational tasks. Such a computational model was inspired by neurobiological systems. A key feature of neural networks is learning, and this refers to the fact that a neural network is trained to perform a task by using examples of data. Such a learning process has clear connections with system identification and parameter estimation. A variety of neural network architectures or models have been employed for constructing representations of complex nonlinear systems from data, and several of these models will be discussed below. From the viewpoint of information flow, there are two classes of neural networks, namely feedforward and recurrent networks. In the former case, the input signal is propagated forward through the network to produce the network output. In the latter case, node outputs can be fed back as node inputs. From the viewpoint of parameter estimation, neural networks can be classified as nonlinear-in-the-parameters or linear-in-the-parameters.

A. Multilayer Perceptrons

A feedforward multilayer perceptron (MLP) is a layered network made up of one or more hidden layers between the input and output layers. Each layer consists of computing nodes, and the nodes in a layer are connected to the nodes in adjacent layers, but there is no connection between the nodes within the same layer and no bridging layer connections. The input layer acts as an input data holder that distributes the inputs to the first hidden layer. The output from the first layer nodes then becomes inputs to the second layer, and so on. The last layer acts as the network output layer. The architecture of an MLP can conveniently be summarized as $n_0 - n_1 - \cdots - n_l$, where n_0 is the number of the network inputs, n_l is the number of the network outputs and n_i, $1 \leq i \leq l - 1$, are the numbers of nodes in the respective hidden layers. Figure 2a depicts the topology of a 4–5–6–3 MLP.

The input–output relationship of a generic node is shown in Fig. 2b. The node computes the weighted sum of the node inputs x_i, $1 \leq i \leq n$, adds a bias weight, and passes the result through a nonlinear activation function f_a to produce the node output y_j:

$$y_j = f_a\left(\sum_{i=0}^{n} w_{j,i} x_i\right),$$
(40)

where $w_{j,i}$ are the node weights and $x_0 = 1$ indicates that $w_{j,0}$ is a bias term. A typical choice of the activation function is the sigmoid function, defined as

$$f_a(v) = \frac{1}{1 + \exp(-v)}.$$
(41)

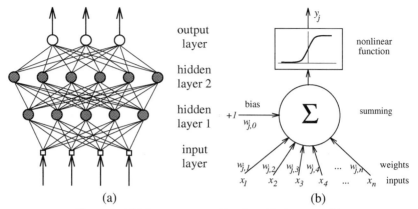

Figure 2   Multilayer perceptron (a) and model of a neuron (b).

For the purpose of function approximation, the output nodes usually do not contain a bias term and the associated activation functions are linear.

An MLP realizes an overall input–output mapping: $f_{\text{MLP}}: R^{n_0} \rightarrow R^{n_1}$. The MLP is a general function approximator, and theoretical works [28, 29] have shown that MLPs with one hidden layer are sufficient to approximate any continuous functions, provided that there are enough hidden nodes. In applications to identifying nonlinear dynamic systems, the number of the output nodes is equal to the number of the system outputs, and the number of the network input is equal to the dimension of $\mathbf{x}(k)$. Collect all of the wieghts of the network model into a vector form Θ. The network input–output mapping $f_{\text{MLP}}(\mathbf{x}(k); \Theta)$ is obviously highly nonlinear in Θ, and training an MLP is equivalent to estimating this parameter vector. After learning, the network mapping can be used as a model of the system dynamics f_{s}.

B. RADIAL BASIS FUNCTION NETWORKS

The RBF network is a processing structure consisting of an input layer, a hidden layer, and an output layer. The hidden layer of a RBF network consists of an array of nodes, and each node contains a parameter vector called a *center*. The node calculates the Euclidean distance between the center and the network input vector, and passes the result through a radially symmetric nonlinear function. The output layer is essentially a set of linear combiners. An example of the RBF network and the model of the Gaussian RBF node are shown in Fig. 3. The overall input–output response of an n_I-input n_O-output RBF network is a mapping

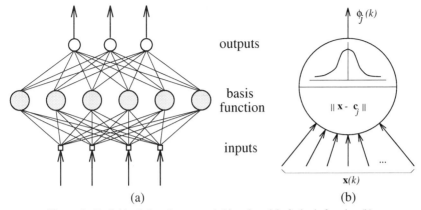

Figure 3   Radial basis function network (a) and model of a basis function (b).

f_{RBF}: $R^{n_{\text{I}}} \to R^{n_{\text{O}}}$. Specifically,

$$f_{\text{RBF}_i} = \sum_{j=1}^{n_{\text{H}}} \theta_{j,i}\phi_j = \sum_{j=1}^{n_{\text{H}}} \theta_{j,i}\phi\big(\|\mathbf{x} - \mathbf{c}_j\|; \rho_j\big), \qquad 1 \leq i \leq n_{\text{O}} \qquad (42)$$

where $\theta_{j,i}$ are the weights of the linear combiners, $\|\cdot\|$ denotes the Euclidean norm, ρ_j are some positive scalars called the widths, $\mathbf{c}_j$ are the RBF centers, $\phi(\cdot; \rho)$ is a function from $R^+ \to R$, and n_{H} is the number of hidden nodes.

The RBF method is a traditional technique for strict interpolation in multidimensional space [49]. On the other hand, using a feedforward neural network to model complex data can also be considered a curve fitting operation in a multidimensional space. Broomhead and Lowe [43] adopted this viewpoint and explicitly revealed the connection between feedforward neural networks and RBF models. The topology of the RBF network is similar to that of the one-hidden-layer perceptron, and the difference lies in the characteristics of the hidden nodes. Two typical choices of $\phi(\cdot)$ for RBF networks are the Gaussian function

$$\phi(v/\rho) = \exp(-v^2/\rho), \qquad (43)$$

and the thin-plate-spline function

$$\phi(v) = v^2 \log(v). \qquad (44)$$

The Gaussian function represents a class of node nonlinearity with the property that $\phi(v/\rho) \to 0$ as $v \to \infty$. The thin-plate-spline function represents another class of node nonlinearity with the property that $\phi(v/\rho) \to \infty$ as $v \to \infty$. These two very different classes of nonlinearities may be referred to as class one and class two, respectively.

Most of the applications of RBF networks can be included in the following framework: use the network input–output mapping f_{RBF} to learn or to approximate some nonlinear mapping from $R^{n_\mathrm{I}} \rightarrow R^{n_\mathrm{O}}$. The RBF network is known to be a general function approximator, and theoretical investigations have concluded that the choice of $\phi(\cdot)$ is not crucial for network approximation capabilities [30, 50, 51]. RBF models based on either class one or class two nonlinearities all have excellent approximation capabilities. Nevertherless, choosing the nonlinearity of the hidden nodes according to the application can often improve performance. Athough each hidden node may have a different width ρ_j, a uniform width is sufficient for universal approximation [30]. All of the widths in the network can therefore be fixed to a value ρ, and this ρ can be derived by using some heuristic rules [52]. For classification applications, however, adjusting individual widths can often improve the generalization properties [53]. Some choices of the nonlinearity, such as (44), do not involve a width.

An extension to the standard RBF network is to replace the Euclidean distance by the Mahalanobis distance in hidden nodes [53, 54]. In this case, the hidden nodes become

$$\phi_j = \phi\big((\mathbf{x} - \mathbf{c}_j)^T V_j^{-1} (\mathbf{x} - \mathbf{c}_j)\big), \qquad 1 \le j \le n_\mathrm{H}, \tag{45}$$

where V_j is an $n_\mathrm{I} \times n_\mathrm{I}$ positive definite matrix. Hartman and Keeler [55] proposed a Gaussian-bar network in which each input dimension is treated differently. In the multi-output case, the hidden nodes are defined by

$$\phi_{j,i} = \sum_{l=1}^{n_\mathrm{I}} \theta_{l,j,i} \exp\big(-(x_l - c_{l,j})^2 / \rho_{l,j}\big), \qquad 1 \le j \le n_\mathrm{H}, \quad 1 \le i \le n_\mathrm{O}. \tag{46}$$

The ith output node is the sum of $\phi_{j,i}$ for $1 \le j \le n_\mathrm{H}$. Hartman and Keeler [55] reported better performance of this Gaussian-bar network over the standard Gaussian network.

In general, the RBF network is nonlinear-in-the-parameters. However, learning can be performed in two phases. First, some learning mechanism is employed to select a suitable set of RBF centers and widths. This effectively determines the hidden layer of the RBF network. Because the output layer is a set of linear combiners, learning the remaining output-layer weights becomes a linear problem. In this sense, the RBF network is often said to have a "linear-in-the-parameters" structure. This is only true after the hidden layer has been fixed separately. Because of this property, learning procedures for the RBF network are simple and reliable.

C. FUZZY BASIS FUNCTION NETWORKS

The architecture of a general n_I-input n_O-output fuzzy system is depicted in Fig. 4. A fuzzy system consists of four basic elements: a fuzzifier, a fuzzy rule base, a fuzzy inference engine, and a defuzzifier. The fuzzifier maps the crisp input space onto the fuzzy sets defined in the input space. The fuzzy rule base consists of a set of n_F linguistic rules in the forms of IF–THEN. The fuzzy inference engine is a decision-making logic that employs fuzzy rules from the rule base to determine a mapping from the fuzzy sets in the input space onto the fuzzy sets in the output space. The defuzzifier performs a mapping from the fuzzy sets in R^{n_O} to the crisp outputs $\mathbf{y} \in R^{n_O}$. A class of fuzzy systems commonly used in practice is constructed based on singleton fuzzification, product inference, and centroid or weighted average defuzzification. Such fuzzy systems can be represented as series expansions of fuzzy basis functions known as fuzzy basis function (FBF) networks or models [31, 56].

An n_I-input n_O-output fuzzy system can be expressed as n_O single-output fuzzy subsystems. The rule base of the ith fuzzy subsystem consists of n_F rules in the following forms:

$$RB_j^i: \text{ IF } x_1 \text{ is } A_{1,j} \text{ AND } \cdots \text{ AND } x_{n_I} \text{ is } A_{n_I,j}, \text{ THEN } y_i \text{ is } B_j^i, \qquad (47)$$

where $1 \leq j \leq n_F$; x_l for $1 \leq l \leq n_I$ are the inputs to the fuzzy system; y_i is the ith output of the fuzzy system, $1 \leq i \leq n_O$; and $A_{l,j}$ and B_j^i are the fuzzy sets characterized by fuzzy membership functions $\mu_{A_{l,j}}(x_l)$ and $\mu_{B_j^i}(y_i)$, respectively.

Under the assumptions of singleton fuzzifier, product inference, and centroid defuzzifier, the input–output mapping of such a fuzzy system, $f_{FBF}: R^{n_I} \to R^{n_O}$, can be shown to have the form [31, 56, 57]

$$f_{FBF_i}(\mathbf{x}) = \frac{\sum_{j=1}^{n_F} \bar{y}_j^i \left(\prod_{l=1}^{n_I} \mu_{A_{l,j}}(x_l) \right)}{\sum_{j=1}^{n_F} \left(\prod_{l=1}^{n_I} \mu_{A_{l,j}}(x_l) \right)}, \qquad 1 \leq i \leq n_O, \qquad (48)$$

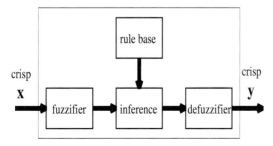

Figure 4   MIMO fuzzy system architecture.

where $\bar{y}_j^i$ is the point at which $\mu_{B_j^i}(y_i)$ achieves its maximum value. Define

$$\phi_j(\mathbf{x}) = \frac{\prod_{l=1}^{n_1} \mu_{A_{l,j}}(x_l)}{\sum_{j=1}^{n_F} \left(\prod_{l=1}^{n_1} \mu_{A_{l,j}}(x_l)\right)}, \qquad 1 \le j \le n_F, \tag{49}$$

which are referred to as FBFs [31, 56]. Then the fuzzy system (48) is equivalent to an FBF expansion,

$$f_{\text{FBF}_i}(\mathbf{x}) = \sum_{j=1}^{n_F} \phi_j(\mathbf{x})\theta_{j,i}, \qquad 1 \le i \le n_O, \tag{50}$$

where $\theta_{j,i}$ are coefficients or parameters of the FBF model.

Fuzzy systems are universal approximators, and theoretical studies have proved that the FBF network (50) can approximate any continuous functions to within any degree of accuracy, provided that a sufficient number of fuzzy rules are used [31, 56, 57]. Although the FBF network is derived within the framework of fuzzy logic, it obviously has many similarities to neural networks such as the RBF network. An advantage of the FBF network is that linguistic information from human experts in the form of the fuzzy IF–THEN rules can be directly incorporated into the model.

Currently fuzzy systems are mainly applied to engineering problems in which the number of inputs and outputs is small. This is because the number of rules or FBFs grows exponentially as the number of inputs and outputs increases. This curse of "rule explosion" limits further applications of fuzzy systems to complex large systems. Research has been directed toward determining the best shape for fuzzy sets that will result in a significant reduction in the number of rules required [58]. Notice that the FBF network (50) has a linear-in-the-parameters structure once the rules have been specificied. Given observed input–output data, the problem of constructing a parsimonious FBF model can be formulated as one of subset model selection. This is the approach adopted in [31], where the OLS algorithm was used to select significant FBFs from a large set of candidate FBFs.

D. RECURRENT NEURAL NETWORKS

Architectures of recurrent neural networks are extremely rich, and it is impossible to give a comprehensive discussion in a short section. For recent advances in the theory and applications of recurrent neural networks, see for example [59]. A straightforward way of obtaining a recurrent neural network is to feed back the network output to the network input. The structure of such feedback or recurrent networks is shown in Fig. 5. Notice that the feedback signal can consist of either the delayed network outputs $\hat{y}(k-1), \ldots, \hat{y}(k-n_d)$ or the delayed prediction

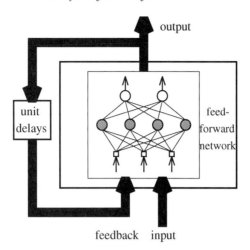

Figure 5 Recurrent network obtained by feeding back the network output to the input.

errors $\epsilon(k-1), \ldots, \epsilon(k-n_d)$, where $\epsilon(k) = y(k) - \hat{y}(k)$ and $y(k)$ is the desired output. The "feedforward network" part of this structure can be a multilayer perceptron, a RBF network, or a FBF network. Although this structure may appear to be similar to the feedforward network except for an "expanded" network input, the design and analysis of such a feedback network are considerably more complex than those of feedforward networks. Advantages gained by using this structure are richer dynamic behaviors and improved representation capabilities. An example in which this approach works well is in adaptive noise cancellation [60].

In a general recurrent network, node outputs can be fed back to node inputs. A class of three-layer recurrent networks with internal feedback is illustrated in Fig. 6a. As in the case of feedforward networks, the input layer simply distributes the network inputs to the hidden-layer nodes, and the output layer is a set of linear combiners. The outputs of the hidden nodes, however, are fed back as part of the inputs to the hidden nodes; the models of the two kinds of connections for hidden-layer neurons are depicted in Fig. 6b. The topology of this three-layer recurrent network can be summarized as $n_0 - n_1 - n_2$, where n_0 is the number of network inputs, n_1 is the number of hidden nodes, and n_2 is the number of network outputs. The output of the jth hidden node is computed as

$$x_{O_j}(k) = f_a\left(\sum_{l=0}^{n_0} w_{F_{j,l}} x_{I_l}(k) + \sum_{l=1}^{n_1} w_{B_{j,l}} x_{O_l}(k-1) \right), \qquad 1 \le j \le n_1 \quad (51)$$

where $x_{I_l}(k)$ for $1 \le l \le n_0$ are the network inputs, $x_{I_0}(k) = 1$ indicates that $w_{F_{j,0}}$ is a bias weight, $x_{O_j}(k)$ are the hidden layer outputs, and $w_{F_{j,l}}$ and $w_{B_{j,l}}$

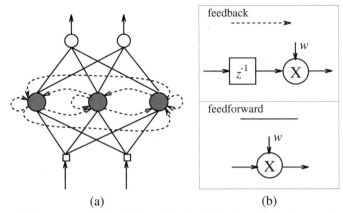

Figure 6   Recurrent network with internal feedback (a) and two kinds of connections (b).

are the feedforward and feedback connection weights, respectively. The single-hidden-layer feedforward network can be viewed as a special case of this recurrent network, where all the feedback connection weights are zeros.

More generally, a recurrent network can include more than one hidden layer, and the network output can also be fed back as part of the network input. A smaller recurrent network with fewer hidden nodes can often achieve the same modeling accuracy as a large feedforward network. In this sense, recurrent networks have better representation capabilities. The learning and analysis for recurrent networks are much more difficult, however. A recurrent network is a highly nonlinear-in-the-parameters and complex dynamic system. Theoretical investigation and practical application of recurrent networks will provide an interesting and challenging research area for many years to come.

V. NONLINEAR-IN-THE-PARAMETERS APPROACH

For artificial neural networks that are nonlinear-in-the-parameters, learning must be based on nonlinear optimization techniques. The optimization criterion is typically chosen to be the mean square error defined in (20), and gradient learning methods remain as the most popular techniques for learning, despite the possible pitfalls that learning can become trapped at a local minimum of the optimization criterion. Global optimization techniques, such as genetic algorithms [61] and simulated annealing [62], although capable of achieving a global minimum, require extensive computation. We describe an efficient gradient learning algorithm called the *parallel prediction error algorithm* (PPEA) [38].

A. PARALLEL PREDICTION ERROR ALGORITHM

The prediction error algorithm [33] discussed in Section III is a general parameter estimation algorithm and can readily be applied to neural networks [9]. Compared with the popular back-propagation algorithm (BPA) [39], which is a steepest-descent gradient algorithm, the prediction error algorithm achieves significantly better performance at a much faster convergence rate, because the Hessian information is utilized in the search. The BPA has computational advantages, however. Consider a generic neural network with a total of p neurons, and let the number of parameters of each node be n_{Θ_j}, where $1 \leq j \leq p$. The total number of parameters in the network is therefore

$$n_{\Theta} = \sum_{j=1}^{p} n_{\Theta_j}. \tag{52}$$

The computational complexity of the BPA is on the order of n_{Θ}, whereas the complexity of the prediction error algorithm can be shown to be on the order of $(n_{\Theta})^2$. Another important advantage of the BPA is that the algorithm is a distributed learning procedure, and weight updating is carried out locally. This is coherent with the massively distributed computing nature and parallel structure of neural networks.

The PPEA is also a distributed learning procedure that is a trade-off between the high performance of the full prediction error algorithm and the low complexity of the BPA. The recursive version of the PPEA is described here. For the generic network of p neurons and n_O outputs, local learning is achieved by using

$$\left.\begin{aligned}
\epsilon(k) &= y(k) - \hat{y}(k) \\
\Psi_j(k) &= \left[\frac{\partial \hat{y}(k)}{\partial \Theta_j}\right]^T \\
P_j(k) &= \frac{1}{\lambda}\Big[P_j(k-1) - P_j(k-1)\Psi_j(k) \\
&\quad \times \left(\lambda I + \Psi_j^T(k)P_j(k-1)\Psi_j(k)\right)^{-1}\Psi_j^T(k)P_j(k-1)\Big] \\
\Theta_j(k) &= \Theta_j(k-1) + P_j(k)\Psi_j(k)\epsilon(k)
\end{aligned}\right\}, \tag{53}$$

where $1 \leq j \leq p$, $\hat{y}(k)$ is the network output vector computed given the previous parameter estimate, Θ_j is the parameter vector of the jth node, I is the $n_O \times n_O$ identity matrix, and λ is the forgetting factor. Essentially, the PPEA consists of

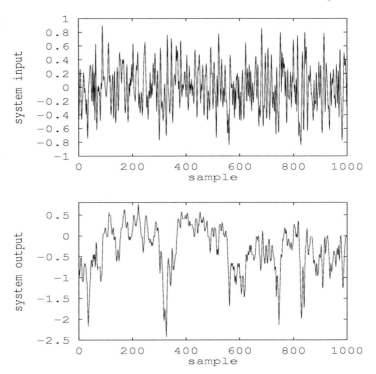

Figure 7   Inputs and outputs of a liquid level system.

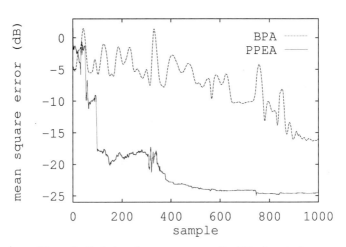

Figure 8   Evolution of mean square error (liquid level system).

many subalgorithms, each one associated with a neuron in the network. The BPA can be derived from this PPEA by replacing all of the P_j matrices with appropriate diagonal matrices [38].

By utilizing "local" or approximate Hessian information contained in the $P_j(k)$ matrices, the PPEA offers significant performance improvements over the BPA. Computational complexity of the PPEA is on the order of $\bar{n}_\Theta$, where

$$\bar{n}_\Theta = \sum_{j=1}^{p} \left(n_{\Theta j}\right)^2. \tag{54}$$

Although this is more complex than the BPA, the increase in computational requirements per iteration can often be offset by a faster convergence rate, so that overall the PPEA is computationally more efficient. We demonstrate better performance and faster convergence of the PPEA over the BPA by considering a real system identification application taken from [11].

EXAMPLE 3.   The data were generated from a single-input single-output nonlinear liquid level system. The system consists of a DC pump feeding a conical flask, which in turn feeds a square tank. The system input $u(k)$, the voltage to the pump motor, and the system output $y(k)$, the water level in the conical flask, are plotted in Fig. 7. A one-hidden-layer feedforward network with five hidden neurons was employed to model this real process, and the network input vector was chosen to be

$$\mathbf{x}(k) = \left[y(k-1)y(k-2)y(k-3)u(k-1)\cdots u(k-5)\right]^T. \tag{55}$$

The network structure was thus specified as 8-5-1, giving rise to $n_\Theta = 50$. The hidden node activation function was the sigmoid function (41), and initial weights were set randomly between -0.3 and 0.3. For the PPEA, all of the P_j matrices were initialized to $100.0I$, where I is identity matrices of appropriate dimensions. A variable forgetting factor,

$$\lambda(k) = \lambda_0 \lambda(k-1) + (1.0 - \lambda_0), \tag{56}$$

was used with $\lambda(0) = 0.95$ and $\lambda_0 = 0.99$. For the BPA, an adaptive gain of 0.01 and a momentum coefficient of 0.9 were found to be appropriate. The evolutions of the mean square error (in dB) obtained by the BPA and the PPEA are depicted in Fig. 8.

B. PRUNING OVERSIZED NETWORK MODELS

Choosing a proper network size is critical for dynamic system modeling using neural networks. If the network is too small, it will not be able to capture the underlying structure of the dynamic system. On the other hand, if the network is too big, it may simply fit to the noise in the training data, resulting in a poor generalization capability. According to the parsimonious principle, the smallest network that fits the data should be chosen to obtain good generalization performance. If no prior knowledge is available, a proper network size must be determined by trial and error. In the training of feedforward networks, one can start with a small network and gradually increase the size by adding more hidden and/or input nodes until the performance stops improving. Using a separate validation data set to test the model performance can often help to determine an appropriate network size.

Even when such a systematic approach is adopted, the final fully connected network model may still contain a large number of "redundant" weights that can be eliminated without affecting the network modeling capability. Pruning refers to the process of deleting redundant or insignificant weights in an oversized network model and is an effective technique for improving generalization properties. The algorithm based on the C-criterion (18), discussed in Section III, is a powerful tool for pruning oversized networks. The algorithm belongs to the approach utilizing complexity regularization, because (18) contains a measure of model complexity. A drawback of this algorithm is its computational cost. When the size of the full network is very large, the computation of the Hessian matrix is expensive, and the stepwise elimination procedure adds further a considerable computational burden. A popular and computationally much simpler method for pruning is known as the "optimal brain damage" (OBD) method [13].

In the OBD method, pruning is based on the saliency measure of network weights. Let J be the cost function used for network learning and θ_i be a generic network weight. The saliency of θ_i defined by the OBD method is

$$S(\theta_i) \approx \frac{\partial^2 J}{\partial \theta_i^2} \theta_i^2. \tag{57}$$

Note that $\partial^2 J / \partial \theta_i^2$ is a diagonal element of the Hessian matrix. The pruning process of the OBD method is as follows. The network is first trained and the saliencies of all the weights are computed using (57). The weights with the smallest saliency are then deleted, and the reduced-size network is retrained. The procedure may have to be repeated several times before a final pruned network model is obtained.

If the standard BPA is used for training, the second derivations $\partial^2 J / \partial \theta_i^2$ must be calculated separately to compute the saliencies of the weights. Notice, however, that if the batch version of the PPEA [38] is employed as the training algorithm, the diagonal elements of the (approximated) Hessian matrix are readily available and the computation of saliency (57) requires little extra cost. The detailed batch PPEA can be found in [11, 38] and will not be repeated here. We use the following example to demonstrate the combined approach of the batch PPEA training and pruning based on the saliency measure.

EXAMPLE 4. This is the same liquid level system as was studied in Example 3. The same 8-5-1 feedforward network was employed, and all 1000 data points were used as the training data. This fully connected 8-5-1 network had a total of 50 weights or parameters. The batch PPEA [38] was used to train the network model with the performance criterion

$$J(\Theta) = \tfrac{1}{2}\log\big(Q(\Theta)\big), \qquad Q(\Theta) = \frac{1}{N}\sum_{k=1}^{N}\epsilon^2(k;\Theta). \qquad (58)$$

After the training, the mean square error $Q(\Theta)$ of the fully connected network model over the training data set was -26.74 dB. Table V summarizes the weight values and the corresponding saliencies of this fully connected network model.

By inspecting the saliency values given in Table V, it was concluded that any saliency smaller than 1.0 could obviously be regarded as insignificant and the corresponding weight could be deleted. The saliency threshold was therefore set to 1.0, and this resulted in 11 weights (underlined in Table V) being eliminated. After retraining, the mean square error of the pruned network model over the training data set was -26.67 dB. To compare the generalization performance of the two network models, we examined the iterative network model output, defined as

$$\hat{y}_d(k) = f_{\text{MLP}}\big(\mathbf{x}_d(k)\big), \qquad (59)$$

where

$$\mathbf{x}_d(k) = \big[\hat{y}_d(k-1)\,\hat{y}_d(k-2)\ \hat{y}_d(k-3)u(k-1)\cdots u(k-5)\big]^T. \qquad (60)$$

Notice that during learning, the network input $\mathbf{x}(k)$ contained past system inputs and outputs. In generating $\hat{y}_d(k)$, the network was driven purely by the system inputs. If the trained model f_{MLP} really captured the system structure, $\hat{y}_d(k)$ should be able to follow the system output $y(k)$ closely. The iterative network outputs of the fully connected and pruned network models are plotted in Fig. 9. The results shown in Fig. 9 clearly demonstrate that the pruned smaller network has superior generalization performance over the fully connected network model.

VI. LINEAR-IN-THE-PARAMETERS APPROACH

Neural networks are never truly linear-in-the-parameters. The class of basis-function neural networks, such as the RBF and FBF networks, only become linear-in-the-parameters after a separate mechanism has been used to fix or to select the basis functions. For this class of neural networks, learning can be "decomposed" into two phases. In the first phase, the hidden layer that is a set of

Table V

Fully Connected 8-5-1 Feedforward Network Model Obtained Using Batch PPEA for Example 4

Node	Link	Weight	Saliency	Node	Link	Weight	Saliency
Output	1	−11.66116	63595.75425	Hidden 3	0	5.20112	6.10191
	2	16.42509	657.95127		1	3.69828	3.66240
	3	0.28607	33.88771		2	−0.30648	0.02250
	4	−28.59449	1152.50990		3	−0.39791	0.03398
	5	26.22748	58284.29225		4	9.21591	3.19361
					5	−8.45146	2.39434
					6	4.75899	0.70737
					7	−3.67219	0.40132
					8	1.65329	0.07686
Hidden 1	0	7.81765	881.21335	Hidden 4	0	−5.39978	10370.91985
	1	− 4.33839	872.64808		1	1.96907	3646.88325
	2	− 6.75760	2511.20255		2	2.99032	9345.46625
	3	12.57424	9719.85827		3	−6.47447	45683.86413
	4	5.54735	100.48206		4	−0.64769	22.94355
	5	− 1.23030	4.72108		5	−1.65110	158.83720
	6	8.51095	246.69947		6	−3.42483	901.78776
	7	− 5.88247	150.03042		7	2.20655	471.78389
	8	1.35375	6.89820		8	−0.17182	2.75034
Hidden 2	0	− 5.08538	5593.24296	Hidden 5	0	−0.26397	1342.97186
	1	1.69406	1548.08633		1	0.10828	88.46429
	2	2.72633	4352.99617		2	0.04182	13.22511
	3	− 6.10927	22012.26445		3	−0.00407	0.12590
	4	0.13381	0.54297		4	0.04258	3.49782
	5	− 2.33698	187.10634		5	−0.00018	0.00006
	6	− 2.94343	399.16156		6	−0.01648	0.52305
	7	2.00734	224.53101		7	0.01038	0.20746
	8	− 0.17158	1.60094		8	−0.00493	0.04687

*The underlined weights are to be eliminated.

basis functions is determined. Often this is done by using some simple and efficient unsupervised method. The second phase of learning is a pure linear learning problem and can easily be solved. The linear-in-the-parameters approach for neural network learning really refers to this two-stage approach. The main advantage of this approach is that it avoids complicated nonlinear gradient-based learning. We will use the RBF network as an example to illustrate this learning approach. The techniques discussed can be applied to other basis-function networks such as the FBF network.

For the RBF network, there are two basic methods of avoiding learning based on complex nonlinear optimization. The first method chooses the RBF centers

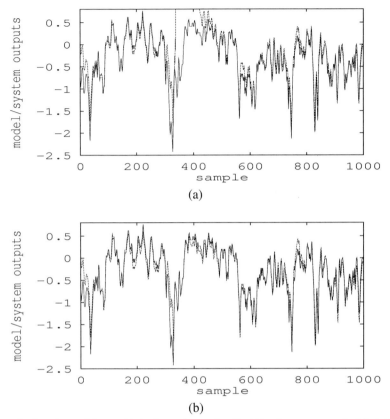

Figure 9 Iterative model ouput (dashed) superimposed on liquid level system output (solid). (a) Full
network model. (b) Pruned network model.

from the training inputs. Learning is formulated as a problem of linear subset
model selection, and the OLS algorithm discussed in Section III provides an efficient tool for constructing parsimonious RBF networks [16]. Other variants of
this theme include the use of mutual information [63] and genetic algorithms [64]
to configure the RBF networks. In the second method, RBF centers are first obtained by using the unsupervised κ-means clustering algorithm, and linear RBF
weights are then learned by using the standard least-squares or least-mean-square
algorithm [52, 65]. If a width parameter is required, it is estimated by using some
heuristic rules [52]. We discuss these two methods here with an emphasis on recent improvements.

A. Regularized Orthogonal Least-Squares Learning

Assume that we have a training set of N samples $\{y(k), \mathbf{x}(k)\}_{k=1}^{N}$. If each input vector $\mathbf{x}(k)$ is used as a RBF center, we have a total of $n_H = N$ basis functions. The desired outputs $y(k)$, the RBF network model outputs, and the errors can be collected together in the matrix form (23). The OLS algorithm can readily be applied to constructing a parsimonious subset network model from this full model [16]. A well-constructed parsimonious model often has better generalization properties and suffers less from overfitting in a noisy environment. A technique for improving generalization of overparameterized neural networks is regularization [21, 22]. For practical purposes, it is highly advantageous to combine regularization techniques with parsimonious construction algorithms, and a regularized OLS (ROLS) algorithm has recently been derived [24, 25].

The usual least-squares criterion in certain circumstances is prone to overfitting. When the data are highly noisy and the model size is large, the problem can be serious. The regularization method improves generalization by adding a penalty function to the cost function. Thus the cost function $NJ_2(\boldsymbol{\Theta}) = \text{trace}(\mathbf{E}^T\mathbf{E})$ of (30) is modified into

$$J_R(\boldsymbol{\Theta}; \xi) = \text{trace}(\mathbf{E}^T\mathbf{E}) + \xi \text{ (penalty function)}, \qquad (61)$$

where ξ is a regularization parameter. The simplest penalty function, known as the zero-order regularization, is $\text{trace}(\boldsymbol{\Theta}^T\boldsymbol{\Theta})$. The zero-order regularization is a technique equivalent to the weight-decaying in gradient descent methods for the MLP [66]. It is also known as the ridge regression in the statistical literature [67].

In the derivation of the ROLS algorithm, the following zero-order regularized error criterion is actually used:

$$J_R(\mathbf{G}; \xi) = \text{trace}(\mathbf{E}^T\mathbf{E} + \xi\mathbf{G}^T\mathbf{G}), \qquad (62)$$

where $\mathbf{G}$ is the orthogonal weight matrix related to $\boldsymbol{\Theta}$ by (29). Notice that the use of $\text{trace}(\mathbf{G}^T\mathbf{G})$ as the penalty is equivalent to the use of $\text{trace}(\boldsymbol{\Theta}^T\boldsymbol{\Theta})$. The choice of the regularized error criterion (62), however, has considerable computational advantages. This is because this cost function can be decomposed into

$$\text{trace}(\mathbf{E}^T\mathbf{E} + \xi\mathbf{G}^T\mathbf{G}) = \text{trace}(\mathbf{Y}^T\mathbf{Y}) - \sum_{j=1}^{n_H}\left(\sum_{i=1}^{m} g_{j,i}^2\right)(\mathbf{w}_j^T\mathbf{w}_j + \xi), \qquad (63)$$

where $\mathbf{w}_j$ are the orthogonal basis vectors. As in the case of the OLS algorithm, we can define the regularized error reduction ratio due to $\mathbf{w}_l$ as

$$[rerr]_l = \left(\sum_{i=1}^{m} g_{l,i}^2\right)(\mathbf{w}_l^T\mathbf{w}_l + \xi)/\text{trace}(\mathbf{Y}^T\mathbf{Y}). \qquad (64)$$

Based on this ratio, significant basis functions can be selected in a forward-selection procedure exactly as in the case of the OLS algorithm [24, 25].

The appropriate value of the regularization parameter ξ depends on the underlying system that generates the training data and the choice of basis function $\phi(\cdot)$. How to choose a good value for ξ has been addressed in the statistical literature [67, 68]. The optimal value of ξ can be determined based on the Bayesian evidence procedure [23]. Applying this Bayesian approach to the ROLS algorithm results in the following iterative procedure for estimating ξ. Given an initial guess of ξ, the algorithm constructs a network model. This in turn allows an updating of ξ by the formula

$$\xi = \frac{\gamma}{N - \gamma} \frac{\text{trace}(\mathbf{E}^T \mathbf{E})}{\text{trace}(\mathbf{G}^T \mathbf{G})}, \tag{65}$$

where

$$\gamma = \sum_{i=1}^{n_S} \frac{\mathbf{w}_i^T \mathbf{w}_i}{\mathbf{w}_i^T \mathbf{w}_i + \xi} \tag{66}$$

is known as the number of good parameter measurements [23], and n_S is the number of basis functions in the selected network model. After a few iterations, an appropriate ξ value can be found. We use two examples taken from [25] to demonstrate the effectiveness of the ROLS algorithm.

EXAMPLE 5.   This simple example was designed to illustrate the problem of overfitting and the power of the regularization technique. In this example, the RBF network with Gaussian basis function and a width $\rho = 0.2$ was used to approximate the scalar function

$$f(x) = \sin(2\pi x), \qquad 0 \le x \le 1. \tag{67}$$

One hundred training data were generated from $f(x) + e$, where x was uniformly distributed in $(0, 1)$ and the noise e had a Gaussian distribution with zero mean and standard deviation 0.4. A separated test data set was also generated for $x = 0, 0.01, \ldots, 0.99, 1.00$. The training data and the function $f(x)$ are plotted in Fig. 10. The training data set was very noisy and highly ill-conditioned.

The ROLS algorithm selected 15 centers from the training set. Figure 11 depicts the mean square error as a function of $\log_{10}(\xi)$ for both the training and testing data sets. The optimal value of ξ for this example was approximately 1.0. However, for a large range of ξ values, the mean square error over the testing set was quite flat, indicating that the performance of the ROLS algorithm was fairly insensitive to the precise value of ξ in this large region. When the evidence formula (65) was used to estimate ξ, ξ converged approximately to 1.0. Figure 12 shows the network mapping constructed by the ROLS algorithm with $\lambda = 1.0$. As a comparison, the network mapping constructed by the OLS algorithm is given

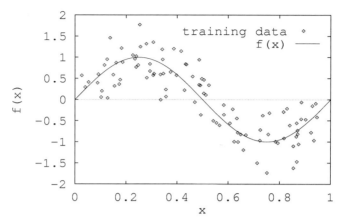

Figure 10   Noisy training data (points) and underlying function (curve) for Example 5.

in Fig. 13, where overfitting can be clearly seen. This example demonstrates that subset selection alone is not immune to overfitting when data are highly noisy, and a combined regularization and subset selection approach is often desirable.

EXAMPLE 6.   This was the time series of annual sunspot numbers. The sunspot time series over the years 1700–1979 is depicted in Fig. 14. The data from 1700 to 1920 were used for training, and the multistep preditions were then computed over the years 1921–1955 and the years 1921–1979, respectively. In

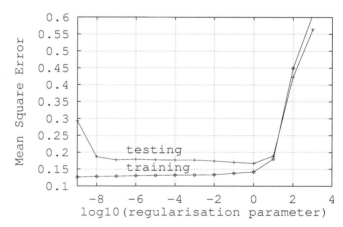

Figure 11   Mean square error as a function of the regularization parameter for Example 5.

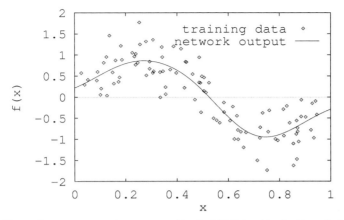

Figure 12 Network mapping constructed by the ROLS algorithm for Example 5.

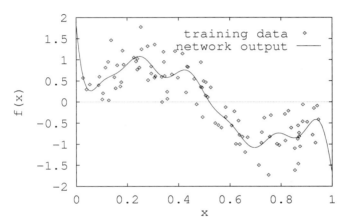

Figure 13 Network mapping constructed by the OLS algorithm for Example 5.

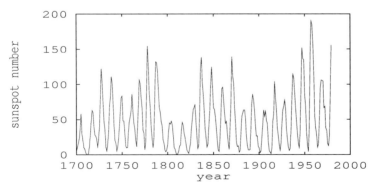

Figure 14 Time series of annual sunspot numbers.

a previous study [69], the OLS algorithm was used to construct a RBF network predictor with thin-plate-spline basis function (44). The algorithm selected 25 centers; the predictive accuracy of the resulting RBF model is shown in Fig. 15. It should be emphasized that the performance of this RBF network constructed with the OLS algorithm is better than some other nonlinear models fitted to the time series [27, 70].

To demonstrate that regularization can further improve performance, the ROLS algorithm was used to construct a RBF network of 25 centers based on the same full network model with $\xi = 10^7$. This value of regularization parameter was found by using the evidence procedure. The predictive accuracy of the network

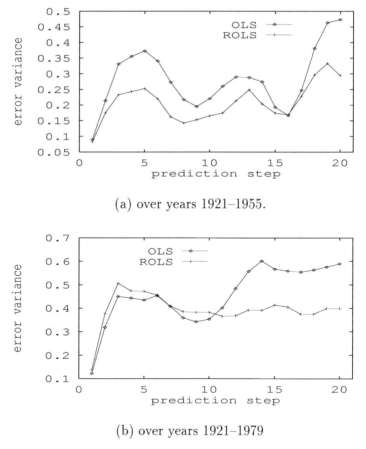

(a) over years 1921–1955.

(b) over years 1921–1979

Figure 15   Normalized variances of multistep prediction errors for sunspot time series.

model obtained by using the ROLS algorithm is also shown in Fig. 15, where it is clearly seen that the ROLS algorithm has better generalization properties.

B. ENHANCED CLUSTERING AND LEAST-SQUARES LEARNING

A popular learning method for RBF networks is the clustering and least-squares learning [52, 65]. The RBF centers are obtained by means of a κ-means clustering algorithm, whereas the network weights are learned by using the least-squares algorithm. The κ-means clustering algorithm is an unsupervised learning method based only on input training samples. It partitions the input data set into n_H clusters and obtains the cluster centers by attempting to minimize the total squared error incurred in representing the data set by the n_H cluster centers [71]. The traditional κ-means clustering algorithm can only achieve a local optimal solution, which depends on the initial locations of cluster centers. A consequence of this local optimality is that some initial centers can become stuck in regions of the input domain with few or no input patterns, and never move to where they are needed. This wastes resources and results in an unnecessarily large network.

An improved κ-means clustering algorithm was recently proposed [72], which overcomes the above-mentioned drawback. By using a cluster variation-weighted measure, the enhanced κ-means partitioning process always achieves an optimal center configuration in the sense that after convergence all clusters have an equal cluster variance. This property ensures that center resources are not wasted and performance does not depend on the initial center locations. This enhanced κ-means clustering algorithm can readily be combined with the least-squares algorithm to provide a powerful learning method for constructing RBF networks [73]. We describe this combined learning method here.

The RBF network structure considered is the normalized Gaussian RBF network,

$$f_{\text{RBF}_i}\big(\mathbf{x}(k)\big) = \sum_{j=1}^{n_H} \theta_{j,i}\phi_j(k), \qquad 1 \le i \le n_O, \tag{68}$$

where

$$\phi_j(k) = \frac{\exp(-\|\mathbf{x}(k) - \mathbf{c}_j\|^2/\sigma_j^2)}{\sum_{l=1}^{n_H} \exp(-\|\mathbf{x}(k) - \mathbf{c}_l\|^2/\sigma_l^2)}. \tag{69}$$

A normalized Gaussian basis function features either localized behavior similar to that of a Gaussian function or nonlocalized behavior similar to that of a sigmoid function, depending on the location of the center [74]. This is often a desired property.

The RBF centers are learned by using the following enhanced κ-means clustering algorithm:

$$\mathbf{c}_j(k+1) = \mathbf{c}_j(k) + M_j(\mathbf{x}(k))(\eta_c(\mathbf{x}(k) - \mathbf{c}_j(k))), \qquad (70)$$

where $0 < \eta_c < 1.0$ is a learning rate, the membership function $M_j(\mathbf{x}(k))$ is defined as

$$M_j(\mathbf{x}) = \begin{cases} 1, & \text{if } v_j \|\mathbf{x} - \mathbf{c}_j\|^2 \leq v_l \|\mathbf{x} - \mathbf{c}_l\|^2 \text{ for all } l \neq j, \\ 0, & \text{otherwise} \end{cases} \qquad (71)$$

and v_j is the variation or "variance" of the jth cluster. To estimate variation v_j, the following updating rule is used:

$$v_j(k+1) = \alpha v_j(k) + (1 - \alpha)(M_j(\mathbf{x}(k))\|\mathbf{x}(k) - \mathbf{c}_j(k)\|^2). \qquad (72)$$

The initial variations, $v_j(0)$, $1 \leq j \leq n_H$, are set to the same small number, and α is a constant slightly less than 1.0.

The learning rate η_c can either be fixed to a small constant or be self-adjusting based on an "entropy" formula [72],

$$\eta_c = 1 - H(\bar{v}_1, \ldots, \bar{v}_{n_H})/\log(n_H), \qquad (73)$$

where

$$H(\bar{v}_1, \ldots, \bar{v}_{n_H}) = \sum_{j=1}^{n_H} -\bar{v}_j \log(\bar{v}_j) \qquad \text{with } \bar{v}_j = v_j \Big/ \sum_{l=1}^{n_H} v_l. \qquad (74)$$

The widths, σ_j^2, $1 \leq j \leq n_H$, can be calculated, after the clustering process has converged, from the variances of the clusters. Because the optimal κ-means clustering distributes the total variation equally among the clusters, a universal width can be used for all of the nodes. The network weights, $\theta_{j,i}$, are then learned by using the usual least squares algorithm.

EXAMPLE 7.   This was a simulated two-dimensional autonomous system taken from [73]. The data were generated by using

$$\begin{aligned} y(k) = {} & (0.8 - 0.5\exp(-y^2(k-1)))y(k-1) \\ & - (0.3 + 0.9\exp(-y^2(k-1)))y(k-2) \\ & + 0.1\sin(\pi y(k-1)) + e(k), \end{aligned} \qquad (75)$$

where the Gaussian white noise $e(k)$ had a zero mean and variance 0.01. Two thousand samples of the time series are depicted in Fig. 16. The first 1000 points were used as the training set and the last 1000 points as the test set. The enhanced clustering and recursive least-squares algorithm was used to construct a RBF net-

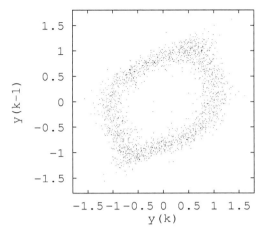

Figure 16 Noisy observations of the two-dimensional time series example.

work model. Figure 17 shows the mean square error as a function of the center number.

The results of Fig. 17 indicated that eight centers were sufficient to model this time series. The noise-free system is a limit cycle depicted in Fig. 18, where the eight center locations obtained by the enhanced κ-means clustering algorithm from noisy data are also plotted. From Fig. 18 it can be seen that an optimal center

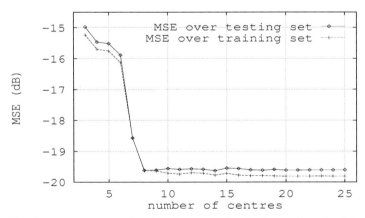

Figure 17 Mean square error as a function of center number for the two-dimensional time series example.

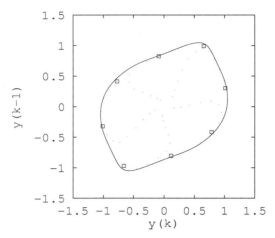

Figure 18   Noise-free two-dimensional system (limit cycle) and RBF center locations (▢).

configuration was obtained. When the network model output was fed back to the input, the iterative network model output,

$$\hat{y}_d(k) = f_{\text{RBF}}\big(\hat{y}_d(k-1), \hat{y}_d(k-2)\big), \tag{76}$$

generated a limit cycle that was indistinguishable from the system limit cycle. The errors between the noise-free system outputs and the iterative network model outputs are given in Fig. 19. The results obtained by using the enhanced clus-

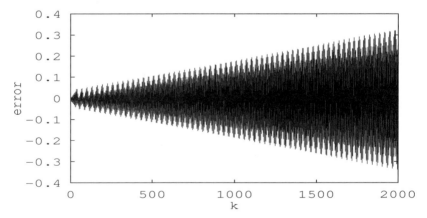

Figure 19   Iterative network model errors for the two-dimensional time series example.

tering and least-squares algorithm are better than some previous results [65, 75]. Furthermore, the present method requires a smaller network size.

There are several alternatives to the κ-means-based clustering algorithms, including the mean-tracking algorithm [76] and fuzzy clustering schemes [77]. All of these provide excellent results and avoid the disadvantages of the basic κ-means method.

C. ADAPTIVE ON-LINE LEARNING

Recursive versions of the OLS algorithm have been derived that update both the model structures and the parameters of nonlinear models on-line. These have been applied to both NARMAX model identification and the training of RBF networks [41, 42, 78, 79]. The new algorithms operate on a window of data and employ a Givens routine to minimize the loss function at every selection step by selecting significant regression variables and computing the parameter estimates in a way that maintains the orthogonality of the vector space. These algorithms can be used as part of an efficient adaptive procedure to track the variations in model structure [41, 79] and network topology [42, 79] and update the associated parameters or weights on-line.

VII. IDENTIFIABILITY AND LOCAL MODEL FITTING

Experiment design for linear system identification is well established [1]. Basically, the input signal chosen for a linear identification experiments should be persistently exciting to ensure identifiability. Persistent excitation in the context of linear system identification means that the input should excite all of the frequencies of interest in the system, and this can be shown to relate to the second-order statistics of the input signal. For nonlinear system identification, however, the second-order statistics of the input are no longer sufficient to determine identifiability, and in general, the input probability density function or higher order statistics are needed. Thus the design of inputs for nonlinear system identification is a very complex problem.

Some useful results have been given in [80]. Roughly speaking, the definition of persistent excitation in the context of nonlinear system identification should be modified as: an input signal is persistently exciting if the input excites all of the frequencies of interest in the system and excites the system over the whole amplitude range of operation. We use a simple example taken from [81] to illustrate the relationship between persistent excitation and identifiability.

EXAMPLE 8.   A nonlinear digital communication channel can be represented
by

$$y(k) = f_s\big(u(k), \ldots, u(k - n_u)\big) + e(k), \tag{77}$$

where the input $u(k)$ is a white sequence taking values from the set $\{\pm 1\}$, and
the noise $e(k)$ is uncorrelated with $u(k)$. Because $u(k)$ is white, it contains all
frequency components, and is an ideal input signal for identifying the linear model
of any order n_u:

$$\hat{y}(k) = \sum_{i=0}^{n_u} \theta_i u(k - i). \tag{78}$$

For nonlinear identification, the input should also excite a sufficient range of amplitudes. The binary nature of $u(k)$ therefore represents a worst scenario, and as
a consequence, parameters in some nonlinear models may not be identifiable. For
example, consider the following channel model:

$$\hat{y}(k) = \sum_{i=0}^{2} \theta_i u(k - i) + \sum_{i=0}^{2}\sum_{j=i}^{2} \theta_{i,j} u(k - i)u(k - j)$$

$$+ \sum_{i=0}^{2}\sum_{j=i}^{2}\sum_{l=j}^{2} \theta_{i,j,l} u(k - i)u(k - j)u(k - l). \tag{79}$$

The rank of the 19×19 autocorrelation matrix of the estimator input vector is
only 8. It is therefore impossible to identify all 19 parameters in (79).

The requirement of exciting a sufficient range of amplitudes is difficult to meet
in practice. Normal operation of an industrial plant is often concerned with controlling the plant close to some operating points. Perturbing signals that the experimenter injects into the plant can have only a small amplitude, so as not to cause
large disturbances in the operation of the plant. Such a small perturbing data set is
bad for nonlinear system identification. If normal operation of the plant includes
several operating levels, several sets of small perturbing data records can be obtained without violating the amplitude constraints for normal operation. These
data records together may cover a sufficient range of amplitudes. This is illustrated in Fig. 20. A "global model" fitting procedure [34] can then be applied to
obtain a nonlinear model that is valid over the whole operating region of interest.

Using a single nonlinear model to represent a nonlinear system obviously has
many advantages. But this depends on whether the system under investigation can
be represented by a single "global" model. Typically, many nonlinear systems require more than one model to capture different dynamic behaviors over different
operating regions. Fitting several local models [82] to a system may be particularly useful for modeling such systems, and this can be included in the framework

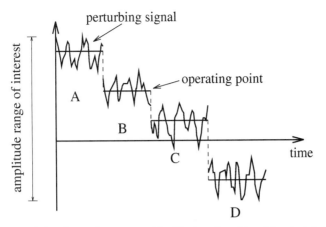

Figure 20 Input design for nonlinear system identification. Perturbing data sets A to D collected together excite the amplitude range of interest.

of the threshold NARMAX (TNARMAX) model [34]. A general TNARMAX model is given by

$$\hat{y}(k) = f^{(l)}(\mathbf{x}(k); \Theta_l), \quad \text{if } \mathbf{x}(k) \in R^{(l)}, \quad 1 \leq l \leq M_{\text{L}}, \tag{80}$$

where $R^{(l)}$ are given regions of the model input space $\mathbf{X}$, $f^{(l)}(\cdot)$ are some nonlinear mappings on $R^{(l)}$, and Θ_l are the parameter vectors of local models $f^{(l)}(\cdot)$, respectively.

A key step in TNARMAX modeling is a proper partition of the model input space $\mathbf{X}$ into M_{L} regions. A simple and effective method is to use the enhanced κ-means clustering algorithm. The data set is divided into M_{L} clusters, as illustrated in Fig. 21. Each data cluster is then modeled. Various models and identification schemes such as those discussed previously can be employed in this local model fitting. To provide a smooth transition from one local model to another, regularization techniques should be adopted in local model identification, and the second-order regularization or curvature-driven smoothing schemes [21] are particularly useful. But care must to be exercised if local models are to provide a representative model of a nonlinear system [82].

VIII. CONCLUSIONS

An artificial neural network that has the ability to learn sophisticated nonlinear relationships provides an ideal means of modeling complex nonlinear systems. In this chapter we have summarized several important results relating to

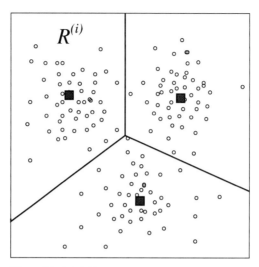

Figure 21   Model input space partition using clustering.

nonlinear dynamic system modeling using neural networks, and highlighted the connections between the traditional system identification approach and the neural network modeling approach. Neural network research has matured considerably during the past decade, and two areas, constructing parsimonious networks and improving generalization properties, have attracted extensive interest. The results of this research are particularly relevant to dynamic system modeling, because what really counts is how well a model captures the underlying system dynamics, not how good it fits the (noisy) observed data.

It is the authors' belief that model structure determination will remain a key future research topic. A crucial issue is finding relevant model inputs. Often inputs to a neural network model

$$\mathbf{x} = [x_1 x_2 \cdots x_{n_l}]^T \tag{81}$$

are overcomplicated in the sense that only a subset of these inputs,

$$\mathbf{x}_S = [x_{l_1} x_{l_2} \cdots x_{l_{n_S}}]^T, \tag{82}$$

are significant and should only be considered as inputs. Development of efficient and reliable methods for detecting relevant model inputs is of fundamental importance because a small reduction in the network input dimension can decrease the network complexity considerably. A useful scheme for determining important inputs is presented in [83]. A second important issue is that many systems are naturally composed of subsystems, and ideally neural network models should be able to reflect this structural information. For example, a nonlinear system of inputs x_1, x_2, x_3, x_4 may simply consist of two additive subsystems,

$$f(x_1, x_2, x_3, x_4) = f_1(x_1, x_4) + f_2(x_2, x_3).\tag{83}$$

If this structural information can be captured, a significant reduction in the network complexity can be achieved. Some neural network models may be inherently better at dealing with this kind of system identification problem than others, and previous research [17, 19, 55] provides a useful starting point for further investigation.

In view of past research results, we can confidently say that both the theory and practice of nonlinear system identification have advanced considerably during the past decade. Future research will involve multidisciplinary approaches, including traditional control engineering, nonlinear dynamical systems theory, and neural networks.

REFERENCES

[1] G. C. Goodwin and R. L. Payne. *Dynamic System Identification: Experiment Design and Data Analysis*. Academic Press, New York, 1977.

[2] L. Ljung and T. Söderström. *Theory and Practice of Recursive Identification*. MIT Press, Cambridge, MA, 1983.

[3] I. J. Leontaritis and S. A. Billings. Input–output parametric models for non-linear systems. Part I. Deterministic non-linear systems. Part II. Stochastic non-linear systems. *Internat. J. Control* 41:303–344, 1985.

[4] I. J. Leontaritis and S. A. Billings. Model selection and validation methods for non-linear systems. *Internat. J. Control* 45:311–341, 1987.

[5] S. Chen and S. A. Billings. Recursive prediction error estimator for non-linear models. *Internat. J. Control* 49:569–594, 1989.

[6] S. Chen and S. A. Billings. Representation of non-linear systems: the NARMAX model. *Internat. J. Control* 49:1013–1032, 1989.

[7] S. Chen, S. A. Billings, and W. Luo. Orthogonal least squares methods and their application to non-linear system identification. *Internat. J. Control* 50:1873–1896, 1989.

[8] K. S. Narendra and K. Parthasarathy. Identification and control of dynamical systems using neural networks. *IEEE Trans. Neural Networks* 1:4–27, 1990.

[9] S. Chen, S. A. Billings, and P. M. Grant. Non-linear systems identification using neural networks. *Internat. J. Control* 51:1191–1214, 1990.

[10] K. J. Hunt, D. Sbarbaro, R. Zbikowski, and P. J. Gawthrop. Neural networks for control systems—a survey. *Automatica* 28:1083–1112, 1992.

[11] S. Chen and S. A. Billings. Neural networks for non-linear dynamic system modelling and identification. *Internat. J. Control* 56:319–346, 1992.

[12] M. Brown and C. J. Harris. *Neurofuzzy Adaptive Modelling and Control*. Prentice Hall, Hemel-Hempstead, England, 1994.

[13] Y. le Cun, J. S. Denker, and S. A. Solla. Optimal brain damage. In *Advances in Neural Information Processing Systems* (D. Touretzky, Ed.), Vol. 2, pp. 598–605. Morgan Kaufmann, San Mateo, CA, 1990.

[14] H. Thodberg. Improving generalization on neural networks through pruning. *Internat. J. Neural Syst.* 1:317–326, 1991.

[15] R. Reed. Pruning algorithms—a survey. *IEEE Trans. Neural Networks* 4:740–747, 1993.

[16] S. Chen, C. F. N. Cowan, and P. M. Grant. Orthogonal least squares learning algorithm for radial basis function networks. *IEEE Trans. Neural Networks* 2:302–309, 1991.

[17] T. D. Sanger. A tree-structured adaptive network for function approximation in high-dimensional space. *IEEE Trans. Neural Networks* 2:285–293, 1991.

[18] J. H. Friedman. Multivariate adaptive regression splines. *Ann. Statist.* 19:1–141, 1991 (with discussion).

[19] T. Kavli. ASMOD—an algorithm for adaptive spline modelling of observation data. *Internat. J. Control* 58:947–967, 1993.

[20] T. Poggio and F. Girosi. Networks for approximation and learning. *Proc. IEEE* 78:1481–1497, 1990.

[21] C. Bishop. Improving the generalization properties of radial basis function neural networks. *Neural Comput.* 3:579–588, 1991.

[22] F. Girosi, M. Jones, and T. Poggio. Regularisation theory and neural networks architectures. *Neural Comput.* 7:219–269, 1995.

[23] D. J. C. MacKay. Bayesian interpolation. *Neural Comput.* 4:415–447, 1992.

[24] S. Chen. Regularised OLS algorithm with fast implementation for training multi-output radial basis function networks. In *Proceedings of the Fourth International Conference on Artificial Neural Networks*, Cambridge, June 26–28, 1995, pp. 290–294.

[25] S. Chen, E. S. Chng, and K. Alkadhimi. Regularised orthogonal least squares algorithm for constructing radial basis function networks. *Internat. J. Control* 64:829–837, 1996.

[26] M. J. L. Orr. Regularisation in the selection of radial basis function centres. *Neural Comput.* 7:606–623, 1995.

[27] S. Chen and S. A. Billings. Modelling and analysis of non-linear time series. *Internat. J. Control* 50:2151–2171, 1989.

[28] G. Cybenko. Approximations by superpositions of a sigmoidal function. *Math. Control Signals Systems* 2:303–314, 1989.

[29] K. Funahashi. On the approximate realization of continuous mappings by neural networks. *Neural Networks* 2:183–192, 1989.

[30] J. Park and I. W. Sandberg. Universal approximation using radial-basis-function networks. *Neural Comput.* 3:246–257, 1991.

[31] L. X. Wang and J. M. Mendel. Fuzzy basis functions, universal approximation, and orthogonal least-squares learning. *IEEE Trans. Neural Networks* 3:807–814, 1992.

[32] D. S. Broomhead and G. P. King. Extracting qualitative dynamics from experimental data. *Physica D* 20:217–236, 1986.

[33] S. A. Billings and S. Chen. Identification of non-linear rational systems using a prediction-error estimation algorithm. *Internat. J. Systems Sci.* 20:467–494, 1989.

[34] S. A. Billings and S. Chen. Extended model set, global data and threshold model identification of severely non-linear systems. *Internat. J. Control* 50:1897–1923, 1989.

[35] Y.-H. Pao, *Adaptive Pattern Recognition and Neural Networks*. Addison-Wesley, Reading, MA, 1989.

[36] S. Chen and S. A. Billings. Prediction-error estimation algorithm for non-linear output-affine systems. *Internat. J. Control* 47:309–332, 1988.

[37] N. R. Draper and H. Smith. *Applied Regression Analysis*. Wiley, New York, 1981.

[38] S. Chen, C. F. N. Cowan, S. A. Billings, and P. M. Grant. Parallel recursive prediction error algorithm for training layered neural networks. *Internat. J. Control* 51:1215–1228, 1990.

[39] D. E. Rumelhart and J. L. McClelland, Eds. *Parallel Distributed Processing: Explorations in the Microstructure of Cognition*. MIT Press, Cambridge, MA, 1986.

[40] G. H. Golub and C. F. Van Loan. *Matrix Computations*. Johns Hopkins Univ. Press, Baltimore, MD, 1989.

[41] W. Luo and S. A. Billings. Adaptive model selection and estimation for nonlinear systems using a sliding data window. *Signal Process.* 46:179–202, 1995.

[42] W. Luo and S. A. Billings. Structure selective updating for nonlinear models and radial basis function neural networks. *Internat. J. Adaptive Control and Signal Processing*, 1997, to appear.

[43] D. S. Broomhead and D. Lowe. Multivariable functional interpolation and adaptive networks. *Complex Systems* 2:321–355, 1988.

[44] S. A. Billings and W. S. F. Voon. Correlation based model validity tests for non-linear models. *Internat. J. Control* 44:235–244, 1986.

[45] S. A. Billings and Q. M. Zhu. Model validation tests for multivariable nonlinear models including neural networks. *Internat. J. Control* 62:749–766, 1995.

[46] T. Bohlin. Maximum power validation of models without higher-order fitting. *Automatica* 14:137–146, 1978.

[47] S. A. Billings, S. Chen, and R. J. Backhouse. The identification of linear and non-linear models of a turbocharged automotive diesel engine. *Mech. Syst. Signal Process.* 3:123–142, 1989.

[48] S. Chen, S. A. Billings, C. F. N. Cowan, and P. M. Grant. Non-linear systems identification using radial basis functions. *Internat. J. Systems Sci.* 21:2513–2539, 1990.

[49] M. J. D. Powell. Radial basis functions for multivariable interpolation: a review. In *Algorithms for Approximation* (J. C. Mason and M. G. Cox, Eds.), pp. 143–167. Oxford, 1987.

[50] M. J. D. Powell. Radial basis function approximations to polynomials. In *Proceedings of the 12th Biennial Numerical Analysis Conference*, Dundee, Scotland, 1987, pp. 223–241.

[51] E. J. Hartman, J. D. Keeler, and J. M. Kowalski. Layered neural networks with Gaussian hidden units as universal approximations. *Neural Comput.* 2:210–215, 1990.

[52] J. Moody and C. J. Darken. Fast learning in networks of locally-tuned processing units. *Neural Comput.* 1:281–294, 1989.

[53] M. T. Musavi, W. Ahmed, K. H. Chan, K. B. Faris, and D. M. Hummels. On the training of radial basis function classifiers. *Neural Networks* 5:595–603, 1992.

[54] S. Lee and R. M. Kil. A Gaussian potential function network with hierarchically self-organizing learning. *Neural Networks* 4:207–224, 1991.

[55] E. J. Hartman and J. D. Keeler. Predicting the future: advantages of semilocal units. *Neural Comput.* 3:566–578, 1991.

[56] L. X. Wang. *Adaptive Fuzzy Systems and Control: Design and Stability Analysis*. Prentice Hall, Englewood Cliffs, NJ, 1994.

[57] B. Kosko. *Neural Networks and Fuzzy Systems*. Prentice Hall, Englewood Cliffs, NJ, 1992.

[58] B. Kosko. *Fuzzy Engineering*. Prentice Hall, Englewood Cliffs, NJ, 1996.

[59] Special Issue on Dynamic Recurrent Neural Networks. *IEEE Trans. Neural Networks* 5(2), 1994.

[60] S. A. Billings and C. Fung. Recurrent radial basis function networks for adaptive noise cancellation. *Neural Networks* 2:273–290, 1995.

[61] D. E. Goldberg. *Genetic Algorithms in Search, Optimization, and Machine Learning*. Addison-Wesley, Reading, MA, 1989.

[62] S. Kirkpatrick, C. Gelatt, and M. Vecchi. Optimization by simulated annealing. *Science* 220:671–680, 1983.

[63] G. L. Zheng and S. A. Billings. Radial basis function network configuration using mutual information and the orthogonal least squares algorithm. *Neural Networks* 9:1619–1637, 1996.

[64] S. A. Billings and G. L. Zheng. Radial basis function network configuration using genetic algorithms. *Neural Networks* 6:877–890, 1995.

[65] S. Chen, S. A. Billings, and P. M. Grant. Recursive hybrid algorithm for non-linear system identification using radial basis function networks. *Internat. J. Control* 55:1051–1070, 1992.

[66] J. Hertz, A. Krough, and R. Palmer. *Introduction to the Theory of Neural Computation*. Addison-Wesley, Redwood City, CA, 1991.

[67] A. E. Hoerl and R. W. Kennard. Ridge regression: biased estimation for nonorthogonal problems. *Technometrics* 12:55–67, 1970.

[68] G. H. Golub, M. Heath, and G. Wahba. Generalized cross-validation as a method for choosing a good ridge parameter. *Technometrics* 21:215–223, 1979.

[69] S. Chen. Radial basis functions for signal prediction and system modelling. *J. Appl. Sci. Comput.* 1, 1994 (Special Issue on the Theory and Applications of Radial Basis Functions).

[70] A. S. Weigend, B. A. Huberman, and D. E. Rumelhart. Predicting the future: a connectionist approach. *Internat. J. Neural Syst.* 1:193–209, 1990.

[71] R. O. Duda and P. E. Hart. *Pattern Classification and Scene Analysis*. Wiley, New York, 1973.

[72] C. Chinrungrueng and C. H. Séquin. Optimal adaptive κ-means algorithm with dynamic adjustment of learning rate. *IEEE Trans. Neural Networks* 6:157–169, 1995.

[73] S. Chen. Nonlinear time series modelling and prediction using Gaussian RBF networks with enhanced clustering and RLS learning. *Electron. Lett.* 31:117–118, 1995.

[74] I. Cha and S. A. Kassam. Interference cancellation using radial basis function networks. *Signal Process.* 47:247–268, 1995.

[75] P. E. An, M. Brown, C. J. Harris, and S. Chen. Comparative aspects of neural network algorithms for on-line modelling of dynamic processes. *Proc. I. Mech. E., Pt. I, J. Systems Control Engrg.* 207:223–241, 1993.

[76] E. L. Sutanto and K. Warwick. Multivariable cluster analysis for high speed industrial machinery. *IEE Proc. Meas. Technol.* 142:417–423, 1995.

[77] G. L. Zheng and S. A. Billings. Radial basis function network training using a fuzzy clustering scheme. Unpublished.

[78] W. Luo, S. A. Billings, and K. M. Tsang. On-line structure detection and parameter estimation with exponential windowing for nonlinear systems. *European J. Control* 2:291–304, 1996.

[79] C. Fung, S. A. Billings, and W. Luo. On-line supervised adaptive training using radial basis function networks. *Neural Networks* 9:1597–1617, 1996.

[80] I. J. Leontaritis and S. A. Billings. Experiment design and identifiability for non-linear systems. *Internat. J. Systems Sci.* 18:189–202, 1987.

[81] S. Chen. Modelling and identification of nonlinear dynamic systems. In *Proceedings SPIE, Advanced Signal Processing: Algorithms, Architectures, and Implementations V*, San Diego, CA, July 24–29, 1994, pp. 270–278.

[82] S. A. Billings and W. S. F. Voon. Piecewise linear identification of nonlinear systems. *Internat. J. Control* 46:215–235, 1987.

[83] H. Pi and C. Peterson. Finding the embedding dimension and variable dependencies in time series. *Neural Comput.* 6:509–520, 1994.

High-Order Neural Network Systems in the Identification of Dynamical Systems

Elias B. Kosmatopoulos
Department of Electrical and
Computer Engineering
University of Victoria
Victoria, British Columbia,
V8W 3P6 Canada

Manolis A. Christodoulou
Department of Electronic and
Computer Engineering
Technical University of Crete
73100 Chania, Crete, Greece

I. INTRODUCTION

In this chapter, the applicability of high-order neural networks (HONNs) and recurrent high-order neural networks (RHONNs) to the identification of unknown nonlinear dynamical systems and related problems is examined. RHONNs have been proved to be capable of approximating, to any degree of accuracy, a quite general class of nonlinear systems. Appropriate learning laws, which are based on well-known adaptive robust parameter estimation techniques, are constructed, and it is shown that these learning laws are globally convergent, stable, and robust. By using these learning laws, the identification error is shown to converge to a ball centered at the origin, and the radius of which can be made arbitrarily small by increasing the number of RHONN high-order connections. A totally novel learning law is proposed that not only is convergent, stable, and robust, but also ensures exponential identification error convergence for any RHONN architecture. In other words, the convergence of the new learning law is independent of the number of RHONN high-order terms, and, moreover, such a convergence is the best that can be achieved. Finally, we mention that the rate of convergence depends on a design parameter γ, and we can make the rate of convergence arbitrarily large by simply increasing γ.

A special class of RHONNs, the so-called gradient RHONNs (g-RHONNs), is also examined; it is shown that the g-RHONNs possess the same approximation capabilities as the RHONNs and that the learning laws that are applicable to RHONNs are also applicable, with appropriate modifications, to the g-RHONNs as well. Moreover, it is shown that g-RHONNs are globally stable dynamical systems, and moreover, they remain stable when either deterministic or stochastic disturbances concatenate their dynamics. The approximation capabilities of g-RHONNs, in connection with their stability and robustness, are shown to have important implications for the fabrication of general-purpose analog hardware, stability analysis of nonlinear systems, etc.

The capabilities of RHONNs and HONNs to solve tasks other than system identification are also examined. A simplified g-RHONN architecture (the so-called Boltzmann g-RHONN) is shown to be capable of estimating unknown probability distributions. RHONNs and g-RHONNs, together with the exponentially convergent learning law, are also shown to be capable of performing spatiotemporal pattern recognition tasks as well as identifying nonlinear systems whose dynamics are corrupted by multiplicative stochastic noise. A simplified HONN structure is shown to be capable of estimating the characteristics of unknown surfaces that are in contact with the end effector of a robot manipulator. Finally, we briefly present some recent results regarding the application of HONNs in the universal stabilization of general nonlinear systems.

We conclude this section by presenting some of the notations that are used throughout this chapter. I denotes the identity matrix. $|\cdot|$ denotes the usual Euclidean norm of a vector. In the case where x is a scalar, $|x|$ denotes its absolute value. If A is a matrix, then $\|A\|$ denotes the square root of the maximum eigenvalue of the matrix A^*A. Now let $f(t)$ be a vector function of time. Then

$$\|f\|_2 := \left(\int_0^\infty |f(\tau)|^2 d\tau \right)^{1/2}$$

and

$$\|f\|_\infty := \sup_{t \geq 0} |f(t)|.$$

We will say that $f \in \mathcal{L}_2$ when $\|f\|_2$ is finite; similarly, we will say that $f \in \mathcal{L}_\infty$ when $\|f\|_\infty$ is finite.

II. RHONNs AND g-RHONNs

Recurrent neural network (RNN) models are characterized by a two-way connectivity between units. This distinguishes them from feedforward neural networks, where the output of one unit is connected only to units in the next layer. In the simple case, the state history of each unit or neuron is determined by a differential equation of the form

$$\dot{x}_i = -a_i x_i + \sum_j w_{ij} y_j, \tag{1}$$

where x_i is the state of the ith neuron, a_i are constants, w_{ij} is the synaptic weight connecting the jth input to the ith neuron, and y_j is the jth input to the above neuron. Each y_j is either an external input or the state of a neuron passed through a sigmoidal function, that is, $y_j = S(x_j)$, where $S(\cdot)$ is a sigmoidal nonlinearity.

The dynamic behavior and stability properties of neural network models of the form (1) have been extensively studied by Hopfield, as well as other researchers [1–4]. These studies showed encouraging results in application areas such as associative memories, but they also revealed limitations of the simple model described by (1).

In a recurrent second-order neural network the total input to the neuron is a linear combination not only of the components y_j, but also of their products $y_j y_k$. Moreover, one can pursue this line and include higher order interactions represented by triplets $y_j y_k y_l$, quadruplets, etc. This class of neural networks forms a recurrent higher order neural network (RHONN).

Consider now a RHONN consisting of n neurons and m inputs. The state of each neuron is governed by a differential equation of the form

$$\dot{x}_i = -a_i x_i + \left[\sum_{k=1}^{L} w_{ik} \prod_{j \in I_k} y_j^{d_j(k)} \right], \tag{2}$$

where $\{I_1, I_2, \ldots, I_L\}$ is a collection of L nonordered subsets of $\{1, 2, \ldots, m + n\}$su, a_i, b_i are real coefficients, w_{ik} are the (adjustable) synaptic weights of the neural network, and $d_j(k)$ are nonnegative integers. The state of the ith neuron is again represented by x_i, and $y = \left[y_1, y_2, \ldots, y_{m+n} \right]^T$ is the vector consisting of inputs to each neuron, defined by

$$
y = \begin{bmatrix} y_1 \\ \vdots \\ y_n \\ y_{n+1} \\ \vdots \\ y_{m+n} \end{bmatrix} = \begin{bmatrix} S(x_1) \\ \vdots \\ S(x_n) \\ u_1 \\ \vdots \\ u_m \end{bmatrix}, \tag{3}
$$

where $u = [u_1, u_2, \ldots, u_m]^T$ is the external input vector to the network. The function $S(\cdot)$ is a monotone increasing, differentiable sigmoidal function of the form

$$
S(x) = \alpha \frac{1}{1 + e^{-\beta x}} - \gamma, \tag{4}
$$

where α, β are positive real numbers and γ is a real number. In the special case that $\alpha = \beta = 1$, $\gamma = 0$, we obtain the logistic function, and by setting $\alpha = \beta = 2$, $\gamma = 1$, we obtain the hyperbolic tangent function; these are the sigmoidal activation functions most commonly used in neural network applications.

We now introduce the L-dimensional vector z, which is defined as

$$
z = \begin{bmatrix} z_1 \\ z_2 \\ \vdots \\ z_L \end{bmatrix} = \begin{bmatrix} \prod_{j \in I_1} y_j^{d_j(1)} \\ \prod_{j \in I_2} y_j^{d_j(2)} \\ \vdots \\ \prod_{j \in I_L} y_j^{d_j(L)} \end{bmatrix}, \tag{5}
$$

and hence the RHONN model (2) is rewritten as

$$
\dot{x}_i = -a_i x_i + \sum_{k=1}^{L} w_{ik} z_k. \tag{6}
$$

Moreover, if we define the adjustable parameter vector as $W_i := [w_{i1} \ w_{i2} \ \ldots \ w_{iL}]^T$, then (2), and hence (6), become

$$
\dot{x}_i = -a_i x_i + W_i^T z. \tag{7}
$$

The vectors $\{W_i: i = 1, \ldots, n\}$ represent the adjustable weights of the network, and the coefficients $\{a_i: i = 1, \ldots, n\}$ are part of the underlying network architecture and are fixed during training. To guarantee that each neuron x_i is bounded-input bounded-output (BIBO) stable, we will assume that each a_i is positive. It is pointed out that in the special case of a continuous-time Hopfield model [1], we have $a_i = 1/R_i C_i$, where $R_i > 0$ and $C_i > 0$ are the resistance and capacitance at the ith node of the network, respectively.

The dynamic behavior of the overall network is described by expressing (7) in vector notation as

$$\dot{x} = Ax + W^T z, \tag{8}$$

where $x = [x_1 \ldots x_n]^T$, $W = [W_1 \ldots W_n] \in \mathcal{R}^{L \times n}$, and $A := \text{diag}\{-a_1, -a_2, \ldots - a_n\}$ is an $n \times n$ diagonal matrix. Because each a_i is positive, A is a stability matrix. Although it is not explicitly written, the vector z is a function of both the network state x and the external input u.

Another class of RHONN models (which will be called *gradient RHONNs* or simply *g-RHONNs* to distinguish them from the RHONN model (8); the reason we call these neural networks *gradient* will be made clear in a later paper) whose dynamics are governed by the following differential equation:

$$\dot{x}_i = -b_i(x_i, \gamma_i) - \sum_{k=1}^{L} w_{ik} \frac{d_i(k)}{S_i} \prod_{j \in I_k} S_j^{d_j(k)} = \mathcal{N}_i(x), \tag{9}$$

where $I_k, w_{ik}, d_j(k)$ are defined as previously. The function $b_i(\cdot)$ is a smooth function of x_i that depends on a vector of parameters $\gamma_i \in \mathfrak{R}^{q_i}$, where q_i are positive integers. The assumptions that the functions $b_i(\cdot)$ must satisfy will be analyzed in the next sections. In our case we have chosen, for simplicity, $a_i(x_i) = 1$. The state of the ith neuron is again represented by x_i, and $S = [S_1, S_2, \ldots, S_n]^T$ is the vector consisting of all inputs to each neuron, defined by

$$S = \begin{bmatrix} S_1 \\ \vdots \\ S_n \end{bmatrix} = \begin{bmatrix} S(x_1) \\ \vdots \\ S(x_n) \end{bmatrix}, \tag{10}$$

where $S(\cdot)$ is given by (4), with $\alpha = \beta = 1$ and $\gamma = 0$. Note that in the g-RHONN case there are no external inputs to the network; however, all of the results presented can be easily extended to the case where there are constant inputs applied to the network. Moreover, in this chapter we will concentrate our attention on the special class of g-RHONNs of the form (9), the synaptic weights of which satisfy the symmetry property, that is, their synaptic weights are restricted to satisfy the following condition:

$$w_{ik} = w_{jk}, \qquad \forall i, j \in \{1, \ldots, n\}. \tag{H1}$$

It can easily be seen that the above assumption is sufficient to guarantee that g-RHONNs are gradient systems. Furthermore, in this chapter we will concentrate on the case where the powers $d_j(k)$ are selected in such a way that the following relation always holds:

$$d_i(k) = d_j(k) \geq 2, \qquad \forall i, j \in \{1, \ldots, n\}, \qquad \forall k \in \{1, \ldots, L\}. \tag{H2}$$

Similar to the RHONN case, we introduce the L-dimensional vectors z and W, defined, respectively, as

$$z^{(i)} = \begin{bmatrix} z_1^{(i)} \\ z_2^{(i)} \\ \vdots \\ z_L^{(i)} \end{bmatrix} = \begin{bmatrix} -\frac{d_i(1)}{y_i} \prod_{j \in I_1} y_j^{d_j(1)} \\ -\frac{d_i(2)}{y_i} \prod_{j \in I_2} y_j^{d_j(2)} \\ \vdots \\ \frac{d_i(L)}{y_i} \prod_{j \in I_L} y_j^{d_j(L)} \end{bmatrix}, \tag{11}$$

$$W = \begin{bmatrix} W_1 \\ W_2 \\ \vdots \\ W_L \end{bmatrix} = \begin{bmatrix} w_{i1} \\ w_{i2} \\ \vdots \\ w_{iL} \end{bmatrix}. \tag{12}$$

Then the g-RHONN model (9) can be written as

$$\dot{x}_i = -b_i(x_i, \gamma_i) + W^T z^{(i)}, \tag{13}$$

or, in vector notation,

$$\dot{x} = B(x, \gamma) + W^T z(x), \tag{14}$$

where $x = [x_1 \ldots x_n]^T$ and $z = [z^{(1)}, \ldots, z^{(n)}]^T \in \Re^{L \times n}$, and

$$B(x, \gamma) := \text{diag}\{-b_1(x_1, \gamma_1), -b_2(x_2, \gamma_2), \ldots, -b_n(x_n, \gamma_n)\}.$$

III. APPROXIMATION AND STABILITY PROPERTIES OF RHONNs AND g-RHONNs

Consider now the problem of approximating a general nonlinear dynamical system by a RHONN. The input–output behavior of the system to be approximated is described by

$$\dot{\chi} = F(\chi, u), \tag{15}$$

where $\chi \in \mathcal{R}^n$ is the state of the system, $u \in \mathcal{R}^m$ is the input to the system, and $F : \mathcal{R}^{n+m} \mapsto \mathcal{R}^n$ is a smooth vector field defined on a compact set $\mathcal{Y} \subset \mathcal{R}^{n+m}$.

The approximation problem consists of determining whether, by allowing enough higher order connections, there exist weights W such that the RHONN model approximates the input–output behavior of an arbitrary dynamical system of the form (15).

To have a well-posed problem, we assume that F is continuous and satisfies a local Lipschitz condition such that (15) has a unique solution (in the sense of Carathèodory [5]) and $(\chi(t), u(t)) \in \mathcal{Y}$ for all t in some time interval $J_T :=$ $\{t: 0 \leq t \leq T\}$, where $\mathcal{Y}$ is a compact subset of $\mathcal{R}^{n+m}$. The interval J_T represents the time period over which the approximation is to be performed. In the sequel, $|\cdot|$ denotes the Euclidean vector norm. Based on the above assumptions, we obtain the following result [6].

THEOREM 1. *Suppose that the system* (15) *and the model* (8) *are initially at the same state* $x(0) = \chi(0)$. *Then for any* $\varepsilon > 0$ *and any finite* $T > 0$, *there exists an integer* L *and a matrix* $W^* \in \mathcal{R}^{L \times n}$ *such that the state* $x(t)$ *of the RHONN model* (8) *with* L *high-order connections and weight values* $W = W^*$ *satisfies*

$$\sup_{0 \leq t \leq T} \left| x(t) - \chi(t) \right| \leq \varepsilon.$$

A similar result has been obtained for g-RHONNs in [7]. It is worth noting that although the proof of Theorem 1 requires the well-known Stone–Weierstrass theorem [8] in the case of g-RHONNs, the g-RHONN vector fields do not satisfy the Stone–Weierstrass theorem; however, as we have proved in [7], the g-RHONN vector fields can be made arbitrarily close to some vector fields that satisfy the Stone–Weierstrass theorem. Finally, it should be remarked that the following theorem is applied only to the case where the unknown system (15) has constant inputs; thus let us assume that $u(t)$ is constant for all t. Then under the same assumptions as in Theorem 1, we have the following result [7].

THEOREM 2. *Assume that the g-RHONN satisfies assumptions* (H1), (H2). *Suppose that the system* (15) *and the g-RHONN* (9) *are initially at the same state* $x(0) = \chi(0)$. *Then for any* $\varepsilon > 0$ *and any finite* $T > 0$, *there exists an integer* L *and a set of weights* w_{ik}^* *satisfying the symmetry property* (H1), *such that the state* $x(t)$ *of the g-RHONN model* (9) *with* L *high-order connections and weight values* $w_{ik} = w_{ik}^*$ *satisfies*

$$\sup_{0 \leq t \leq T} \left| x(t) - \chi(t) \right| \leq \varepsilon.$$

The above two theorems prove that if sufficiently a large number of higher order connections are allowed in the RHONN or g-RHONN models, then it is possible to approximate any dynamical system to any degree of accuracy. This is strictly an existence result; it does not provide any constructive method for obtaining the correct weights W^*. In the next section we consider the learning problem of adaptively adjusting the weights such that the RHONN (g-RHONN) model identifies general dynamical systems.

A. STABILITY AND ROBUSTNESS PROPERTIES OF g-RHONNs

In this subsection we examine the stability and robustness properties of the g-RHONN (9). Following the same guidelines as in [4], we select the following Lyapunov function for the g-RHONN (9):

$$V = \sum_{k=1}^{L} w_{ik} \prod_{j \in I_k} S_j^{d_j(k)} + \sum_{i=1}^{n} \int_0^{x_i} b_i(\eta, \gamma_i) S'(\eta) \, d\eta. \tag{16}$$

Taking the time derivative of V along the solutions of (9) and assuming that the antiderivative of $b_i(x_i, \gamma_i) S'(\eta)$ is zero at $x_i = 0$, we readily obtain

$$\dot{V} = -\sum_{i=1}^{n} S_i' \left[b_i(x_i, \gamma_i) + \sum_{k=1}^{L} w_{ik} \frac{d_i(k)}{S_i} \prod_{j \in I_k} S_j^{d_j(k)} \right]^2 = -\sum_{i=1}^{n} S_i' \mathcal{N}_i^2(x) \le 0.$$

The equilibrium points of the g-RHONN are these points x^e in the state space for which

$$b_i(x_i^e, \gamma_i) + \sum_{k=1}^{L} w_{ik} \frac{d_i(k)}{S(x_i^e)} \prod_{j \in I_k} S(x_i^e)^{d_j(k)} = \mathcal{N}_i(x^e) = 0.$$

Let $\mathcal{E}$ denote the set of equilibrium points. Also let

$$\mathcal{D}(x) = \sum_{i=1}^{n} S_i' \left[b_i(x_i, \gamma_i) + \sum_{k=1}^{L} w_{ik} \frac{d_i(k)}{S_i} \prod_{j \in I_k} S_j^{d_j(k)} \right]^2. \tag{17}$$

Before we proceed to the analysis of the stability and robustness analysis of the g-RHONN, we need the following assumptions about the functions $b_i(\cdot)$:

S1. $\int_0^{x_i} b_i(\eta, \gamma_i) S'(\eta) d\eta \to +\infty$ as $|x_i| \to \infty$.
S2. $|b_i(x_i, \gamma_i) S'(x_i)| \to \infty$ as $|x_i| \to \infty$.
S3. For any positive constants a_i, d_i, there exist γ_i such $\frac{1}{2} b_i^2(x_i, \gamma_i) > |b_i'(x_i, \gamma_i)| a_i + |b_i(x_i, \gamma_i)| d_i$, where $b_i'(\cdot) = (\partial b_i / \partial x_i)(\cdot)$.
S4. There exists a positive number c_1 such that
$c_1 \sum_{i=1}^{n} \int_0^{x_i} b_i(\eta, \gamma_i) S'(\eta) d\eta \ge \frac{1}{2} \sum_{i=1}^{n} S_i' b_i^2(x_i, \gamma_i)$.

Note that from assumption S2 we directly obtain $|b_i(x_i, \gamma_i)| \to \infty$ as $|x_i| \to \infty$. Consider now the next lemma [7].

LEMMA 1. *The following statements are true:*

(a) $V(x) \to +\infty$ iff $|x| \to +\infty$.
(b) $\mathcal{D}(x) \to +\infty$ iff $|x| \to +\infty$.
(c) $\mathcal{D}(x) = 0 \Leftrightarrow x \in \mathcal{E}$.

It is not difficult, using the fact that $\dot{V} \leq 0$ and the results of the above lemma, to establish the stability of the g-RHONN dynamics, in the sense that all of the so-lutions of the g-RHONN remain bounded, and moreover, they converge asymptot-ically to an equilibrium point. We will avoid such an analysis here. The interested reader is referred to [4] for such an analysis. However, we must mention that the g-RHONNs used in [4] are slightly different from the ones analyzed here. For in-stance, in the g-RHONNs of [4], the g-RHONN solutions are always nonnegative. Obviously this is not the case here.

Now let ν be a positive scalar. Then $\mathcal{B}(\nu)$ denotes the ball centered at the origin with radius ν. The next proposition [7] shows that the g-RHONNs remain stable in the case where bounded deterministic signals disturb their dynamics.

PROPOSITION 1. *Let the g-RHONN dynamics be perturbed as follows:*

$$\dot{x}_i = -b_i(x_i, \gamma_i) - \sum_{k=1}^{L} w_{ik} \frac{d_i(k)}{S_i} \prod_{j \in I_k} S_j^{d_j(k)} + \xi_i(t) = \mathcal{N}_i(x) + \xi_i(t), \quad (18)$$

where $\xi \in \Re^n$ is a bounded signal. Let $\bar{\xi} = \sup_t |\xi(t)|^2$. Then, if $b_i(\cdot)$ satisfies assumptions S1, S2, then

(a) *The solutions of the perturbed g-RHONN (18) are bounded for each t, and moreover the term $\mathcal{D}(x)$ converges to $\mathcal{B}(\bar{\xi})$.*
(b) *There exists a positive constant λ such that at each t,*

$$\int_{t_0}^{t} \tfrac{1}{2} \mathcal{D}(\tau) \, d\tau \leq \lambda + \tfrac{1}{2} \int_{t_0}^{t} |\xi(\tau)|^2 \, d\tau.$$

(c) *In the special case where $\xi \in \mathcal{L}_2$, the solutions of the g-RHONN converge to an equilibrium point asymptotically.*

The above proposition proves the stability of solutions of the g-RHONN in the case where bounded signals disturb its dynamics.[1] However, in many cases we are interested in examining the stability properties of g-RHONNs whose dynamics are disturbed by signals that are not bounded. For instance, in the case where the signals $\xi(t)$ in (18) are Gaussian white noise processes, the above proposition is no longer valid. For this reason we investigate the following perturbed version of the g-RHONN (9):

$$dx_i = -b_i(x_i, \gamma_i)dt - \sum_{k=1}^{L} w_{ik} \frac{d_i(k)}{S_i} \prod_{j \in I_k} S_j^{d_j(k)} dt + \sum_{j=1}^{m} f_{ij}(x) \, d\zeta_j, \quad (19)$$

where $f_{ij}(\cdot)$ are bounded and smooth functions, and ζ is a standard m-dimensional Wiener process with $E\{\dot{\zeta}(t)\} = 0$ and $E\{\dot{\zeta}(t)\dot{\zeta}(\tau)\} = \sqrt{2}I\delta(t-\tau)$,

---

[1]It is mentioned that part (c) of the above proposition is a corollary of Theorem 2 of [9].

where I is the $L \times L$ identity matrix. The stochastic differential equation (19) can be written in a form more familiar to engineers:

$$\dot{x}_i = -b_i(x_i, \gamma_i) - \sum_{k=1}^{L} w_{ik} \frac{d_i(k)}{S_i} \prod_{j \in I_k} S_j^{d_j(k)} + \sum_{j=1}^{m} f_{ij}(x)\dot{\zeta}_j, \qquad (20)$$

where $\dot{\zeta}$ is an m-dimensional white noise process. The next proposition [7] establishes the stability of the perturbed g-RHONN (19).

PROPOSITION 2.   *Consider the perturbed g-RHONN* (19). *Assume that $f_{ij}(\cdot)$ are bounded functions, and that $b_i(\cdot)$ satisfy assumptions* S1–S4. *Then there exist positive numbers c_i, $i = 1, 2$, such that*

$$\sup_{0 \le t < \infty} \Pr\{\mathcal{V}(t) \ge C\} < \frac{\mathcal{V}(0) + (c_2/c_1)}{C}.$$

Several remarks are in order:

• g-RHONNs are gradient stable dynamical systems. Moreover, they remain stable when either deterministic or stochastic disturbances concatenate their dynamics. This fact is very important, especially when one wishes to implement g-RHONNs in analog hardware. It is well known that stable dynamical systems may lose their stability properties, or even more so that their trajectories may escape to infinity, even when small disturbances perturb their dynamics. Therefore, when stable systems are implemented in hardware, the unavoidable disturbances and perturbations (e.g., thermal noise, component malfunction) that are inherent in hardware implementations may distort or destroy the hardware. Fortunately, this is not the case for the g-RHONNs; their robustness under deterministic and stochastic disturbances guarantee the efficient behavior of their hardware implementation.

• The fact that a large class of dynamical systems can be approximated arbitrarily closely by a stable and robust gradient g-RHONN is a very important result. From a theoretical point of view, this result states that for every dynamical system belonging to this class there is another one very close to it, which is not simply a stable one, but also is a gradient dynamical system; gradient dynamical systems possess a certain Lyapunov function. Moreover, their stability properties (equilibria, regions of attraction, etc.) can be easily detected because of the special properties that such systems possess. On the other hand, the stability properties of general dynamical systems are very difficult to analyze, even in the case where these systems possess a certain Lyapunov function (see, e.g., [10] and the references therein). In fact, the methods for analyzing the stability properties of dynamical systems require both symbolic manipulations and computational procedures of high computational burden (see, e.g., [10]). Therefore, the results of this section can be utilized to detect and analyze the stability properties of either known or unknown dynamical systems as follows: (i) approximate the dynamical system by a g-RHONN system and (ii) detect the stability properties of the g-RHONN (this can be easily done by using the gradient system properties); the stability properties of the g-RHONN will be approximately equal to those of the system under study. Of course, such a procedure will be successful only in the case where the dynamical system possesses no periodic or unbounded solutions.

• From a nonlinear circuit-theoretical point of view, the results of this section can be utilized to give a definitive answer to the following problem [11]: Given a state equation $\dot{\chi} = F(\chi)$, does there exist a stable and robust nonlinear circuit that realizes this equation? Moreover, the results of this section seem to give a definitive answer to another more general question: Does there exist a general-purpose nonlinear circuit, which for every state equation $\dot{\chi} = F(\chi)$, is able to realize this equation? This question seems to be answered if one simply implements in hardware the g-RHONN together with an appropriate learning law like the ones proposed in the next section. Finally, we mention that the g-RHONN possesses certain advantages over other general-purpose architectures, like the pseudoreciprocal or reciprocal circuits, in the sense that the later could exhibit complicated dynamics, including chaos [11], whereas g-RHONNs guarantee stable and robust performance.

IV. CONVERGENT LEARNING LAWS

Consider now the problem of identifying a general nonlinear dynamical system of the form (15) by a RHONN (g-RHONN). The identification problem (once we have selected the high-order terms for the neural network) consists of constructing an appropriate learning algorithm for adjusting the weights such that, after the termination of the algorithm, the RHONN (g-RHONN) model approximates the behavior of an arbitrary dynamical system of the form (15). For simplicity we analyze here only the learning laws for the RHONN case.[2] The learning laws for the g-RHONN case are very similar and are not described here; the interested reader is referred to [7] for more details.

To have a well-posed problem, we assume that F is continuous and satisfies a local Lipschitz condition such that (15) has a unique solution (in the sense of Carathèodory [5]) and $\chi(t) \in \mathcal{Y}$ for all t in some time interval $J_T := \{t: 0 \leq t \leq T\}$, where $\mathcal{Y}$ is a compact subset of $\mathcal{R}^n$. The interval J_T represents the time period over which the approximation is to be performed.

In formulating the problem it is noted that by adding and subtracting $a_i \chi_i + W_i^{*T} z(\chi, u)$, the dynamic behavior of each state of the system (15) can be ex-

---

[2]In fact, the learning laws for the g-RHONN case are very similar to those for the RHONN; the only difference is that, in the g-RHONN case, condition (H1) must not be violated.

pressed by a differential equation of the form

$$\dot{\chi}_i = -a_i \chi_i + W_i^{*T} z(\chi, u) + v_i(t), \tag{21}$$

where the modeling error $v_i(t)$ is given by

$$v_i(t) := F_i\big(\chi(t), u(t)\big) + a_i \chi(t) - W_i^{*T} z\big(\chi(t), u(t)\big). \tag{22}$$

The function $F_i(\chi, u)$ denotes the ith component of the vector field $F(\chi, u)$, and the unknown optimal weight vector W_i^* is defined as the value of the weight vector W_i that minimizes the $\mathcal{L}_\infty$-norm difference between $F_i(\chi, u) + a_i \chi$ and $W_i^T z(\chi, u)$ for all $(\chi, u) \in \mathcal{Y} \subset \mathcal{R}^{n+m}$, subject to the constraint that $|W_i| \le M_i$, where M_i is a large design constant. The region $\mathcal{Y}$ denotes the smallest compact subset of $\mathcal{R}^{n+m}$ that includes all of the values that (χ, u) can take, that is, $(\chi(t), u(t)) \in \mathcal{Y}$ for all $t \ge 0$. Because by assumption $u(t)$ is uniformly bounded and the dynamical system to be identified is BIBO stable, the existence of such $\mathcal{Y}$ is ensured. It should be pointed out that in our analysis we do not require knowledge of the region $\mathcal{Y}$, or upper bounds for the modeling error $v_i(t)$.

In summary, for $i = 1, \ldots, n$, the optimal weight vector W_i^* is defined as

$$W_i^* := \arg \min_{|W_i| \le M_i} \left\{ \sup_{(\chi, u) \in \mathcal{Y}} \left| F_i(\chi, u) + a_i \chi - W_i^T z(\chi, u) \right| \right\}. \tag{23}$$

We have developed two different types of learning algorithms for adjusting the weights of RHONNs (g-RHONNs). The first type of these algorithms is based on classical robust adaptive algorithms, which are well-known in the parameter estimation and adaptive control literature. The second type is a new class of learning algorithms which, as we will see in the sequel, possess many interesting properties that classical robust adaptive algorithms do not possess.

A. ROBUST ADAPTIVE LEARNING LAWS

The following lemma is useful in the development of the adaptive identification scheme presented in this subsection.

LEMMA 2.   *The system described by*

$$\dot{\chi}_i = -a_i \chi_i + W_i^{*T} z(\chi, u) \qquad \chi_i(0) = \chi_i^0 \tag{24}$$

can be expressed as

$$\dot{\zeta}_i = -a_i \zeta_i + z \qquad \zeta_i(0) = 0 \tag{25}$$

$$\chi_i = W_i^{*T} \zeta_i + e^{-a_i t} \chi_i^0. \tag{26}$$

Using Lemma 2, the dynamical system described by (24) is rewritten as

$$\chi_i = W_i^{*T} \zeta_i + \epsilon_i \qquad i = 1, 2, \ldots, n, \tag{27}$$

where ζ_i is a filtered version of the vector z (as described by (25)), and $\epsilon_i :=$ $e^{a_i t} \chi_i^0$ is an exponentially decaying term that appears if the system is in a nonzero initial state. By replacing the unknown weight vector W_i^* in (27), by its estimate W_i and ignoring the exponentially decaying term ϵ_i, we obtain the RHONN model (filtered regressor RHONN),

$$x_i = W_i^T \zeta_i \qquad i = 1, 2, \ldots, n. \tag{28}$$

The exponentially decaying term $\epsilon_i(t)$ can be omitted from (28) because, as we will see later, it does not affect the convergence properties of the scheme.

Although there are many different algorithms from the robust adaptive parameter estimation literature that may be used, we chose to use a gradient algorithm, switching σ-modification as follows:

$$\dot{W}_i = \begin{cases} -\Gamma_i \zeta_i e_i & \text{if } |W_i| \leq M_i \\ -\Gamma_i \zeta_i e_i - \sigma_i \Gamma_i W_i & \text{if } |W_i| > M_i \end{cases}, \tag{29}$$

where σ_i is a positive constant chosen by the designer, Γ_i are symmetrical positive definite matrices, and $e_i := \chi_i - x_i$ denotes the identification error. The above weight adjustment law is a standard gradient learning law if W_i belongs to a ball of radius M_i. In the case where the weights leave this ball, the weight adjustment law is modified by the addition of the leakage term $\sigma_i \Gamma_i W_i$, the objective of which is to prevent the weight values from drifting to infinity. This modification is known as the *switching σ-modification* [12].

In the following theorem we use the vector notation $v := [v_1 \ldots v_n]^T$, $e :=$ $[e_1 \ldots e_n]^T$, and $\tilde{W}_i := W_i^* - W_i$ denotes the parameter estimation error, that is, the difference between the optimal ith weight vector and its actual value.

THEOREM 3.   *Consider the unknown system* (15) *and the filtered regressor RHONN model given by* (28), *the weights of which are adjusted according to* (29). *Then for $i = 1, \ldots, n$,*

  (a) $e_i, \tilde{W}_i \in \mathcal{L}_\infty$
  (b) *There exist constants λ, μ such that*

$$\int_0^t |e(\tau)|^2 \, d\tau \leq \lambda + \mu \int_0^t |h(s)v(\tau)|^2 \, d\tau,$$

  where $h(s)$ is a stable linear filter.

(c) *Moreover, if the regressor vector ζ_i is persistently exciting, that is, there exist positive scalars $\mathbf{a_j}$, $\mathbf{b_j}$ and T such that for all $t \geq 0$,*

$$\mathbf{a}_i I \leq \int_t^{t+T} \zeta_i(\tau)\zeta_i^T(\tau)\,d\tau \leq \mathbf{b}_i I, \tag{30}$$

then the parameter error $\widetilde{W}_i$ converges exponentially to the residual set

$$\underline{\mathcal{D}}_i = \left\{ \widetilde{W}_i \mid |\tilde{W}_i| \leq c\bar{\varrho}_i \right\}$$

for some $c \in \Re^+$, where $\bar{\varrho}_i = \sup_{t \geq 0} |\varrho_i(t)| =: \|\varrho_i\|_\infty$ and ϱ_i is the L-dimensional signal whose ℓth entry is defined as

$$\varrho_{i\ell} = h_{i\ell}(s)v_i,$$

where $h_{i\ell}$ is an asymptotically stable filter.

In simple words, the above theorem states that the weight adaptive law (29) guarantees that e_i and $\widetilde{W}_i$ remain bounded for all $i = 1, \ldots, n$, and furthermore, the "energy" of the state error $e(t)$ is proportional to the "energy" of the modeling error $v(t)$. In the special case where the modeling error is square integrable, that is, $v \in \mathcal{L}_2$, then $e(t)$ converges to zero asymptotically. Moreover, if some persistence of excitation conditions hold, then the synaptic weights converge very closely to their optimal weights. It is worth noting that persistence of excitation conditions are both necessary and sufficient for parameter convergence [13].

B. Learning Laws That Guarantee Exponential Error Convergence

Classical adaptive and robust adaptive schemes (like the one we presented in the previous section) are unable to ensure convergence of the identification error to zero, in the case of modeling errors. Therefore, the usage of such schemes in "black box" identification of nonlinear systems ensures, in the best case, bounded identification error. In this subsection, a new learning (adaptive) law is presented, which when applied to RHONNs, ensures that the identification error converges to zero exponentially fast, and moreover, in the case where the identification error is initially zero, it remains equal to zero during the whole identification process. This learning law has been initially proposed in [14]; the interested reader is referred to [14] for more details and for variants and modifications of the learning law presented in this subsection.

The learning law is as follows:

$$w_{ik}(t) = \vartheta_{ik}(t) + \varphi_{ik}(t) \tag{31}$$

$$\dot{\vartheta}_{ik} = -\gamma \frac{e_i n_{ik}}{\zeta_{ik}} - \sigma_i w_{ik} + \sigma_i \frac{\xi_i n_{ik}}{\zeta_{ik}} \tag{32}$$

$$\varphi_{ik}(t) = \frac{\xi_i(t) n_{ik}}{\zeta_{ik}(t)} + \eta_{ik}(t) \tag{33}$$

$$\dot{\eta}_{ik} = -\xi_i \frac{d}{dt}\left(\frac{n_{ik}}{\zeta_{ik}}\right) - \sum_{\ell=1}^{L} \frac{w_{i\ell}\dot{\zeta}_{i\ell}}{\zeta_{ik}} n_{ik}, \tag{34}$$

where

$$\sigma_i = \begin{cases} 0 & \text{if } |w_i| \le M_i \\ \left(\frac{|w_i|}{M_i} - 1\right)^q \sigma_{i0} & \text{if } M_i < |w_i| \le 2M_i \\ \sigma_{i0} & \text{if } |w_i| > 2M_i \end{cases} \tag{35}$$

n_{ik} are design constants that satisfy

$$\sum_{k=1}^{L} n_{ik} = 1, \qquad \forall i, \tag{36}$$

and γ, σ_{i0} are positive design constants. The variables ϑ_{ik}, φ_{ik}, ξ_i, η_i are auxiliary one-dimensional signals.

The next theorem [14] establishes the properties of the learning law (31)–(35):

THEOREM 4. *Consider the unknown system (15) and the filtered regressor RHONN model given by (28) the weights of which are adjusted according to (31)–(35). Moreover, assume that the sigmoidal $S(\cdot)$ is such that $S(x) > 0$ for all x.[3] Then,*

(a) *The identification error e_i converges to zero exponentially fast.*
(b) $\widetilde{W}_i$, $W_i \in \mathcal{L}_\infty$.
(c) *Moreover, if ζ_i is persistently exciting, ϕ_i, $\dot{\phi}_i$ are bounded, and M_i is chosen to be sufficiently large, then the parameter error $\widetilde{W}_i$ converges exponentially to the residual set*

$$\underline{\mathcal{D}}_{\sigma_i} = \left\{ \widetilde{W}_i \mid |\widetilde{W}_i| \le c\bar{\varrho}_i \right\}$$

---

[3]This assumption is made only for implementation reasons; in fact, it can easily be seen that such an assumption does not affect the approximation properties of the RHONN.

for some $c \in \Re^+$, where $\bar{\varrho}_i = \sup_{t \geq 0} |\varrho_i(t)| =: \|\varrho_i\|_\infty$ and ϱ_i is the L-dimensional signal whose ℓth entry is defined as

$$\varrho_{i\ell} = h_{i\ell}(s)v_i + \sigma_i \frac{\xi_i}{\zeta_{i\ell}},$$

where $h_{i\ell}$ is an asymptotically stable filter.

In fact, the above theorem simply states that the proposed learning law guarantees exponential convergence to zero (in fact, if the identification error is initially zero, then it remains zero forever), independently of the number of high-order connections. A very interesting fact is that this rapid error convergence does not destroy the learning properties of the learning scheme; the parameter convergence properties of the above learning law are very similar to the ones of the learning law presented in the previous subsection. In fact, for sufficiently small σ_i, the convergence properties of the two learning laws are almost the same.

V. THE BOLTZMANN g-RHONN

The Boltzmann g-RHONN [15] is a simplified version of the g-RHONN presented previously. Its dynamics are given by the following differential equation:

$$\dot{z}_i = -\sum_{k=1}^{L} w_{ik} \frac{d_i(k)}{z_i} \prod_{j \in I_k} z_j^{d_j(k)}, \tag{37}$$

where

- $\{I_1, I_2, \ldots, I_L\}$ is a collection of L nonordered subsets of $\{1, 2, \ldots, n\}$,
- w_{ik} are the synaptic weights of the neural network,
- $d_j(k)$ are integers, $d_j(k) \geq 0$,
- z_i is the state of the ith neuron, $i = 1, 2, \ldots, n$.

One can easily see that in the case of the Boltzmann g-RHONN, the function $S(\cdot)$ is selected to be the identity function $S(x) = x$.

As in to the g-RHONN case, we can easily see that, in the case where the synaptic weights satisfy the following identity:

$$w_{ik} = w_{jk} \overset{\triangle}{=} w_k, \qquad \forall i, j, \tag{38}$$

the neural network (37) becomes a gradient dynamical system. More precisely, if we define

$$L(z) = \sum_{k=1}^{L} w_k \prod_{j \in I_k} z_j^{d_j(k)}, \tag{39}$$

then the neural network (37) can be written as

$$\dot{z} = \nabla L(z) \Leftrightarrow dz = \nabla L(z)\,dt.$$

Note now that if we define

$$\phi_i(z) := \begin{bmatrix} \dfrac{d_i(1)}{z_i}\,\prod_{j\in I_1} z_j^{d_j(1)} \\[2ex] \dfrac{d_i(2)}{z_i}\,\prod_{j\in I_2} z_j^{d_j(2)} \\[2ex] \vdots \\[2ex] \dfrac{d_i(L)}{z_i}\,\prod_{j\in I_L} z_j^{d_j(L)} \end{bmatrix}, \tag{40}$$

and

$$w_i := \begin{bmatrix} w_{i1} \\ w_{i2} \\ \vdots \\ w_{iL} \end{bmatrix}, \tag{41}$$

then the neural network dynamics can be rewritten as

$$\dot{z}_i = -w_i^\tau \phi_i(z). \tag{42}$$

Consider now an unknown ergodic probability distribution with density $p(x)$. Then we can approximate the logarithm of $p(x)$ by a polynomial as follows:

$$\log\big(p(x)\big) = -\sum_{k=1}^{L} w_k^* \prod_{j\in I_k} x_j^{d_j(k)} + \nu(x), \tag{43}$$

where $w^* := [w_1^*, \ldots, w_L^*]^\tau$ are the optimal weights, defined by

$$w^* := \arg\min_{w}\left\{\sup_{x\in\mathcal{Y}}\left|\log(p(x)) + \sum_{k=1}^{L} w_k \prod_{j\in I_k} x_j^{d_j(k)}\right|\right\}, \tag{44}$$

where $\mathcal{Y}$ is any compact subset of $\mathcal{X}$. The term $\nu(\cdot)$ represents the modeling error term and is defined by

$$\nu(x) := \log\big(p(x)\big) + \sum_{k=1}^{L} w_k \prod_{j\in I_k} x_j^{d_j(k)}. \tag{45}$$

We note here that from the approximation properties of polynomials, we have that the modeling error term ν can be made arbitrarily small.

Consider now that the weight vectors w_i are adjusted according to the following least-squares method

$$\dot{w}_i = P_i \psi_i (x_i - w_i^\tau \psi_i), \qquad w_i(0) = 0 \tag{46}$$
$$\dot{P}_i = -P_i \psi_i \psi_i^\tau P_i, \qquad P(0) = L \cdot \mathbf{I}, \tag{47}$$

where ψ_i is a filtered version of ϕ_i and is computed as follows:

$$\dot{\psi}_i = \phi_i(x), \qquad \psi(0) = 0. \tag{48}$$

Note that the matrix ϕ is a function of the observations x and not a function of the neural network states z.

The next theorem [15] establishes the convergence properties of the learning law (46)–(48).

THEOREM 5.   *Consider a signal source that randomly transmits signals x according to an ergodic but unknown distribution with density $p(x)$. Moreover, consider the neural network (42) and the learning laws (46)–(48). Define the neural network energy functions:*

$$L_i(w_i, z) = \sum_{k=1}^{L} w_{ik} \prod_{j \in I_k} z_j^{d_j(k)}, \qquad i \in \{1, \ldots, n\}. \tag{49}$$

Moreover assume that

(A2) *There exist real numbers $\varepsilon_i \geq 0$ such that*

$$\int_0^t \|\mu_i(s)\| ds \leq \varepsilon_i r_i(t), \qquad t \geq 0, \tag{50}$$

where $\mu_i := \partial v / \partial x_i$,

$$r_i(t) = e + \int_0^t \|\phi_i(x(s))\|^2 ds, \qquad e = 2.718282\ldots. \tag{51}$$

Then $-L_i(t)$ converges to $\log(p(x))$ in the sense that[4]

$$\limsup_{t \to \infty} \left\{ \sup_{x \in \mathcal{Y}} \|L_i(w_i(t), x) + \log(p(x))\| \right\} \leq A_i k_i \varepsilon_i + B_i, \qquad \text{a.s.,} \tag{52}$$

---

[4]We say that $\limsup_{t \to \infty} \|\chi(t)\| < \epsilon$ almost surely (a.s.), where $\chi(t)$ is a stochastic process, if $\Pr\{\limsup_{t \to \infty} \|\chi(t)\| < \epsilon\} = 1$. We mention that the a.s. convergence criterion is one of the strongest criteria for stochastic convergence.

where A_i, B_i are bounded nonnegative constants, and

$$k_i = \limsup_{t \to \infty} \frac{r_i(t)}{\lambda_i^{\min}(t)} < \infty, \tag{53}$$

where $\lambda_i^{\min}(t)$ denotes the minimum eigenvalue of $P_i^{-1}(t)$.

In fact, what the above theorem states is that in the case where the modeling error term ν is small enough, and the signals ϕ_i are persistently exciting, the energy functions of the neural network will converge very closely to the natural logarithm of the unknown probability density; the notion of "persistence of excitation" (see, e.g., [13]) is well-known in adaptive algorithms. Intuitively speaking, the signals ϕ_i are persistently exciting whenever their "energy" $\int_0^t \|\phi_i(s)\|^2 \, ds$ is not negligible; however, in the case where the energy of those signals is negligible, the learning problem does not make sense. In other words, whenever the signals ϕ_i are not persistently exciting, any learning algorithm that may be used will fail.

In simple words, if we provide the g-RHONN with a sufficiently large number of high-order connections, and we train it by using the learning laws (46)–(48) then, provided that the persistence of excitation assumption holds, after a large training period the following relation will be valid:

$$-L_i(w_i, x) \approx \log\left(p(x)\right)$$

for all of the energy functions of the neural network and for all x in $\mathcal{Y}$, or, equivalently,

$$p(x) \approx e^{-L_i(w_i, x)}.$$

Remark 1. Note that each of the vectors w_i consists of an estimate of the optimal vector w^*. Therefore, we do not have to update all of the vectors w_i to obtain an estimation of the vector w^*; it suffices to update only one of the vectors w_i by using the learning laws (46)–(48) to obtain a reliable estimation of the vector w^*. On the other hand, in the case where we update all of the vectors w_i, it is natural to ask which of the w_i is the best estimation of the vector w^*. Of course, in the simplest case we may consider the average $\bar{w} = 1/n \sum_{i=1}^n w_i$ as an estimation of the vector w^*. It is not difficult to see that in this case the limsup of the $\|\bar{w}(t) - w^*\|$ will be a.s. less than or equal to $1/n \sum_{i=1}^n \alpha_i k_i \varepsilon_i$. Another approach is the following: by assuming that the modeling error term $\nu(x)$ is negligible, we can see, using the results of [16], that[5]

$$\|w_i(t) - w_k^*\| = O\left(\frac{\log(r_i(t))}{\lambda_i^{\min}(t)}\right).$$

[5]We say that the positive function $x(t)$ is $O(y(t))$, where $y(t)$ is a positive function, if there is a constant K such that $x(t) < Ky(t)$ for all t.

In this case, at each time we can use for the estimation of w^* the $w_j(t)$ for which

$$\frac{\log(r_j(t))}{\lambda_j^{\min}(t)} = \arg \min_{i \in \{1,\ldots,n\}} \left(\frac{\log(r_i(t))}{\lambda_i^{\min}(t)} \right).$$

VI. OTHER APPLICATIONS

A. ESTIMATION OF ROBOT CONTACT SURFACES

When it is desired to design robot manipulators that will operate in uncertain and unknown environments, it is necessary to provide them with appropriate devices and algorithms that are capable of estimating and learning unknown surfaces and objects that are in contact with the robots' end-effectors. One solution to the surface estimation problem is to use computer vision systems. However, the usage of vision systems considerably increases the cost and the complexity of the application, and, on the other hand, the current vision systems do not perform efficiently when either the environment is very noisy or the unknown surfaces are complex enough. Another possible solution is to use force or tactile sensors to estimate the contact force that is applied between the end-effector and the constraint surface. In fact, it can be shown that the contact force F is given by $F = \Lambda(\theta, \dot{\theta}, \tau, \vartheta)$, where $\theta, \dot{\theta}$ are the vectors of robot joint angular positions and velocities, respectively; τ is the vector of joint torques; ϑ is the vector of the unknown surface shape parameters; and $\Lambda(\cdot)$ is a nonlinear function. Therefore, if $\theta, \dot{\theta}, \tau$, and F are available for measurement, we can apply an extended Kalman filter or any other nonlinear parameter estimation method to estimate the unknown surface parameters ϑ. In the simplest case, the parameter vector ϑ can be estimated through the following gradient estimation algorithm:

$$\dot{\hat{\vartheta}} = -\Pi \frac{\partial \Lambda}{\partial \vartheta}\bigg|_{\vartheta=\hat{\vartheta}} \left(F - \Lambda(\theta, \dot{\theta}, \tau, \hat{\vartheta}) \right),$$

where $\hat{\vartheta}$ denotes the estimated value of ϑ and Π is a positive definite (possibly time-varying) matrix. The above methodology, slightly modified, has been followed by Bay and Hemami [17] to solve the unknown surface estimation problem. However, it is a well-known fact that nonlinear parameter estimation algorithms do not guarantee the convergence of $\hat{\vartheta}$ to the actual parameter vector ϑ; this is due to the fact that, because of the nonlinear dependence of the function $\Lambda(\cdot)$ on the parameter vector ϑ, there are many local minima of the error functional that the parameter estimation algorithm minimizes. Therefore the nonlinear parameter estimation algorithm might become trapped in a local minimum, and thus, the estimation procedure will fail. Even worse, nonlinear parameter estimation algorithms may become unstable, which means that the estimated values $\hat{\vartheta}$ will reach unacceptably large values (theoretically infinite).

In [18] we showed that, using HONN approximators, we can formulate the unknown surface estimation problem as a linear parameter estimation problem, and therefore a linear parameter estimation algorithm can be applied. The linear parameter estimation algorithms, contrary to nonlinear ones, ensure stability and convergence. Furthermore, we proposed a new learning architecture that is capable of estimating the unknown surface shape parameters, even in the case where the contact force F is not available for measurement. This learning architecture consists of a linear parameter estimation algorithm, as in the case where the contact force is available for measurement, and an appropriate approximator, which approximates (estimates) the unmeasured force. As we showed, the whole scheme is globally stable and convergent. In fact, the learning algorithms for training those neural networks are very similar to the ones presented in Section IV.

The key idea used in [18] is that during the constrained motion, the second time derivative of the constraint equation is zero. In other words, if $\phi(x) = $ constant is the mathematical description of the constraint surface, then during the constrained motion, the following relation must be valid:

$$\ddot{\phi}(x) = 0.$$

If we approximate the constraint surface $\phi(x)$ by using a HONN, and if we assume that the constraint forces are available for measurement, then the above differential equation reduces to an algebraic equation that is linear with respect to the (unknown) parameters of the surface. Therefore a linear parameter estimation algorithm can be directly applied. When the contact forces are not available for measurement, the robot dynamics (which are assumed to be completely known) are utilized to obtain, via another neural network approximator, a reliable estimation of the constraint forces, which in turn is used in the linear parameter estimation algorithm. It is worth noting, that when the forces are not available for measurement, the proposed learning strategy requires the end-effector to be provided with a cheap sensor, which detects if the end-effector is contact with the surface.

Whereas the approach of [17] is applicable to surfaces that are modeled by quadratic functions in Cartesian coordinates, our methodology is applicable to any surface that can be described by a smooth function of the Cartesian coordinates. Thus our approach includes as special cases the quadratic function approach of [17] as well as simple surface objects such as ellipsoids, spheres, etc. It is worth noting that to be capable of estimating the unknown surface, an appropriate control procedure must be applied that will be capable of keeping the end-effector in contact with the surface for as long a time as possible and bringing the end-effector back to the surface in the case where contact has been lost. The problem of constructing such a control procedure as well as other practical aspects are discussed in [18], and the interested reader is referred there for further details.

B. RHONNs FOR SPATIOTEMPORAL PATTERN RECOGNITION AND IDENTIFICATION OF STOCHASTIC DYNAMICAL SYSTEMS

In [19] the authors have shown that RHONNs (or g-RHONNs), the weights of which are adjusted according to the exponential error convergent learning law presented in Section IV.B, are capable of identifying stochastic nonlinear systems of the form

$$d\chi = F(\chi)\,dt + G(\chi)\,d\xi, \qquad (\Sigma)$$

where $\chi \in \Re^n$ denotes the n-dimensional incoming spatiotemporal pattern; $F(\cdot)$ and $G(\cdot)$ are smooth, bounded and locally Lipschitz, but otherwise unknown functions; and ξ is an m-dimensional standard Wiener process. The stochastic dynamical system (Σ) can be written in a more convenient form:

$$\dot{\chi} = F(\chi) + G(\chi)\dot{\xi}, \qquad (\Sigma')$$

where $\dot{\xi}$ denotes an m-dimensional white Gaussian process, with $E\{\dot{\xi}(t)\} = 0$ and $E\{\dot{\xi}(t)\dot{\xi}^T(\tau)\} = \sqrt{2}I\delta(t - \tau)$, where I is the identity matrix. The stochastic dynamical system (Σ) (known as the Ito process) is quite general. In fact, the linear time invariant additive noise model $\dot{\chi} = A\chi + B\dot{\xi}$, as well as the nonlinear additive noise model $\dot{\chi} = F(\chi) + B\dot{\xi}$, which are used in many engineering problems, are simple cases of the model (Σ). The main difficulty that arises when one tries to extend the deterministic calculus to the theory of stochastic systems of the form (Σ) is that the well-known chain differentiation rule is no longer applicable. Instead, the so-called Ito formula must be applied in those cases.

In this [19] we showed that the learning law of Section IV.B preserves the same properties as those of the deterministic case. Among these properties, the most important is the following: the identification (or prediction) error between the estimated (by the neural network) and the actual signal value converges to zero exponentially fast. This property, combined with the fact that the identification error converges exponentially to zero for any preselected RHONN or g-RHONN (i.e., the convergence is independent of the number of high-order terms), has a threefold significance:

- Despite the conventional parameter estimation and adaptive algorithms, the proposed architecture guarantees exponential error convergence, which is the best convergence that can be achieved.

- In the existing approaches for system identification using neural networks, the number of neurons or high-order connections [6, 7] must be sufficiently large to guarantee convergence of the error to zero. If this is not the case, the error will never converge to zero (except in some extreme cases). Contrary to these methods, in the proposed architecture, the error converges to zero for any selection of high-order connections.
- At last, we mention that an exact (in the sense that the identification error converges to zero) solution to the problem of identification of unknown systems of the form (Σ) is, to the best of our knowledge, proposed for first time in the signal processing, information theory and adaptive and learning systems theory literature.

Finally, we mention that RHONNs and g-RHONNs whose weights are adjusted according to the learning law of Section IV.B can also be used for spatiotemporal pattern recognition problems because, as we argued in [19], the spatiotemporal pattern recognition problem is a special class of the problem of identification of systems whose dynamics are concatenated by multiplicative stochastic disturbances (i.e., systems of the form (Σ)).

C. UNIVERSAL STABILIZATION USING HIGH-ORDER NEURAL NETWORKS

In the recent work of one of the authors [20], HONNs have been used for the solution of the stabilization problem of unknown nonlinear systems of the form

$$\dot{x} = f(x) + g(x)u, \tag{54}$$

where $x \in \mathfrak{R}^n$ denotes the state vector of the system, u is the scalar control input, and f and g are at least C^1 vector fields, with $f(0) = 0$. It can be shown that a quite large class of dynamical systems can be represented by a differential equation of the form of (54). Moreover, systems belonging to the general class $\dot{z} = F(z, v)$ can be rendered in the form of (54) by simply applying a prefeedback $\dot{v} = k(z) + l(z)u$; in this case, the resulting system can be rewritten in the form of (54) with $x = [z^\tau, v^\tau]^\tau$, $f(\cdot) = [F^\tau(\cdot), k^\tau(\cdot)]^\tau$, $g(\cdot) = [0, \ldots, 0, l^\tau(\cdot)]^\tau$.

The asymptotic stabilization problem (or simply stabilization problem) of system (54) consists of constructing an appropriate feedback $u = q(x, t)$, such that the closed-loop system $\dot{x} = f(x) + g(x)q(x, t)$ is stable and, moreover, the solution $x(t)$ converges asymptotically to (or very close to) the equilibrium $x = 0$. The stabilization problem plays a central role in the Automatic Control theory, because of its practical importance on the one hand, and because of the fact that the solution of the stabilization problem is a prerequiste for the solution of other more general control problems, such as set-point regulation, asymptotic tracking, robust control, etc.

Contrary to the linear case $f(x) = Ax$, $g(x) = B$, where the stabilization problem has found satisfactory solutions even for the case where A and B are uncertain or unknown, the nonlinear case has not yet been solved, even for the case where $f(\cdot)$ and $g(\cdot)$ are known exactly. Of course, there are many proposed solutions for certain classes of systems; for instance, there are stabilizing controllers for systems whose vector fields satisfy certain boundedness, sector, or Lipschitz conditions; for feedback-linearizable systems; and for systems for which a known Lyapunov function exists.

Recently, adaptive control methods have also been proposed for the stabilization of nonlinear systems whose dynamics are not explicitly known (see, e.g., [21, 22], and the references therein). Such adaptive schemes are applicable to systems whose dynamics depend on known nonlinearities and unknown parameters. Most of these stabilizing controllers are adaptive extensions of existing stabilizing controllers for the case where the system dynamics are explicitly known; thus the existing adaptive stabilizing controllers are applicable at most to the class of systems for which a known stabilizer exists for the case where the system dynamics are explicitly known. Moreover, there are certain classes of systems for which a known stabilizer exists for the case where the system dynamics are explicitly known, but there exists no known adaptive stabilizer for the case where some of the parameters of the system are unknown.

A generalization of adaptive stabilizers has also appeared recently, where neural networks are used for the stabilization of unknown systems (see [23, 24], and the references therein). These neural network stabilizers are extensions of the existing adaptive ones; in fact, by using neural nets, we can sufficiently approximate the system dynamics with neural networks. Thus the trick is to consider the system vector fields $f(\cdot)$ and $g(\cdot)$ as neural networks with unknown parameters (plus some small modeling error terms). In that sense, the stabilization problem for the case where $f(\cdot)$ and $g(\cdot)$ are unknown reduces to the case where $f(\cdot)$ and $g(\cdot)$ are known functions that depend on unknown parameters; hence the adaptive stabilization methods are directly applicable to this case as well. However, as in the adaptive case, the neural network approach is applicable only to a small class of nonlinear systems.

In [20] we solved the so-called universal stabilization (US) problem, which can be stated as follows: given any family of nonlinear systems of the form of (54) satisfying the assumptions stated before, does there exist a unique stabilizer that stabilizes any (stabilizable) system of the family? Pomet [25] attempted to solve the US problem for the general class of systems of the form $\dot{x} = f(x, u)$, where $f(\cdot)$ is a smooth vector field and $u \in \Re^m$, and he conjectured that information of the type "$f(\cdot)$ is smooth and the system is stabilizable" is not sufficient for constructing a stabilizer for the aforementioned system, in the sense that some additional information must be provided to make us capable of designing a stabilizer for the above system. Moreover, Pomet [25] found "universal" stabilizers for certain classes of systems; the problem with Pomet's stabilizers is that a parameterized control Lyapunov function for the systems to be controlled must be known *a priori*. In [20] we gave a positive answer to the US problem as stated above. More precisely, we extended the switching adaptive controller of [22] (proposed by the author for the adaptive control of feedback-linearizable systems) by using control Lyapunov functions (CLFs) [26] and the approximation properties of high-order neural networks. By making use of Lyapunov stability arguments, we showed that the proposed adaptive/neural network stabilizer is capable of stabilizing any system of the form of (54), provided that the neural networks of the stabilizer are "large enough" to approximate the unknown nonlinearities. The usage of the switching adaptive technique is critical for overcoming the problem where the identification model (that is, the model that is used as an estimation of the actual system) becomes "uncontrollable"; on the other hand, the neural network approximators make it possible to overcome the problem of dealing with unknown nonlinearities.

It is worth noting that the only assumptions about the systems (54) is that their order n is known (such an assumption can be relaxed to: an upper bound of the order n is known), that the controlled system is stabilizable, and that the vector fields $f(\cdot)$ and $g(\cdot)$ are at least C^1. No other information about the vector fields $f(\cdot)$ and $g(\cdot)$ is required, with the exception of some information regarding the growth properties of $f(\cdot)$ and $g(\cdot)$; however, even in the case where this information cannot be provided, we can still design the universal stabilizer, but the price paid for that will be a more conservative stabilizer and the need to operate some preliminary experiments.

The results of [20] have both theoretical and practical implications. From a theoretical point of view, it is the first time that a universal stabilizer has been proposed; moreover, it seems that the proposed controller can be the starting point for the solution of more general but similar problems like universal stabilization of more general classes of systems than that of (54), universal controllers that achieve not only stabilization but fix-point regulation and asymptotic tracking as well, etc. Another interesting fact is that the existing methods, in many cases, were incapable of designing a stabilizer, even for the case where the system's (54) dynamics are completely known; thus our method can be also applied in the case where the system's dynamics are known, but there exists no method for constructing a stabilizer for this system. On the other hand, the proposed stabilizer can be used in uncertain and rapidly changing environments where neither nonadaptive nor adaptive controllers can be used, because they require some *a priori* knowledge about the system dynamics. However, it must be mentioned that the results of [20] are valid, provided that the neural networks of the stabilizer are "large enough" to capture the unknown nonlinearities; thus in many cases the proposed stabilizer may be very complicated, although the stabilization problem can be solved by using a very simple stabilizer. Thus, in practice, the application of the proposed stabilizer must be appropriately combined with any *a priori* information for the resulted stabilizer to be as simple as possible.

VII. CONCLUSIONS

In this chapter we have summarized our recent results regarding the application of various high-order neural network structures in system identification and related problems. We have shown that the proposed neural networks are capable of approximating and identifying a very large class of dynamical systems. Moreover, we have shown that there exist learning laws that are globally convergent, stable, and robust. It is worth noting that the proposed neural network structures are linear-in-the-weights, and thus there is one and only one minimum in the error functional; this is contrary to the multilayer neural network, where there are many local minima.

The proposed architectures have also been shown to possess other interesting properties such as stability, robustness, capability of solving other than system identification problems (e.g., estimation of robot contact surfaces, estimation of unknown probability distributions, universal stabilization of unknown nonlinear dynamical systems), etc. It is worth noting that we have successfully applied RHONNs to the problem of fault detection and identification of TV cable amplifier plants [27].

REFERENCES

[1] J. J. Hopfield. Neurons with graded response have collective computational properties like those of two-state neurons. *Proc. Nat. Acad. Sci. U.S.A.* 81:3088–3092, 1984.
[2] M. A. Cohen and S. Grossberg. Absolute stability of global pattern formation and parallel memory storage by competitive neural networks. *IEEE Trans. Systems Man Cybernet.* SMC-13:815–826, 1983.
[3] Y. Kamp and M. Hasler. *Recursive Neural Networks for Associative Memory.* Wiley, New York, 1990.
[4] A. Dempo, O. Farotimi, and T. Kailath. High-order absolutely stable neural networks. *IEEE Trans. Circuits Systems* 38:??, 1991.
[5] J. K. Hale. *Ordinary Differential Equations.* Wiley-Intersci. Publ., New York, 1969.
[6] E. B. Kosmatopoulos, M. M. Polycarpou, M. A. Christodoulou, and P. A. Ioannou. High-order neural network structures for identification of dynamical systems. *IEEE Trans. Neural Networks* 6:422–431, 1995.
[7] E. B. Kosmatopoulos and M. A. Christodoulou. Structural properties of gradient recurrent high-order neural networks. *IEEE Trans. Circuits Syst. II Analog Digital Signal Process.* 42:592–603, 1995.
[8] L. V. Kantorovich and G. P. Akilov. *Functional Analysis,* 2nd ed. Pergamon, Oxford, 1982.

[9] M. A. Cohen. The construction of arbitrary stable dynamics in nonlinear neural networks. *Neural Networks* 5:83–103, 1992.

[10] H.-D. Chiang and J. S. Thorp. Stability regions of nonlinear dynamical systems: a constructive methodology. *IEEE Trans. Automat. Control* 34:1229–1241, 1989.

[11] R. Lum and L. O. Chua. The identification of general pseudoreciprocal vector fields. *IEEE Trans. Circuits Systems* 39:102–122, 1992.

[12] P. A. Ioannou and A. Datta. Robust adaptive control: a unified approach. *Proc. IEEE* 1991.

[13] P. A. Ioannou and J. Sun. *Stable and Robust Adaptive Control.* Prentice Hall, Englewood Cliffs, NJ, 1995.

[14] E. B. Kosmatopoulos, M. A. Christodoulou, and P. A. Ioannou. Dynamical neural networks that ensure exponential error convergence. *Neural Networks*, to appear.

[15] E. B. Kosmatopoulos and M. A. Christodoulou. The Boltzmann g-RHONN: a learning machine for estimating unknown probability distributions. *Neural Networks* 7:271–278, 1994.

[16] H.-F. Chen and L. Guo. Continuous-time stochastic adaptive tracking—robustness and asymptotic properties. *SIAM J. Control Optim.* 28:513–527, 1990.

[17] J. S. Bay and H. Hemami. Dynamics of a learning controller for surface tracking robots on unknown surfaces. *IEEE Trans. Automat. Control* 35:1051–1054, 1990.

[18] E. B. Kosmatopoulos and M. A. Christodoulou. High-order neural networks for learning of robot contact surface shape. *IEEE Trans. Robotics Automation*, to appear.

[19] E. B. Kosmatopoulos and M. A. Christodoulou. Filtering, prediction, and learning properties of ECE neural networks. *IEEE Trans. Systems Man Cybernet.* 24:971–981, 1994.

[20] E. B. Kosmatopoulos. Universal stabilization using control Lyapunov functions, adaptive derivative feedback and neural network approximators. Unpublished (also to be presented in *IEEE Conference on Decision and Control*, 1996).

[21] I. Kanellakopoulos, P. V. Kokotovic, and A. S. Morse. Systematic design of adaptive controllers for feedback linearizable systems. *IEEE Trans. Automat. Control* 36:1241–1253, 1991.

[22] E. B. Kosmatopoulos. A switching adaptive controller for feedback linearizable systems. Unpublished (also to be presented in *IEEE Conference on Decision and Control*, 1996).

[23] R. M. Sanner and J.-J. E. Slotine. Gaussian networks for direct adaptive control. *IEEE Trans. Neural Networks* 3:837–863, 1992.

[24] G. A. Rovithakis and M. A. Christodoulou. Adaptive control of unknown plants using dynamical neural networks. *IEEE Trans. Systems Man Cybernet.* 24:400–412, 1994.

[25] J.-B. Pomet. Remarks on sufficient information for adaptive nonlinear regulation. In *Proceedings of the 31st IEEE Conference on Decision and Control*, Tuscon, Arizona, December 1992, Vol. 2, pp. 1737–1741.

[26] E. D. Sontag. A "universal" construction of Arstein's theorem on nonlinear stabilization. *Systems Control Lett.* 13:117–123, 1989.

[27] J. T. Dorocic, E. B. Kosmatopoulos, S. Neville, and N. J. Dimopoulos. Recurrent neural networks for fault detection. In *Proceedings of the 4th IEEE Mediterranean Symposium on New Directions in Control and Automation*, Chania, Greece, June 1996, pp. 17–22.

Neurocontrols for Systems with Unknown Dynamics*

William A. Porter
Department of Electrical
and Computer Engineering
University of Alabama
Huntsville, Alabama 35899

Wie Liu
Department of Electrical
and Computer Engineering
University of Alabama
Huntsville, Alabama 35899

Luis Trevino
EP-13
Bldg. 4666
Marshall Space Flight Center
MSFC, Alabama 35812

I. INTRODUCTION

In recent years considerable attention has been paid to computational models based, in part, on natural neurological behavior. The promise of such models arises from an implicit parallelism, a compatibility with distributive memory, and design procedures that do not rely on analytic models of the computational process. Technological developments in analog VLSI, optical computation, etc., have made nonstandard computing structures feasible and thereby reinforced interest in artificial neural networks and their application to a broad spectrum of problems.

Our interest here is the possibility of realizing feedback controllers for discrete-time dynamical systems using neural designs. This topic has been addressed by other researchers. Many prior studies have used the capability of neural networks to synthesize complex nonlinear functions. One such use is to implement state space control laws for partially known nonlinear dynamical systems [1–5]. Most of these designs use iterative off-line techniques to develop the neural controller. These controllers are then inserted in a conventional feedback structure.

The iterative adjustment of parameters can produce effective designs. However, such methods are computationally slow and come equipped with no assurance of convergence. For on-line adjustment of controller parameters, the iterative methods are too slow and, in addition, introduce the danger of coupling with process dynamics [6, 7]. Indeed, the potential for instability arising from on-line gra-

*Reprinted in part with permission from *IEEE Trans. Aerospace Electronic Systems* 31:1331–1340, 1995 (©1995 IEEE).

307

dient adjustment of parameters was one reason adaptive control theorists turned to other methods some years ago.

In another line of development, researchers [5, 8] have explored the decomposition of compact regions in the state space. The neural network is designed to map a subregion into a choice of control, which ensures stability and/or movement along a reference trajectory. Such methods can produce acceptable designs. However, at least partial knowledge of the plant dynamics and access to the plant state variables are necessary.

The design procedure of the present study presumes no model information about the plant. In particular, no knowledge concerning state dimension, linearity, or stability is presumed. The total available information is assumed to consist of a finite-duration input/output (i/o) histogram (e.g., the system response, where the input is white noise). We note in advance that the quality of the design is dependent in part on the degree to which the i/o histogram accurately reflects the behavior of the system.

By using an algorithm developed elsewhere [9–11], an expanded form of the i/o histogram is scanned for critical exemplars. The set of critical exemplars, referred to as a *kernel*, specifies the neural controller. Although we refer to this specification as "training," it is noniterative and immediate. Thus complete designs, including specification of the architecture, are available at the computational latency of the scan operation. Moreover, partial scans produce valid partial designs; hence the possibility of on-line training, using i/o data from a sliding window, is a possibility.

Because the design format uses minimal *a priori* information, it is broadly applicable to the control of plants that are not precisely modeled. To reflect such applications we consider here output trajectory tracking. To evaluate the neural controller in this context it is connected, in a closed loop feedback fashion, to the system; its inputs include the system outputs. The neural controller also accepts a job description in the form of a goal trajectory that the system outputs are to be driven along. The neural controller then causally determines the input signals, which stimulate the system to track the desired output trajectory. We remark that the design uses no *a priori* information about the reference trajectory.

A study with a similar motivation, but with more restrictive goals and a different neural network, has been reported [12]. That study focused on disturbance rejection rather than tracking and gave example applications to systems different from those used here. We will remark further on connections between the two controller designs after some technical details are in hand. However, we note in advance that the present design, even without training, exhibits an implicit capability for rejecting such disturbances.

Although most neural network researchers draw tacit inspiration from neurological phenomena, very few neural network designs actually reflect such origins. Our architecture can be viewed as a single-layer perceptron architecture

[9, 13, 14]. However, we use higher order moment information and have no sigmoidal activation function. Because of this, we make no claims concerning resemblance to natural neural calculations.

The use of higher order moments provides a capability to store and distinguish information from the i/o histogram far in excess of the vector space dimension. Thus, from the most recent i/o values and the target trajectory values, the Higher Order Moment Neural Array, HOMNA, [9–11] can extrapolate to determine a choice of input value. Intuitively, one might view the HOMNA controller as a very fast look-up and extrapolation device, with its table implicit in the synaptic interconnects. Such a perspective makes plausible the possibly superfluous nature of a detailed dynamic model and the potential for adaptivity via an update of this implicit table.

In applying the HOMNA design technique for a select domain, it is demonstrated to be applicable to specifying a propulsion pressurization system controller. Propulsion systems are classified as highly nonlinear and time-varying systems, particularly due to the cryogenics and pressurization gases involved. As a result, controlling the inlet pressure for rocket engine test-bed testing gives rise to problems in fine tuning, disturbance accommodation, and determining control gains for new profile operating regions [15].

II. THE TEST CASES

In our study controller designs were evaluated for discrete systems of order ≤ 8. Both multivariate and single-variate systems were tested. The results we present here are more focused but nonetheless are typical of the broader study. It is convenient to catalog these results within the context of the following difference equation:

$$
\begin{aligned}
y(j + 1) = {} & ay(j) + by(j - 1) + cy(j - 2) + d\mathrm{sat}[y(j - 1)] \\
& + ey(j - 1)u(j) + u(j),
\end{aligned}
\tag{1}
$$

where $u(\cdot)$ is the input sequence, $y(\cdot)$ is the output sequence, and $\mathrm{sat}[r] = r^2/[1 + r^2]$. By choosing the coefficients in Eq. (1), a variety of models result. We shall discuss simulations for the five cases identified in Table I.

To collect the i/o histogram we follow a standard format. First a random sequence is constructed by draws from a uniform distribution over $[-1, 1]$. This random sequence is used as an input to the model in question. The model is initially at rest. We disregard the first 10 samples, which provides a random initialization. The histogram is then taken as $\{(y(j + 1), u(j))\}$ for the ensuing 100 samples. The i/o histogram is used to design the neural controller. We shall provide more details in Section III on this design procedure. For the moment it suffices to know

Table I
Typical Designs Tested[a]

Case	a	b	c	d	e	Type
I	0.6,	0.6,	0,	0,	0	Linear, 2nd order, unstable
II	0.5,	−0.5,	0.5,	0,	0	Linear, 3rd order, stable
III	0.5,	−0.5,	0,	−0.5,	0	Nonlinear, 2nd order
IV	0.5,	−0,	0,	0,	−0.5	Nonseparable, 2nd order
V	0.5,	0,	0,	0.5,	−0.5	Nonseparable, nonlinear

[a]Reprinted with permission from W. A. Porter and W. Liu, *IEEE Trans. Aerospace Electron. Systems* 31:1331–1339 (©1995 IEEE).

that the design procedure is fast and direct. In fact, it can be implemented on-line as the i/o histogram is generated.

The neural controller is then connected in a closed-loop feedback fashion to the dynamic process (see Fig. 1). Its inputs include the process output, $y(\cdot)$. The neural controller also accepts a job description in the form of a goal trajectory, $\bar{y}(\cdot)$, that the process output is to be driven along. The neural controller then generates the signal, $u(\cdot)$, which attempts to drive the model along the desired output trajectory. The neural controller performance is then evaluated via simulation. At the risk of seeming repetitious, we note that the models of Eq. (1) are used only to generate the i/o histogram and in the test evaluation. The neural design is not privy to this information.

The neural controller design methodology was tested for a variety of target output profiles. However, for brevity, we discuss here only the target profile of Fig. 2. The fact that this function is composed of straight line segments is immaterial to the performance results. Indeed, performance on smooth periodic trajectories exceeds substantially the results reported herein. We note that the neural design

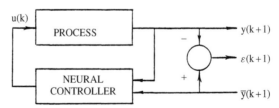

Figure 1 The design evaluation. Reprinted with permission from W. A. Porter and W. Liu, *IEEE Trans. Aerospace Electron. Systems* 31:1331–1339 (©1995 IEEE).

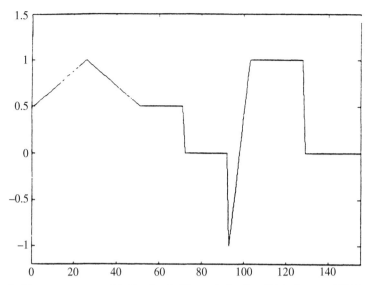

Figure 2   The target output profile. Reprinted with permission from W. A. Porter and W. Liu, *IEEE Trans. Aerospace Electron. Systems* 31:1331–1339 (©1995 IEEE).

procedure does not use any information about the testing profile. The test profile is also used causally by the neural controller, that is, no look ahead is allowed. Of course, if a test profile is selected that takes on values outside the range of the initial experimentation on the system, then the neural design will be extrapolating beyond its domain of information about the system. Our tests have shown the design to be robust relative to extrapolation outside the original i/o range, however, this robustness is certainly model dependent.

For two of the cases of Table I, namely cases II, and III, the controller design (of Section III) provided excellent tracking of the reference profile. To illustrate, in Fig. 3 we display the superposition of the reference and actual output for these two cases (solid line for $\bar{y}(\cdot)$; dashed line for $y(\cdot)$). Several other such linear/nonlinear models of varying order ≤ 8 were also tested with similar results. Select simulations of linear multivariate systems also substantiated excellent performance. Because cases II and III did not prove challenging, we shall not mention them further.

The remaining three cases provided a more substantial challenge to the neural design. We note that systems IV and V have a response-dependent control gain. Nonseparable systems such as these constitute a classically difficult problem for

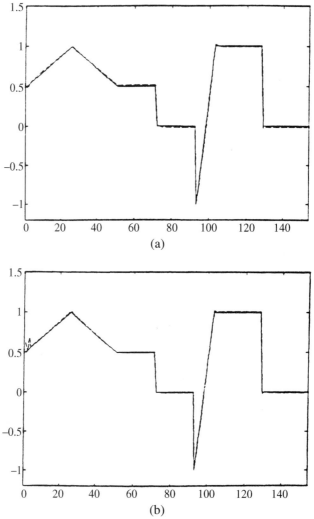

Figure 3 (a) Case II performance. (b) Case III performance. Reprinted with permission from W. A. Porter and W. Liu, *IEEE Trans. Aerospace Electron. Systems* 31:1331–1339 (©1995 IEEE).

controller designs. We note that system I is unstable. The difficulty presented here is not in the controller design, but rather in collecting an i/o histogram. In Section IV we shall reopen the discussion of controller performance for systems I, IV, and V. We consider first, however, the technical features of the design.

III. THE DESIGN PROCEDURE

For each system i/o histogram, a training set is constructed. This construction is characterized by the map, E_{pq}, where

$$E_{pq}\big(y(j+1), u(j)\big) = \big(y(j+1), y(j), y(j-1), \ldots,$$
$$\big(y(j-p), u(j), \ldots, u(j-q)\big)\big).$$

We refer to the map E_{pq} as an *expansion map* and the range of this map as an *expanded histogram*. It is noted that E_{pq} is an easy computation on the i/o data sets, provided enough data are taken to avoid endpoint ambiguities. In all cases, we collected i/o data of sufficient length to unambiguously define expanded histograms of cardinality 100.

The expansion map, E_{pq}, gives data sets that are natural for training neural system identifiers (see [11, 16]). However, for training a neural controller, a slight modification is useful. In this modification we delete $u(j)$ from $E_{pq}(y(j+1), u(j))$ and treat this as a companion, but separate, vector. That is

$$x(j) = E'_{pq}\big(y(j+1), u(j)\big)$$
$$= \big(y(j+1), \ldots, \big(y(j-p), u(j-1), \ldots, u(j-q)\big), \tag{2}$$

and finally,

$$\big\{(x(j); u(j)), \ j = 1, \ldots, 100\big\}$$

identifies the set of training input/output pairs.

With these definitions in hand, our design strategy is quite straightforward. We train a neural network on the i/o pairs of Eq. (2). As the input to this network, we let $(y(j), \ldots, y(j-p), u(j-1), \ldots, u(j-q))$ be taken as the actual prior data from the plant and controller. We take $y(j+1)$ from the reference trajectory. The next network output is an estimate for the proper control, $u(j)$, which is then fed back to the plant.

Concerning the structure of the neural controller, this may be characterized as a feedforward single hidden layer network that implements the calculation

$$u(j) = K\Psi\big(Sx(j)\big). \tag{3}$$

Here $S: R^{p+q+2} \to R^m$ is a matrix connecting the input, $x(j)$, to a single layer of m neurons. The map Ψ is nonlinear and acts on R^m. We assume that Ψ is decoupled and that all neurons have the same activation function, ψ. In short, $(\Psi\lambda)_j = \psi(\lambda_j), j = 1, \ldots, m$. The matrix $K: R^m \to R$ is the output map.

In the present design we use $\psi(r) = e^r$; however, finite polynomials and sigmoidal functions may also be used [9, 10]. The HOMNA algorithm, of course, must be synchronized to the choice of ψ.

In a manner identified in Sections II.A–D, the HOMNA algorithm scans through the expanded training set. The number of neurons (i.e., m) and, in fact, the rows of the matrix, S, are selected by the algorithm. The matrix K is also a prescribed function of the selected i/o exemplar pairs [10, 11, 16].

Because S and K are prescribed functions of the kernel, it is possible to determine the evolution of these matrices as the kernel expands. In fact, the HOMNA algorithm provides this information [10, 16]. As a result, the controller design evolves as the kernel, and hence the canonical network, expands. Each step in this evolvement is a valid controller based on the information then available to the kernel algorithm.

A. Using Higher Order Moments

For perspective, we note that if the neural net was to be linear, then the cardinality of the memory is bounded by the dimension of the range of E'_{pq}, namely $p + q + 2$. Such a network might typically project $(x(j), u(j))$ onto the memory set to determine $u(j)$ by linear combination. However, the memory limitation would prohibit accurate modeling of many system conditions.

This observation about a linear solution to the posed design problem and its inherent limitations is more than just idle speculation. Indeed, the HOMNA design can be best understood by analogy with the linear case, the principal departure being that the design takes place in a companion tensor space. This transition in setting utilizes the higher order moments and effectively removes the original constraint on memory size.

To introduce this transition let $H = R^n$ denote the Euclidean space, equipped with the usual inner product denoted by $\langle x, y \rangle$, $x, y \in H$. The space H also has a tensor product [12], which we denote by $x \otimes y$. In R^n it is often convenient to utilize the representation $x \otimes y = xy^*$. Here x, y are column vectors. y^* is the transpose (row vector) and xy^* is the indicated rank 1 matrix. Taking the closure over linear combinations, we arrive at the space $H \otimes H$. It follows easily that $H \otimes H$ is isomorphic to $R^{n \times n}$, the space of $n \times n$ matrices.

The dot product on H has a natural extension to $H \otimes H$. For this, the definition

$$\left\langle\!\!\left\langle \sum_i x_i \otimes y_i, \sum_j u_j \otimes v_j \right\rangle\!\!\right\rangle = \sum_{ij} \langle x_i, u_j \rangle \langle y_i, v_j \rangle \tag{4}$$

suffices. For $H = R^n$ it is readily verified that

$$\langle\!\langle x \otimes y, u \otimes v \rangle\!\rangle = tr\{(xy^*)(uv^*)^*\}.$$

We refer to the inner product of Eq. (4) as the natural inner product on $H \otimes H$.

Now let $v = \{0, 1, \ldots, m\}$ denote an index set. Recall the definition of a Cartesian product $K = K_0 \times K_1 \times \cdots \times K_m$, where the K_j are linear spaces. Let $K_j = \otimes^j H$ be used in this Cartesian product. The resultant K, equipped with the natural inner product, is called a *tensor scale* (over H, v) and is denoted by $H(v)$.

It is convenient to explicitly identify the embedding map $H \to H(v)$. For $x \in H$, consider the definition

$$\tau_m(x) = (a_0 x_0, a_1 x_1, \ldots, a_m x_m), \tag{5}$$

where $a_i \in R$ and $x_0 = 1, x_1 = x, x_2 = x \otimes x, \ldots, x_m = \otimes^m x$.
For example, with $v = \{1, 2, 3\}$, we have

$$\tau_3(x) = (a_0, a_1 x_1, a_2 x \otimes x, a_3 x \otimes x \otimes x).$$

For convenience we think of the polynomial $\psi_m(r) = \sum a_j r^j$ as a companion to τ_m.

It follows easily that

$$\langle\langle \tau_m(x), \tau_m(y) \rangle\rangle = \sum_{j=0}^{m} a_j \langle x, y \rangle^j = \psi_m(\langle x, y \rangle). \tag{6}$$

Our tensor embedding definitions extend easily to analytic functions. For example, with $a_j = 1/j!$ we have

$$\langle\langle \tau(x), \tau(y) \rangle\rangle = \exp(\langle x, y \rangle). \tag{7}$$

The exponential embedding was, in fact, used for the simulations reported herein.

B. EMBEDDING THE CONTROLLER DESIGN IN $H(v)$

We return now to the neural design. Suppose now that $\{x(j): j = 1, \ldots, 100\} \subset R^m$ is the extended histogram. This set has a well-defined tensor image, namely, $\{\tau(x(j)): j = 1, \ldots, 100\}$. Whereas the vector space set has rank $\leq p + q + 2$, the tensor space image has no such restraint (see [10, 17, 18]) and, in fact, may be linearly independent.

A subset $\mathcal{K}$ of $\{\tau(x_j)\}$ is said to be a "kernel" for $\{\tau(x_j)\}$ provided: (i) $\{\tau(x_j)\}$ is tightly clustered about the manifold span $\{\mathcal{K}\}$ and (ii) $\mathcal{K}$ is not redundant. The phraseology "tightly clustered" and "nonredundant" is explained in part by the following algorithm for identifying a kernel:

(a) Select any initial point $\tau(x(a))$, form the manifold span $\tau(x(a))$.
(b) Select any $\tau(x(b))$. If $\tau(x(b))$ is sufficiently close to span $\{\tau(x(a))\}$, reject it. If not, form span $\{\tau(x(a)), \tau(x(b))\}$.
(c) Select any $\tau(x(c))$ and continue until the set is exhausted.

For a measure of closeness $\|\tau(x) - P\tau(x)\|^2$, where P is the orthogonal projector on the span, is useful. However, since $\|\tau(x) - P\tau(x)\|^2 / \|\tau(x)\|^2$ lies in the range $[0, 1]$, this measure is often more convenient. The "sufficiently close" criteria then reduce to a threshold test.

It is apparent that the kernel is scan order dependent and hence not unique; however, for a fixed threshold level the kernel size is almost invariant. It is also apparent that the kernel size varies inversely with the threshold level. For thresholds of 0 (respectively 1) the kernel is all inclusive (respectively orthonormal).

Suppose then that a kernel for the embedded histogram set $\{\tau(x_j)\}$ has been selected. The affiliated manifold and orthogonal projector generated by the kernel are then well defined. If $x \in R^m \times \varepsilon\mathbb{R}^m$ is an arbitrary data point, then $P\tau(x)$ is a unique linear combination of the generating subset. Each vector of E'_{pq}, however, has an associated $u(\cdot)$ value. We form the same linear combination of these values and identify this combination as the input value affiliated with x.

The above overview is a quite accurate picture of the HOMNA design. The HOMNA design, however, does not do any tensor space calculations. In view of the computational complexity of such tensor operations, this is indeed fortunate. The key to this is that all of the above operations, in particular, forming the orthogonal projection on span $\{\tau(x(k)), \ k = a, b, \ldots\}$, principally involve dot products. Equations (6) and (7), however, note that dot products can be computed in the underlying space. This Hilbert space congruence is the mechanism for bringing the tensor space calculations back down to the underlying vector space.

C. HOMNA TRAINING ALGORITHMS

The HOMNA training algorithm has been discussed elsewhere [9–11, 18], and our summary here will be concise. Our objective is to expose the interrelationship between the tensor space calculations and their vector space counterparts.

Consider now a matrix $S \in R^{m \times n}$. Let S^* denote the matrix transpose. The matrix SS^* is nonnegative symmetric. Let ψ denote an arbitrary function on R. The matrix G is formed by letting ψ act on the individual entries of SS^*, that is,

$$G_{ij} = \psi\big([SS^*]_{ij}\big) \qquad i, j = 1, \ldots, m. \tag{8}$$

It is well known that the rank of SS^* is equal to the dimension of the row space of S. The rank of G, however, is determined by both S and ψ. In particular, for polynomial ψ and when the rows of S are distinct, G can be nonsingular, even when SS^* has rank 1.

Suppose now that ψ is fixed and that $\{x_1, x_2, \ldots\}$ denotes the rows of S. For clarity let S_m denote the matrix S with m rows. Similarly, let G_m denote the matrix G of Eq. (8) when $S = S_m$. The matrices $\{G_m\}$, although of distinct dimensions, do have a definitive relationship to each other. Using Eq. (8), it can be verified that

$$G_{m+1} = \left\{ \begin{array}{cc} G_m & \phi_m \\ \phi_m^* & r_m \end{array} \right\}, \tag{9}$$

where $\phi_m \in R^m$ denotes a vector, ϕ_m^* its transpose, and r_m denotes a scalar. Equation (9) uses the obvious block partitioning notation. Suppose then that G_m is invertible. The following lemma identifies the conditions under which G_{m+1} shares this invertibility.

LEMMA 1.   *The matrix G_{m+1} is invertible if and only if G_m is invertible and*

$$m_m = r_m - f_m^* G_m^{-1} f_m \neq 0. \tag{10}$$

The inverse matrix G_{m+1}^{-1} has the block partitioned form

$$G_{m+1}^{-1} = \left\{ \begin{array}{cc} G_m^{-1} + \mu_m^{-1} G_m^{-1} \phi_m \phi_m^* G_m^{-1} & -\mu_m^{-1} G_m^{-1} \phi_m \\ -\mu_m^{-1} \phi^* G_m^{-1} & \mu_m^{-1} \end{array} \right\}. \tag{11}$$

The content of this lemma is well known [9, 10]. The identity of Eq. (11) may be verified by inspection. Further details on these identities are available. For this note that the matrix S_m has the training set vectors $\{x_1, \ldots, x_m\}$ as rows. Thus,

$$S_{m+1} = \left[S_m^* \vdots x_{m+1} \right]^*$$

describes the evolution of S_m with m. Using the partitioned form of S_{m+1}, it follows that

$$S_{m+1} S_{m+1}^* = \left\{ \begin{array}{cc} S_m S_m^* & S_m x_{m+1} \\ x_{m+1} S_m^* & x_{m+1}^* x_{m+1} \end{array} \right\}.$$

Since ψ acts componentwise, it follows that $G_m = G(S_m)$ has the block partitioned form of Eq. (9), where

$$\begin{array}{cc} (\phi_m)_i = \psi[(S_m x_{m+1})_i] \\ r_m = \psi[x_{m+1}^* x_{m+1}] \end{array}, \qquad i = 1, \ldots, m. \tag{12}$$

Hence, the inversion identify of Eq. (10) is related back to the training set.

The specific form of the training algorithm is now considered. We refer to the notation of Eqs. (9–11). We also continue the notations G_m, S_m introduced above.

Let $\{y_1 \ldots, y_m\}$ denote the desired signatures of the training vectors $\{x_1, \ldots, x_m\}$ and Y_m the matrix formed using the signature vectors as columns. In many applications $Y_m = S_m^*$. The synaptic matrix K of Eq. (7) is given in general by

$$K_m - Y_m G_m^{-1}, \tag{13}$$

where the invertibility of G_m is ensured. Here H_m is the obvious mth step form of K.

D. Tensor Space Matchups
with the HOMNA Calculations

We now relate Section III.A to the calculations of Sections III.B and III.C. For this, suppose that $\{\tau(x_i): i = 1, 2, \ldots, m\}$ is a linearly independent set in $H(\nu)$. Then the Grammian matrix, G, which has $G_{ij} = \langle\langle \tau(x_i), \tau(y_i)\rangle\rangle$, is invertible. However, using Eq. (7), we see that this is exactly the matrix of Eq. (8). Indeed, if T is the matrix whose columns are the $\tau(x_i)$, then also $G = T^*T$.

Concerning the tensor space project, P, of Section III.B, we note that $P = TG^{-1}T^*$. Because P acts on $\tau(x)$ we examine $T^*\tau(x)$. The ith entry of this vector is apparently $\langle\langle \tau(x_i), \tau(x)\rangle\rangle = \psi(\langle x_i, x\rangle)$. However, this is identical to $\psi(Sx)$, where the matrix S has the set $\{x_i\}$ as rows. In short, $(T^*T)^{-1}T^*\tau(x) = G^{-1}\psi(Sx)$ for all $X \in R^m$.

Proceeding in this manner, it can also be shown that

$$\left\|\tau(x) - P\tau(x)\right\|^2 = \mu(x),$$

where μ is given by Eqs. (10) and (12). Similarly,

$$\left\|\tau(x)\right\|^2 = r(x).$$

For the controller application we note that $(T^*T)^{-1}T^*\tau(x)$ consists of the coefficients in the linear expansion of $P\tau(x)$ along the basis of the underlying manifold. Thus if Y is taken as the matrix whose columns are the values $\{u(j)\}$ for the selected vectors, then $\hat{\mu} = YG^{-1}\psi(Sx)$ is exactly the linear interpolation identified in Section III.A.

We remarked above that the invertibility of G_m was ensured. Because G_m is the Grammian of the affiliated kernel and because the kernel selection algorithm guarantees linear independence, the invertibility of G_m follows.

We also note that the matrices $\{K_m, S_m\}$ are explicitly defined from the kernel. Hence a completely trained neural network evolves with the kernel. This provides a potential for on-line adjustment of the network as the plant characteristics vary.

IV. MORE DETAILS ON THE
CONTROLLER DESIGN

Reconsider now the discussion of Section III. If this design were to be characterized from a conventional control design perspective, two analogies could be justified. The first analogy is that of an inverse system design. The second is that of a discrete PID controller design with the integral component absent.

During the several simulations of our research, design performance did show some characteristics appropriate to both analogies. In particular, designs based on small kernels occasionally exhibit small constant offset errors to step and/or ramp inputs. Increasing kernel size is one cure; however, intrigued by the possibility of blending a conventional controller design cure (integral-type feedback), we explored a gain adjustment. This adjustment computes a windowed average,

$$\Delta = \frac{1}{N} \sum [y(j) - \bar{y}(j)],$$

and then multiplies the neural controller output, namely $u(\cdot)$, by $1 + \Delta$. This refinement had no connection to the neural design. We discuss it here only to illustrate that the basic neural design can be fine-tuned with conventional controller design insight.

In all designs reported herein, the parameters $p = 3$, $q = 0$ were used to define E'_{pq}. However, simulations were run for several p,q combinations over the range $0 \le p \le 8, 0 \le q \le 5$ without discovering an unpredictable difference in performance.

With regard to the unstable plant (Case I), we employed a simple artifice to facilitate obtaining a i/o histogram. The Case I model was first embedded in a stabilizing feedback loop, namely, $v(j) = 0.4z(j - 1)$; $u(j) = w(j) - v(j)$. The histogram was then taken off the plant, that is, $\{(u, z)(\cdot)\}$, whereas $w(\cdot)$ was chosen randomly, as in the other experiments. For testing, the feedback loop is removed and the neural controller is connected as in Fig. 1. We shall see shortly that the neural controller provides stabilizing as well as controlling capabilities.

For the two cases II and III, the neural controller gave excellent performance, to the extent that a plot of the system response is almost indistinguishable from the target trajectory (recall Fig. 3a, b). Hence, as a space-saving measure, we shall forgo further remarks on these cases.

As a final point of this section we take note of the number of neurons necessary to implement the controller designs. The entries of Table II display the kernel cardinality, which equals the neuron cardinality, for all five system cases.

Table II
Kernel Cardinality[a]

Case	I	II	III	IV	V
Kernel no.	21	27	26	27	20

[a]Reprinted with permission from W. A. Porter and W. Liu, *IEEE Trans. Aerospace Electron. Systems* 31:1331–1339 (©1995 IEEE).

V. MORE ON PERFORMANCE

With regard to the unstable system of Case I, we take note of Fig. 4a, which displays the performance of the basic neural controller design. The controller obviously stabilizes the system and drives the plant, with some success, along the

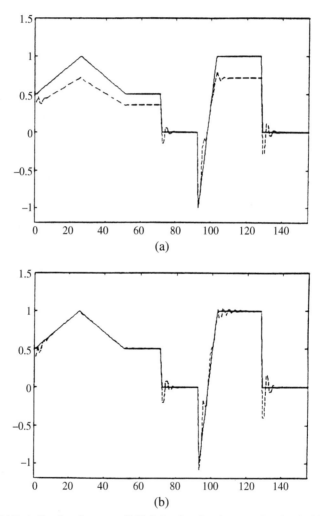

(a)

(b)

Figure 4   (a) Basic Case I performance. (b) Variation Case I performance. Reprinted with permission from W. A. Porter and W. Liu, *IEEE Trans. Aerospace Electron. Systems* 31:1331–1339 (©1995 IEEE).

reference trajectory. The step and ramp offsets apparently arise from an i/o histogram that is distorted by the stabilizing loop necessary for its collection.

In Fig. 4b we display the same controller with the averaging gain adjustment described in Section IV. We note that the step and ramp offsets are driven to zero. The system tends to "ring" a little more, which reflects a slight interaction between the gain adjustment and plant dynamics.

In the system of Case IV, the value of the control gain is response dependent. Figure 5a and b displays the basic controller performance and the controller performance with the average gain adjustment in place. The basic design performs very well; the variation design gives the expected improvement. We note a slight tendency to ringing in the variation design performance.

The Case V system contains both the bilinear term and the nonlinear term. Initial simulations for a design at kernel no. = 20 showed little difference from the plots of Fig. 5a and b. To sample the effects of kernel size, we raised the threshold on the kernels algorithm to secure a kernel of cardinality 18. The resultant performances are displayed in Fig. 6a and b.

A comparison of Figs. 5 and 6 is indicative of design robustness as a function of kernel size.

A. DISTURBANCE REJECTION

In a recent study [12] the ability of neural controllers to reject disturbance inputs was considered. Although [12] used a different neural architecture and training (back-propagation), there remain enough similarities that a few comparative remarks are helpful in placing the two studies in perspective.

The most general format of [12] is summarized by the equation set

$$
\begin{aligned}
y(k+1) &= f\big[Y(k), U(k) + V(k)\big] \\
v(k+1) &= g\big[V(k)\big],
\end{aligned}
\tag{14}
$$

where (in the notation of the present study)

$$
\begin{aligned}
Y(k) &= \big[y(k)\cdots y(k-p)\big] \\
V(k) &= \big[v(k)\cdots v(k-q)\big].
\end{aligned}
\tag{15}
$$

As is apparent from Eqs. (14) and (15), the disturbance $v(\cdot)$ has the form of unmodeled dynamics, as contrasted to an external disturbance arising from noise. The i/o histogram for the neural design is collected with the disturbance in place and using uniformly distributed noise as the input sequence $v(\cdot)$. The nominal trajectory was derived from a smooth input, normally a sum of two sinusoidal functions. The disturbance model was contrived to be cyclic, so that long-term

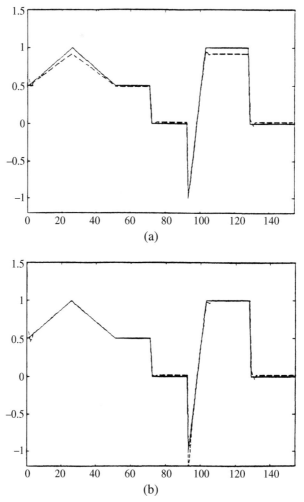

Figure 5   (a) Basic Case IV performance. (b) Alternate Case IV performance. Reprinted with permission from W. A. Porter and W. Liu, *IEEE Trans. Aerospace Electron. Systems* 31:1331–1339 (©1995 IEEE).

suppression could be studied. Because of the difference in the specific equations studied and affiliated distinct assumptions, a direct comparison with the present study is not meaningful. It is of interest, however, to note that disturbance reduction (as defined in [12]) is a tacit capability of the present design. This arises from the smooth cyclic nature of the disturbance, which the neural net treats as a

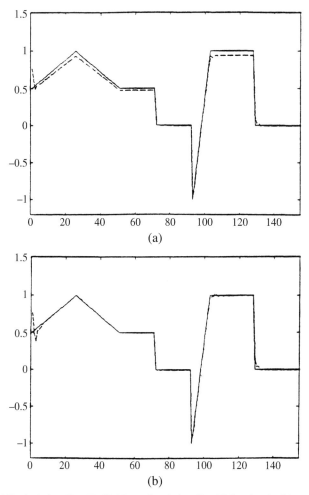

Figure 6 (a) Basic design Case V. (b) Alternative design Case V. Reprinted with permission from W. A. Porter and W. Liu, *IEEE Trans. Aerospace Electron. Systems* 31:1331–1339 (©1995 IEEE).

slightly erroneous previous calculation and hence corrects during the next calculation.

To illustrate, we selected the cyclic disturbance

$$x_1(k+1) = x_1(k) + 0.2x_2(k)$$
$$x_2(k+1) = 2x_1(k) + x_2(k) - 0.1[x_1^2(k) - 1]x_2(k) \tag{16}$$
$$v(k) = x_1(k),$$

introduced in example 4 of [12]. This disturbance was simply added at the input of the plant, that is, $u(k) \to u(k) + v(k)$. The neural design was not otherwise adjusted. Our testing included all five systems of Table I; however, for brevity we display here only the results for system V.

Two reference trajectories were tested, namely, the reference trajectory of Fig. 2 and a reference trajectory used in [12]. This latter trajectory is periodic and is given by the solution of the equation

$$\bar{y}(k+1) = 0.6\bar{y}(k) + r(k)$$
$$r(k) = \sin[2\pi k/10] + \sin[(2\pi k)/25]. \tag{17}$$

Because the HOMNA is trajectory independent, the results were qualitatively identical. We display here only the results from Eq. (17).

In Fig. 7 we display the open loop response for system V. The solid curve is the perturbation of Eq. (16), and the dashed curve is the open-loop response to this signal. We note that system V provides an open-loop gain of ≥ 4 for periodic signals in the frequency range of the cyclic disturbance.

Our next experiment utilized the controller for system V reported on earlier. The controller was used to drive the plant along the trajectory generated by Eq. (17). The controller performance gave a response that was numerically indistinguishable from the target trajectory. Because of considerations of space we shall forgo plotting the two visually identical curves.

Our next experiment was to add the disturbance at the input. In Fig. 8 we plot the target trajectory (solid line) and the closed-loop response (dashed line).

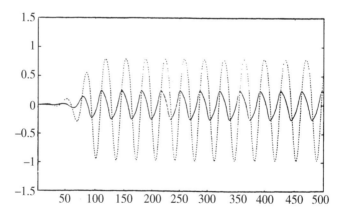

Figure 7 Disturbance and open-loop response. Reprinted with permission from W. A. Porter and W. Liu, *IEEE Trans. Aerospace Electron. Systems* 31:1331–1339 (©1995 IEEE).

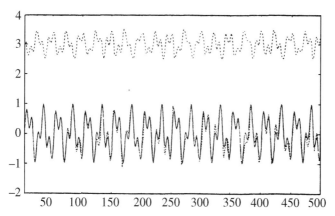

Figure 8 Suppressing the disturbance. Reprinted with permission from W. A. Porter and W. Liu, *IEEE Trans. Aerospace Electron. Systems* 31:1331–1339 (©1995 IEEE).

Although the controller was designed without information on either the reference trajectory or the disturbance, its ability to suppress the disturbance is evident.

The third curve in Fig. 8 is a plot of the coefficient of the control, namely, $1 - 0.5y(j - 1)$. The curve has been offset, by adding 2.0, to avoid plot overlap. We note that the neural design has to cope with a system with substantial variation in its i/o incremental gain.

B. PROPULSION SYSTEM APPLICATION

Using the presented design procedure and the integration technique of Sections III and IV, respectively, a neural controller is synthesized for a propulsion inlet pressure control system. The existing system is represented in Fig. 9 [15]. The 23,000 gallon LOX run tank is the plant considered for this design. An electrohydraulic (servo) valve serves as the primary controlling element. The valve position feedback system serves as a minor loop, and the pressure feedback system serves as the outer or major loop. The existing controller is a standard PID servocontroller. The reference command trajectory is provided by a programmable logic controller. The linearized state equations for the system are

$$x_1 = x_2 - (0.8\,kg + c)x_1 \tag{18}$$

$$x_2 = 5\,kg\,au - (0.8\,kg\,c + d)x_1 + x_3 \tag{19}$$

$$x_3 = 5\,abkg\,u - (0.8\,kg\,d + f)x_1 + x_4 \tag{20}$$

$$x_4 = -0.8\,kg\,f x_1, \tag{21}$$

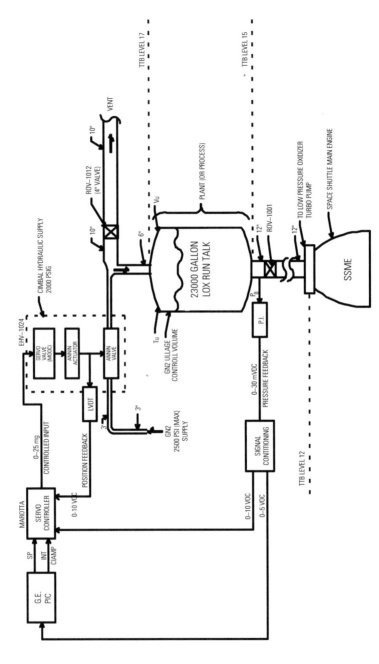

Figure 9 Propellant LOX run tank pressurization system.

where

$kg = 1$, servo valve minimum gain
$x_1 =$ bottom tank pressure
$x_2 =$ ullage temperature
$x_3 =$ ullage volume
$x_4 =$ valve stem stroke length

From the linearization technique of [19] and the assigned average operating values for the state variable, the coefficients are determined to be

a = 120.05
b = 89.19
c = 214.30
d = 5995.44
e = 14.70

Figure 10 further illustrates the controller design approach using the HOMNA technique.

1. Using a developed PID-system model, an i/o histogram is established in the required format for a select cardinality of 300. A ramp control setpoint signal (from 0 to 120 psig) served as the reference trajectory. The input portion of the

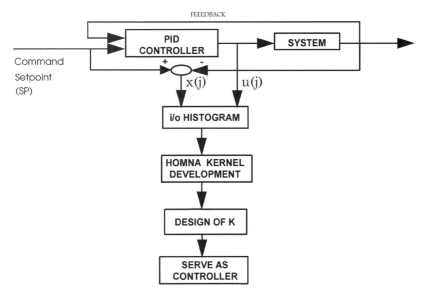

Figure 10 System and HOMNA design procedure.

histogram is selected to be a five-dimensional (300×5) matrix: the current sample and four successive delayed samples. The output portion is a one-dimensional (300×1) vector. Therefore, the i/o histogram is simply represented by a 300×6 matrix.

2. Using the HOMNA design equations of Sections III and IV, a reduced training i/o set is specified ("S"). The input portion of the set provides the mapping of real-time system inputs to the neural net controller (NNC). The output segment of the set is represented by the last column vector of the reduced i/o set.

3. After configuring the reduced i/o set into the required formats, using MAT-LAB, the gain equation, K, as described by (8) and (13), is developed. For convenience, this is represented by

$$K_m = Y_m G_m^{-1} = Y_m \psi(SS^*), \tag{22}$$

where

K_m = neural gains (row vector) for single neural layer
Y_m = NNC controller output signature row vector
S = established matrix set of step 2
ψ = any (decoupled) operation: exponential, etc.

4. For this application, the exponential function served as the operator ψ. The vector K can be viewed as a mapping function of the input, by way of S, to the neural net controller output, $u(j)$. The simulation approach is similar to that presented by Fig. 1. For convenience, this is represented by Fig. 11.

5. For select cases, the integration technique of Section IV was applied to reduce system offsets.

The simulation plots for a typical trajectory profile are shown in Figs. 12–15. The offsets and integration effects are evident in Figs. 12 and 13. Figures 14 and 15 compare the responses of the HOMNA-based controller with those of the PID controller (with the application of noise).

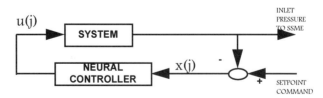

Figure 11   Simulation scheme for NNC and system.

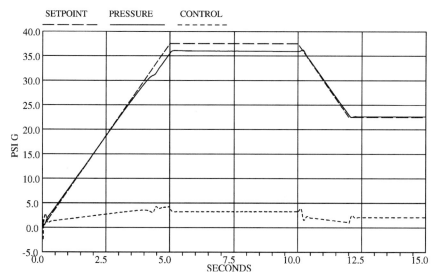

Figure 12   HOMNA-based controller response, no integration.

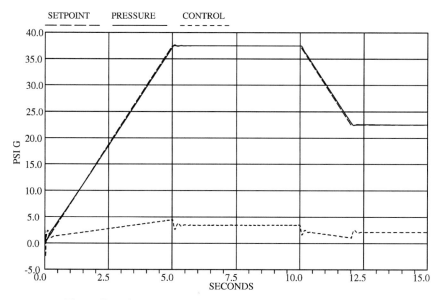

Figure 13   HOMNA-based controller response, using integration scheme.

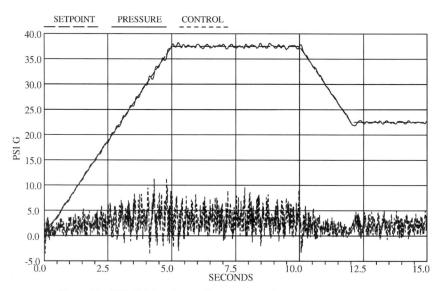

Figure 14 HOMNA-based controller response, noise and integration applied.

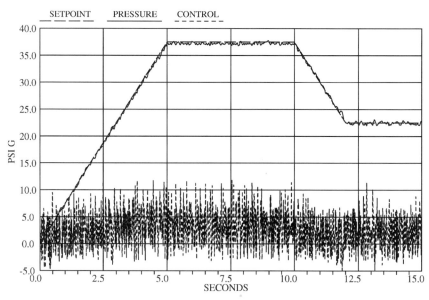

Figure 15 PID controller response, noise applied.

VI. CLOSURE

The HOMNA design of neural controllers has been demonstrated for a variety of nonlinear and nonseparable system models. Such designs show good fidelity in tracking a broad range of reference trajectories. These designs also appear to suppress smooth disturbances at either the plant input or output. They have the capability of simultaneously providing stability and tracking. For the presented propulsion system application, the HOMNA-based neural controller has proved itself viable for propulsion system applications and is capable of serving as a stand-alone or parallel backup to an existing PID controller. The HOMNA controller has proved itself capable of accurately tracking a typical pressure profile. In comparison to the PID controller response in the presence of noise, the minute differences are negligible. However, in comparisons of each controller output, the magnitude and oscillatory behavior of the PID controller are more significant than the HOMNA controller. Therefore, in terms of the life cycle of the hydraulic servo valve, the choice of the HOMNA controller would be a better choice.

Concerning the training time for the HOMNA design, this is obviously the once-through scan time of the HOMNA kernel selection algorithm. Our experience shows that this is normally some fraction of a single epoch of back-propagation training for a perceptron network with the same number of neurons and using the extended histogram as the training set.

In view of the short training time and Eqs. (11)–(13), it is not difficult to discern that the kernels algorithm can potentially compute in real time. This implies that the HOMNA design can potentially update to control plants with nonstationary characteristics. A study is in progress to determine what limits might exist and how to best use this capability for adaptive designs.

REFERENCES

[1] M. Chatty, K. Furukawa, and R. Suzuki. A hierarchical neural-network model for control and learning of voluntary movement. *Biol. Cybernet.* 57:169–185, 1987.

[2] W. Li and J. J. E. Slotine. Neural network control of unknown nonlinear systems. In *Proceedings of the 1989 ACC*, 1989, pp. 1136–1141.

[3] W. T. Miller, F. H. Glanz, and L. G. Kraft. Application of a general learning algorithm to the control of robotic manipulators. *Internat. J. Rob. Res.* 6:84–98, 1987.

[4] K. S. Narendra and K. Pathasarathy. Gradient methods for the optimization of dynamical systems containing neural networks. *IEEE Trans. Neural Networks* 2:252–262, 1991.

[5] D. H. Nguyen and B. Widrow. Neural networks for self-learning control systems. *IEEE Control System Magazine*, April 1990.

[6] P. C. Parks. Lyapunov redesign of model reference adaptive control systems. *IEEE Trans. Automat. Control* 11:362–367, 1966.

[7] K. J. Astrom and B. Wittenmark. *Adaptive Control*. Addison-Wesley, Reading, MA, 1989.

[8] R. M. Spanner and J. E. Slotine. Gaussian networks for direct adaptive control. *IEEE Trans. Neural Networks* 3: 1992.

[9] W. A. Porter. Using polynomic embedding for neural network design. *IEEE Trans. Circuits Systems* 39: 1992.

[10] W. A. Porter. Recent advances in neural arrays. *Cir. Sys. Signal Process.* 12: 1993.

[11] W. A. Porter and W. Liu. Object recognition by a massively parallel 2-D neural architecture. *Multidimens. Systems Signal Process.* 5:179–201, 1994.

[12] S. Mukhopadhyay and K. S. Narendra. Disturbance rejection in nonlinear systems using neural networks. *IEEE Trans. Neural Networks* 4: 1993.

[13] M. Minsky and S. Papert. *Perceptrons*. MIT Press, Cambridge, MA, 1969.

[14] W. A. Porter. Neuromic arrays: design and performance. In *Proceedings of WNN-AIND91*, Auburn University, February 22–13, 1991.

[15] L. C. Trevino. Neural net controller for inlet pressure control of rocket engine testing. *Third CLIPS Conference Proceedings*, Houston, TX, September 1994.

[16] W. A. Porter and S. X. Zhang. Neural identifiers for systems with unknown models. In *Transactions of the International Conference on Fuzzy Theory and Technology*, Durham, NC, October 1993.

[17] W. A. Porter and S. Ligade. Training the higher order moment neural array. *IEEE Trans. Signal Process.* 42: 1994.

[18] W. A. Porter and S. X. Zheng. A nonbinary neural network design. *IEEE Trans. Computers* 42: 1993.

[19] L. C. Trevino. *Modeling, simulation, and applied fuzzy logic for inlet pressure control for a space shuttle main engine at technology test bed*. Masters Thesis, University of Alabama Huntsville, Huntsville, AL, 1993.

On-Line Learning Neural Networks for Aircraft Autopilot and Command Augmentation Systems

Marcello Napolitano

Department of Mechanical and
Aerospace Engineering
West Virginia University
Morgantown, West Virginia 26506-6106

Michael Kincheloe

Lockheed Martin Electronics and Missiles
Orlando, Florida 32800

I. INTRODUCTION

Flight control system design for high-performance aircraft is an area in which there is an increasing need for better control system performance. The current research and development efforts involving unmanned aerial vehicles (UAVs) and remote piloted vehicles (RPVs) promise to further expand the operational flight envelope in which aircraft flight control systems must perform. The design of suitable flight control systems for these vehicles will be challenging because of the coupled, nonlinear, and time-varying dynamics that will lead directly to uncertainties in modeling. Classical and modern control law design methods rely on linearized models to compute the controller gains and interpolation algorithms to schedule the gains in an attempt to meet performance specifications throughout the flight envelope. A better approach could be implemented through a flight control system capable of "learning" throughout the flight envelope. That is, the controller updates its free parameters based on a function approximation used to map the current flight condition to an appropriate controller response. At first glance, this appears similar to gain scheduling, but there is a critical distinction: the learning controller is adapting on-line, based on the actual aircraft system rather than a model.

Several classic and modern approaches have been proposed for designing longitudinal and lateral-directional control laws for these aircraft. Typically, these techniques are either time-domain or frequency-domain based and are applied to linearized, time-invariant aircraft models. These control system design methods have a number of disadvantages. These designs are performed off-line for a limited number of linearized, time-invariant models representing different conditions in the flight envelope, requiring intensive gain scheduling computations in an effort to meet performance specifications at any point in the flight envelope. Although some approaches may be capable of handling mild nonlinearities, these approaches are not suitable for highly nonlinear problems or inconsistencies between the actual aircraft dynamics and its mathematical model. According to Baker and Farrell [1], conventional and modern control system design techniques for high-performance nonlinear systems are suitable where there is little or no modeling uncertainty. These controllers require extensive tuning in real-world implementation when the physical models used for design are not representative of the actual system. A completely new approach to the design and implementation of control laws can be found in artificial neural network (ANN) technology. To give the reader a better understanding of ANNs, a brief discussion of their evolution will be presented. The fundamental concept of ANNs dates back to the work of Frank Rosenblatt (Cornell University) in the late 1950s. Rosenblatt was concerned with typed-character recognition. His idea was to simulate the processing structure of the human brain with sets of parallel, fully interconnected processing elements (neurons) containing a binary, nonlinear, hard-limiting activation function [2]. The human brain is a massively parallel network of neurons with a processing speed on the order of milliseconds. Each neuron has between 10^{11} and 10^{14} associated connections. A neuron is the basic processing unit of the human nervous system. The neurons receive input signals from other neurons. If the combined input signals are sufficient to exceed some threshold, an output is generated [3]. This process can be simulated with hardware or software. Rosenblatt implemented his network via hardware and called it the perceptron. The perceptron had the capability to determine whether an input belonged to one of two classes. Rosenblatt "trained" his network, using what he called the perceptron convergence procedure. He was able to prove leaning convergence for data sets that were separable into two classes. The perceptron learning procedure exhibited oscillatory behavior when inputs were not distinctly separable.

Working at the same time as Rosenblatt were Bernard Widrow and Marcian Hoff (Stanford Electronic Laboratories). Widrow and Hoff introduced the adaptive linear neuron (ADALINE) and multiple adaptive linear neuron (MADALINE). These differed from the perceptron in that they used a least mean square (LMS) training procedure and a continuous nonlinear activation function. The MADALINE has seen applications in adaptive nulling of radar jammers, adaptive modems, and adaptive equalizers in telephone lines [4]. The problem with all of these ANNs is that they are single-layered networks that are only capable of classifying linearly separable patterns. No suitable training algorithm was available to train multilayered networks with a nonlinear activation function. The interest in ANN technology was limited until the back-propagation training paradigm was introduced. Back-propagation (BP) is an algorithm based on method of steepest descent that minimizes the mean squared error between the desired network output and the actual network output. Learning techniques of this type (where the desired output is known *a priori*) are called *supervised learning*. It should be noted that another class of ANNs exists where the training is unsupervised, which allows the networks to recognize patterns in massive input sets that are not readily obvious. Because this type of network is notdirectly applicable to a control-type problem, it will not be discussed. The BP training algorithm revitalized interest in ANNs by removing the limitation on the number of layers and providing for a continuous (rather than binary) output range [5].

More detailed descriptions of neural network theory as well as discussions of the various training algorithms are available in [6] and [7]. The critical point of these neural architectures is the adaptability inherent in their parallel processing structure. This structure provides excellent generalization capability and allows the network, given sufficient training, to map

- Simple or complex dynamic systems
- Time-invariant and/or time-varying systems
- Noise-free or noise-corrupted systems
- Linear and nonlinear systems

Recently, several investigations have proposed the implementation of neural networks in flight control systems, both as controllers and as estimators. In 1991, C. M. Ha (General Dynamics) [8] applied ANN technology to the lateral-directional control laws in a high-performance aircraft model with six degrees of freedom (DOF). Ha used a linearized state–space representation of the 6-DOF model and a hybrid approach coupling a feedback neural controller and a feedforward neural filter. The networks were trained off-line with the linear model. Once a suitable degree of learning had occurred, the network architectures were frozen and the controllers were tested on-line in a linear, time-domain simulation. Ha was able to demonstrate reasonable tracking of pilot command inputs using this model. Ha was also able to use a similar scheme to design and demonstrate off-line trained neural controller in a reconfiguration strategy for the pitch-rate controller in a longitudinal flight control system with the same linearized model [9]. Baker and Farrell [1] provide an excellent and complete discussion of the issues related to the design and implementation of learning adaptive (neural) controllers in flight control systems and the associated performance expectations relative to gain scheduling-based modern control strategies. Milligan *et al.* [10] successfully introduced a "hybrid" adaptive learning controller for the pitch-rate command augmentation system in a high-performance aircraft model. All of these investigations fall short of proposing on-line learning neural network (OLNNET) controllers as replacements for gain-scheduling-based control schemes, as they describe the successful implementation of neural controllers trained off-line with linearized mathematical models.

OLNNET controllers have been proposed for specific flight control problems such as sensor failure detection, identification, and accommodation (SFDIA) [11], and actuator failure detection, identification, and accommodation (AFDIA) [12, 13]. These studies illustrated the successful implementation of OLNNET controllers for flight control applications and particularly demonstrated the OLNNET controller's capacity for adapting to sudden changes in the aircraft model. The controller described in these studies were all of the multiple input–single output (MISO) configuration. The purpose was to demonstrate the OLNNET effectiveness in regaining control of the aircraft after a failure mode or battle damage. No investigations into the model following performance or into the multiple input–multiple output (MIMO) capability were made.

This text attempts to expand the current data base of neural network research. OLNNET controllers were designed and implemented on-line for five typical autopilot functions using the NASA nonlinear six-degree-of-freedom simulation code distributed for the 1991 AIAA Controls Design Challenge. The OLNNET autopilot performance was evaluated under both linear and nonlinear conditions, and at multiple points in the aircraft flight envelope. These controllers are MISO, and the goal was to investigate their global learning performance. In the second part of the investigation, OLNNET controllers with MIMO configurations were implemented for lateral directional command augmentation systems (CASs) in the same aircraft model. The purpose was to illustrate that a single neural controller architecture is able to exhibit excellent learning and adaptation performance, not only throughout the flight envelope, but while tracking multiple models. The remainder of this chapter is organized as follows. Section II provides a detailed description of the standard back-propagation and extended back-propagation algorithms used. Section III gives a brief description of the NASA 6-DOF model used to simulate closed-loop performance. Section IV discusses OLNNET autopilots and associated results. Section V covers the investigations related to the MIMO OLNNET CAS controllers. Finally, Section VI provides a summary of the topics covered as well as recommendations for further study.

II. THE NEURAL NETWORK ALGORITHMS

The back-propagation (BP) algorithm is a gradient-based method for the updating of network free parameters (weights and thresholds for each layer) to minimize some performance index. In general, the network consists of input and output sets fully connected through one or more layers of processing elements (neurons), as shown in Fig. 1.

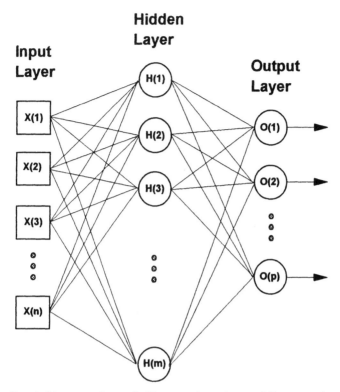

Figure 1   Generic 3-layer neural network. There are *n* input elements fully connected to *m* hidden-layer neurons. These hidden-layer neurons are fully connected to *p* output layer neurons.

The input to each processing element is the weighted sum of the outputs of the previous layer plus a bias term. This weighted sum serves as the input signal to a nonlinear activation function, the output of which becomes the output signal for the processing element. Figure 2 shows the structure of a typical processing element.

A traditional BP network has two phases of operation:

- The forward phase
- The update (training) phase

In the forward phase the activation signal for each of the hidden layer neurons is computed as the weighted sum of the output signals from the previous layer. For the first hidden layer the input signal is provided by the input pattern. For each successive layer the activation signal is the weighted sum of the output signals of

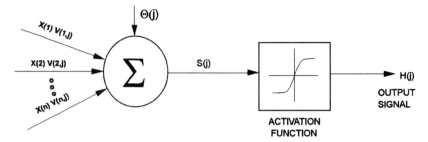

Figure 2 Internal structure of a typical processing element (neuron). The weighted sum of all input signals serves as an input signal to a nonlinear activation function which generates the network output.

all processing elements of the previous layer:

$$S_j = \sum_{i=1}^{L} (v_{ij} X_i + \theta_i). \tag{1}$$

The activation signal in each processing element is then sent to a nonlinear activation function, in this case a sigmoid function, which can be modified to map the activation signal into any desired output range by setting an upper bound (U) and lower bound (L), and over any activation signal range by setting the slope (T). In this study, the standard back-propagation (SBP) output range was set to ± 1 with a slope of 1:

$$H_j = f(S_j) = \frac{U - L}{1 + e^{-S_j/T}} + L. \tag{2}$$

Similarly, the output layer activation signals are provided by the weighted sums of the outputs of the previous hidden-layer neurons. Again, this weighted sum is sent to the modified sigmoid activation function to give the output of each output layer node:

$$O_k = f(S_k) = \frac{U - L}{1 + e^{-s_k/T}} + L \tag{3}$$

$$S_k = \sum_{j=1}^{M} (w_{jk} H_j + \Gamma_k). \tag{4}$$

The training algorithm for the SBP network is well documented and is available in a number of standard texts on neural networks [6, 7] and therefore will not be presented here. The extended back-propagation (EBP) algorithm, which was first proposed by Nutter and Chen (West Virginia University) [12], is not as readily available and will be discussed.

Although the SBP algorithm has renewed interest in the study of artificial neural systems for many applications, it does have limitations. The learning speed can be slow, especially for large order systems, and because of the complex mapping of the error gradient, there can be local minima problems. The first issue can be improved by increasing the learning speed (i.e., a larger learning rate), but the increase in learning speed comes at the expense of an increase in the likelihood of the network becoming "trapped" in a local minimum, thus being unable to find a suitable set of weights that minimize the performance index. These problems are therefore coupled. Many modifications to the BP learning rules, such as using a "momentum" term, changing the sigmoid function, using sigmoid functions with adaptive slopes, or allowing for adaptive learning rates, have been proposed to address these issues [7]. The EBP algorithm was proposed as an approach to solving the SBP difficulties. The extended back-propagation network is a heterogeneous network, which means that each neuron in the hidden and output layers has the capacity to update not only the incoming weights but also the parameters of its activation function. Specifically, each neuron can change the upper (U) and lower (L) bounds as well as the activation function slope (T).

A. EXTENDED BACK-PROPAGATION TRAINING ALGORITHM

In the training mode the input pattern is presented and the network output is computed. The output of each output layer processing element is then compared to the corresponding target output, and an error signal (δ_k) is computed for each activation function parameter (S, U, L, and T) for each output neuron:

$$\delta_{S_k} = f'_{S_k}(Y_k - O_k) \quad \text{where } k = 1 \text{ to no. of output layer neurons} \tag{5}$$

$$\delta_{U_k} = f'_{U_k}(Y_k - O_k) \tag{6}$$

$$\delta_{L_k} = f'_{L_k}(Y_k - O_k) \tag{7}$$

$$\delta_{T_k} = f'_{T_k}(Y_k - O_k). \tag{8}$$

It is important to note that only the output layer processing element has desired (or target) outputs; the hidden-layer free parameters are updated based on the "backward-propagated" error signal from the output (or next hidden) layer. The error signals for each successive hidden layer processing element can be computed from Eqs. (9)–(12):

$$\delta_{S_j} = f'_{S_j}\left[\sum_{k=1}^{N}(\delta_{S_k} w_{jk})\right] \tag{9}$$

$$\delta_{U_j} = f'_{U_j}\left[\sum_{k=1}^{N}(\delta_{U_k}w_{jk})\right] \tag{10}$$

$$\delta_{L_j} = f'_{L_j}\left[\sum_{k=1}^{N}(\delta_{L_k}w_{jk})\right] \tag{11}$$

$$\delta_{T_j} = f'_{T_j}\left[\sum_{k=1}^{N}(\delta_{T_j}w_{jk})\right]. \tag{12}$$

The activation function derivatives with respect to each of the updatable parameters are given by Eqs. (13–16):

$$f'_S = \frac{(O_k - U_k)(O_k - L_k)}{T_k(U_k - L_k)} \tag{13}$$

$$f'_U = \frac{1}{1 + e^{-S_k/T_k}} \tag{14}$$

$$f'_L = 1 - f'_U \tag{15}$$

$$f'_T = -\frac{S_k f'_{S_k}}{T_k}. \tag{16}$$

Similarly, the activation function derivatives for the hidden-layer free parameters are given by Eqs. (17–20):

$$f'_S = \frac{(O_j - U_j)(O_j - L_j)}{T_j(U_j - L_j)} \tag{17}$$

$$f'_U = \frac{1}{1 + e^{-S_j/T_j}} \tag{18}$$

$$f'_L = 1 - f'_U \tag{19}$$

$$f'_T = -\frac{S_j f'_{S_j}}{T_j}. \tag{20}$$

From Eqs. (5–8) the weight, threshold, and activation function parameter update equations for each neuron of the output layer become, respectively,

$$w_{jk_{NEW}} = w_{jk_{OLD}} + \eta\delta_{S_k}H_j + \alpha\Delta w_{jk_{OLD}} \tag{21}$$

$$\Gamma_{k_{NEW}} = \Gamma_{k_{OLD}} + \eta\delta_{S_k} + \alpha\delta\Gamma_{k_{OLD}} \tag{22}$$

$$U_{k_{NEW}} = U_{k_{OLD}} + \eta\delta_{S_k} + \alpha\Delta U_{k_{OLD}} \tag{23}$$

$$L_{k_{NEW}} = L_{k_{OLD}} + \eta\delta_{S_k} + \alpha\Delta L_{k_{OLD}} \tag{24}$$

$$T_{k_{NEW}} = T_{k_{OLD}} + \eta\delta_{S_k} + \alpha\Delta T_{k_{OLD}}. \tag{25}$$

Similarly, the hidden-layer weight, threshold, and activation function parameter updates are given by Eqs. (26–30), respectively:

$$v_{ij_{NEW}} = v_{ij_{OLD}} + \eta \delta_{S_j} X_i + \alpha \Delta v_{ij_{OLD}} \tag{26}$$

$$\Theta_{j_{NEW}} = \Theta_{j_{OLD}} + \eta \delta_{S_j} + \alpha \Delta \Theta_{j_{OLD}} \tag{27}$$

$$U_{j_{NEW}} = U_{j_{OLD}} + \eta \delta_{U_j} + \alpha \Delta U_{j_{OLD}} \tag{28}$$

$$L_{j_{NEW}} = L_{j_{OLD}} + \eta \delta_{L_k} + \alpha \Delta L_{j_{OLD}} \tag{29}$$

$$T_{j_{NEW}} = T_{j_{OLD}} + \eta \delta_{T_j} + \alpha \Delta T_{j_{OLD}}, \tag{30}$$

where η is the learning rate, and α is the momentum rate. The learning rate governs the magnitude of the parameter updates at each successive training step. The momentum rate weights the influence of the previous weight update on the current update. It should be noted that additional degrees of freedom can be obtained be setting different learning and momentum rates for each of the free parameters. It should also be noted that in a typical derivation of the BP learning rules [6], the momentum term is not included. Several studies [7, 14] have demonstrated that the momentum term can substantially decrease the required training iterations and thereby the training time. A theoretical derivation of the momentum term is given in [15].

III. AIRCRAFT MODEL

The aircraft model used in this study was a nonlinear, six-degree-of-freedom model representative of a modern high-performance supersonic fighter with full flight envelope nonlinear aerodynamics and full-envelope thrust with first-order engine response data. This is the FORTRAN software model distributed for the 1991 AIAA Control Design Challenge [16]. The model can receive control inputs from conventional ailerons, two horizontal elevators capable of symmetric or differential movement, a single vertical rudder, and a throttle. All control surfaces have identical actuators with rate limits of 24 degrees per second. The maximum deflections for the elevators and the rudder are ± 30 degrees. The maximum deflection for the ailerons is ± 21.5 degrees.

In the original configuration the software did not allow the user to input specific maneuvers or access the control surfaces directly. The model was trimmed via four conventional autopilots (altitude hold, velocity hold, roll angle hold, and sideslip suppression) to a preset condition. These autopilots were designed by the NASA model developers with classical control theory (PI and PID) with gains computed for a single flight condition. The conventional autopilots were included with the software to provide simple, basic autopilot functions and to give the user a template to interface future flight controller development with the core simulation. It should be mentioned here that no attempt was made by the authors to enhance the performance of these controllers; performance plots are included throughout this text to provide the reader with a point of reference in assessing the performance of the OLNNET controllers. The authors want to emphasize here that these conventional controllers do not represent the state of the art in modern control theory. The reader should keep in mind that powerful control schemes exist that could significantly improve the performance in a conventional configuration.

The model was modified to allow a user to "fly" the simulation manually through keyboard commands. This provided a primitive pilot-in-the-loop interface for parts of the investigation. Also added was the capability to toggle the original conventional autopilots and the new neural flight control systems via the computer keyboard. Finally, the model was changed to allow the user to fly both longitudinal and lateral-directional preprogrammed maneuvers. This capability was used extensively in the evaluation of the neural command augmentation systems.

IV. NEURAL NETWORK AUTOPILOTS

This chapter presents the results of a study that was conducted with the goal of demonstrating the suitability of on-line learning neural network (OLNNET) controllers for the control laws in the autopilot functions of a high-performance aircraft. OLNNET controllers were designed to replace the conventional control laws for five autopilot functions. The neural controllers were trained on-line with two different learning algorithms: the standard back-propagation (SBP) algorithm and the extended back-propagation (EBP) algorithm. The objective was to show that these neural-based control laws, which do not require gain scheduling, exhibit high levels of performance throughout the flight envelope and are generally robust to system nonlinearities.

OLNNET controllers were designed for the autopilot functions of

- Altitude hold
- Pitch angle hold
- Airspeed hold
- Roll angle hold
- Direction hold

The design of OLNNET controllers can appear complex because of the multiple degrees of freedom available in the network architectures. Previous studies [11–13] have demonstrated that OLNNET controllers are highly robust in the following areas:

- Number of hidden-layer processing elements
- Number of hidden layers: single hidden layer (three layer) networks perform as well as more complex multiple hidden layer networks
- Selection of the input data set and pattern

Therefore, the emphasis of the design process for the OLNNET autopilots was placed on the selection of the performance index to be minimized and the selection of the learning rates.

The most critical element in the design of neural controllers is the selection of the performance index to be minimized. An interesting and important result of this investigation is that the parameters in the performance index for each of the five neural autopilots are the same parameters as required in the feedback loops of conventional or "classic" autopilots. The performance indices for the neural controllers have typical PI, PD, and PID configurations. The architectures for each of the neural controllers, along with their associated performance indices, are shown in Table I. It should be noted that some classic design specifications such as overshoot and settling time for the on-line learning neural controllers can be influenced by the selection of the weight of each term in the performance index. These effects, however, need to be better understood and are beyond the scope of this text.

The second critical area of the design of a neural controller is the selection of the learning rate. Because the OLNNET controller is trained on-line in real time, it is obvious that most of the training must take place in the first few seconds after activation of the controller. This implies that a large learning rate must be used. A common problem with large learning rates (> 0.2) is that as the learning rate increases, so does the likelihood of the network becoming trapped in a local minimum. The training of OLNNET controllers is, however, a local problem. It is not necessary, for example, that the altitude hold autopilot "learn" the system dynamics throughout the flight envelope, but only at the current flight condition. As the flight condition changes, the controller can adapt its free parameters to accommodate the change and perform the required function while only gaining a local knowledge of the system. A brief parametric study of the effects of the learning rates was performed, and a learning rate of 0.5 for all neural controllers was selected. This rate provided fast learning convergence and showed no adverse local minima effects.

A. PHASE I: FLIGHT ENVELOPE PERFORMANCE

In phase I of this study, OLNNET controllers trained with both the SBP and the EBP algorithms were compared to the single set-point designed conventional controllers contained in the original model. It should be noted here that the comparison of the performance of the OLNNET controllers to that of the conventional controllers is intended to provide the reader with a point of reference by which to gauge the OLNNET performance. It is not intended to be a comparison of neural versus conventional design. The on-line learning capabilities of two of the

Table I

Architectures for Neural Network Autopilots Showing Performance Indices and Weighting Terms

Altitude hold system
NN input: θ, q, h, $\overset{\circ}{h}$, δ_E
NN input window: 3
NN output: δ_E
Number of hidden-layer neurons: 18
Learning rate: 0.5
Neural architecture: 15, 18, 1
Performance index: $J = 0.25(h - h_{ref})$
$+0.7(\overset{\circ}{h} - \overset{\circ}{h}_{ref}) + 0.05(q - q_{ref})$
h_{ref} is user defined: $h_{ref} = q_{ref} = 0.0$

Airspeed hold system
NN inputs: U, a_X, δ_T
NN input window: 3
Number of hidden-layer neurons: 12
Learning rate: 0.5
Neural architecture: 9, 12, 1
Performance index: $J = 0.3(U - U_{ref})$
$+0.7(a - a_{X, ref})$
U_{ref} is user defined: $a_{X, ref} = 0.0$

Roll-angle control system
NN input: Φ, p, δ_A, δ_R
NN input window: 3
Network output: δ_A
Number of hidden-layer neurons: 15
Learning rate: 0.5
Neural architecture: 12, 15, 1
Performance index: $J = 0.5(\Phi - \Phi_{ref})$
$+0.5(p - p_{ref})$
Φ_{ref} is user defined: $p_{ref} = 0.0$

Direction hold system
NN inputs: r, r, β, δ_R
NN input window: 3
NN output: δ_R
Number of hidden-layer neurons: 15
Learning rate: 0.5
Neural architecture: 12, 15, 1
Performance index: $J = 0.5(r - r_{ref})$
$+0.5(r - r_{ref})$
$r_{ref} = r_{ref} = 0.0$

Pitch hold system
NN input: θ, q, δ_E
NN input window: 3
NN output: δ_E
Number of hidden-layer neurons: 12
Learning rate: 0.5
Neural architecture: 9, 12, 1
Performance index: $J = 0.5(\theta - \theta_{ref})$
$+0.05(q - q_{ref})$
θ_{ref} is user defined: $q_{ref} = 0.0$

neural autopilot systems (altitude hold and airspeed hold) were evaluated in a dynamic simulation over 18 different flight conditions. These flight conditions and their corresponding locations in the model flight envelope are shown in Fig. 3. Because the controller performance is based on the speed of the gradient-based minimization of the performance index, the OLNNET controllers cannot be designed for performance in terms of the classic time-domain specifications such as damping, natural frequency, overshoot, settling time, rise time, etc. To evaluate

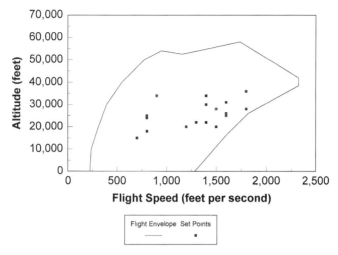

Figure 3 NASA 6-DOF aircraft model flight envelope showing 18 set points used in the autopilot investigation.

the OLNNET controller capability, a performance measure had to be defined. It was decided to use a modified definition of settling time (T_S) where

- $T_{S, \text{ALT}}$ = time required to go from ±100 feet of the target altitude settling to an oscillation around the target of ±6 feet.
- $T_{S, \text{VEL}}$ = time required to go from ±50 feet/sec of the target airspeed to an oscillation around the target of ±3 feet/sec.

The settling times were computed for the three controllers (SBP, EBP, and conventional) for the two autopilot systems (altitude hold and airspeed hold) over the 18 flight conditions. They are shown in Table II. It should be noted that because of the propulsive characteristics as modeled in the original NASA simulation, the airspeed hold autopilot systems experienced some model-related difficulties at lower flight speeds (< 700 feet/sec). This affected both the OLNNET and the conventional systems. Figures 4 and 5 show plots of the settling times for the altitude and airspeed autopilots, respectively, for each of the three controllers at each flight condition.

It can be seen that both EBP and SBP performed well compared to the conventional controllers under all flight conditions. It can also be noted that the EBP generally exhibited performance superior to that of the SBP controller over the range of flight conditions, as was expected.

Table II

Complete Listing of the 18 Flight Conditions Along with the Settling Time Performance of the SBP, EBP, and Conventional Altitude and Airspeed Autopilots

Flight condition	Altitude (ft)	Airspeed (ft/sec)	Ts, Altitude (sec) EBPA	Ts, Altitude (sec) SBPA	Ts, Altitude (sec) CONV	Ts, Airspeed (sec) EBPA	Ts, Airspeed (sec) SBPA	Ts, Airspeed (sec) CONV
1	15,000	700	26.42	28.68	73.59	5.94	38.10	31.52
2	18,000	800	15.04	17.72	12.56	7.08	7.44	26.01
3	20,000	1,200	12.54	13.54	19.24	8.26	9.26	119.21
4	22,000	1,400	48.53	71.10	42.19	18.89	20.88	121.44
5	24,000	800	12.67	20.15	23.14	23.26	29.86	145.00
6	22,000	1,300	4.19	13.61	19.93	14.25	15.61	75.61
7	26,000	1,600	11.03	12.21	16.31	19.13	18.73	119.15
8	30,000	1,400	4.60	4.04	19.90	6.46	5.92	51.03
9	28,000	1,800	15.81	18.61	22.02	28.52	32.15	96.45
10	31,000	1,600	12.79	13.01	16.45	7.20	9.69	7.97
11	34,000	900	12.55	12.13	15.07	15.31	21.86	178.78
12	28,000	1,500	12.49	12.99	17.16	14.05	15.25	172.53
13	34,000	1,400	10.87	12.61	24.74	5.94	6.18	8.38
14	36,000	1,800	11.23	11.55	25.78	25.88	25.90	110.33
15	25,000	1,600	11.69	11.77	16.14	6.46	7.00	55.01
16	22,000	1,300	13.31	13.40	19.60	5.90	6.69	151.33
17	25,000	800	11.45	18.29	12.63	18.77	24.26	98.94
18	20,000	1,500	10.49	10.83	15.88	17.46	17.81	81.76

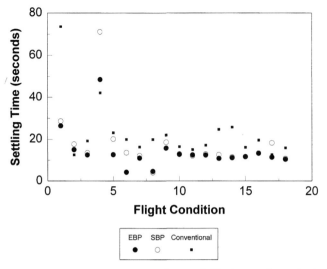

Figure 4 Graphical comparison of settling times for neural altitude controllers trained with the standard back-propagation (SBP) and the extended back-propagation (EBP) algorithms.

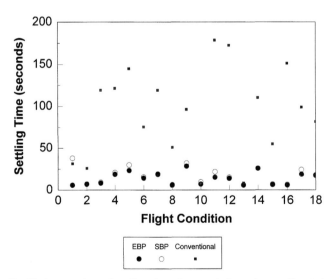

Figure 5 Graphical comparison of settling times for neural airspeed controllers trained with the standard back-propagation (SBP) and the extended back-propagation (EBP) algorithms.

B. PHASE II: NEURAL AUTOPILOT CONTROLLERS
UNDER LINEAR AND NONLINEAR CONDITIONS

As stated earlier, an important feature of neural controllers is their robustness to system nonlinearities. The three remaining autopilot systems (pitch angle hold, roll angle hold, and direction hold) were studied under both linear and nonlinear conditions. For this phase of the study, the original FORTRAN simulation was modified to allow user command inputs from the computer keyboard. This allowed the user to fly the simulation and manually activate the autopilots in any flight configuration. Although it may not be realistic to activate the autopilot systems under nonlinear flight conditions, it is important to determine the OLNNET controller performance and evaluate the on-line learning capabilities in these regions.

For evaluation of the neural roll angle hold autopilot, the simulation was manually "flown" into a linear configuration (low angle of attack, low roll rate) and rolled to both small ($< 45°$) and large ($> 45°$) bank angles. The neural roll autopilot was then activated in each roll situation. The neural roll autopilot was instructed to hold the aircraft at a bank angle of $0°$. The settling time, defined as the time required to stabilize the aircraft with $±0.5°$ of the target roll angle, was computed for each test. Figure 6 shows the results of the roll angle hold autopilot performance linear conditions. As can clearly be seen, the neural roll autopilot performance was very good. Next, the aircraft was flown into a nonlinear condition (high angle of attack, moderate roll rate) and rolled to both large and small roll angles, with the neural roll autopilot activated in each situation. The settling time was computed with the same criteria as in the linear test. Figure 7 shows the results of the nonlinear test. Again, the neural autopilot exhibited very good performance. It should be noted that control over the directional dynamics was maintained through the directional neural autopilot in both the linear and nonlinear phases. It is likely that, given the conditions at the time the system was activated, some dynamic coupling effects (induced by the simultaneous high values of p, q, and r) were present. Nevertheless, wing level conditions were regained in a reasonable time.

The neural directional autopilot was evaluated under both linear and nonlinear conditions similarly. The aircraft was manually flown into both linear and nonlinear conditions, and the neural directional autopilot was instructed to cancel any yaw rate and acceleration. The settling time, defined as the time required to stabilize the aircraft within $±0.1$ degree/sec of the target yaw rate, was computed for each test. Figures 8 and 9 show the results of the linear and nonlinear simulations, respectively. The performance under both conditions, even with the high initial angular velocities, is excellent. Again, it should be noted that control over the lateral (rolling) dynamics was maintained through the neural roll autopilot.

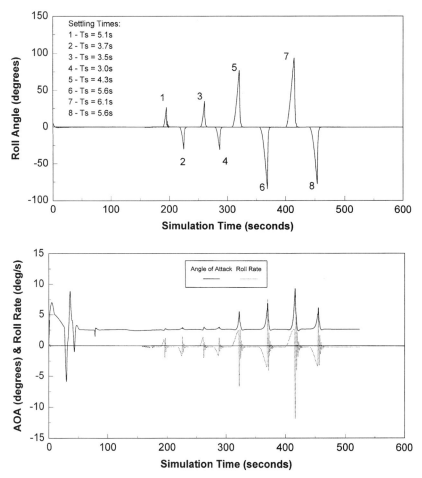

Figure 6 Neural roll autopilot, linear conditions. Top: Roll angle versus time for eight linear test runs (four at low roll angles and four at larger roll angles). Bottom: Angle of attack and roll rate during runs.

In the final evaluation for phase II, the neural pitch controller was tested at low and high angles of attack. Unlike the previous autopilot functions, when the pitch controller is activated it is instructed to hold a user-defined pitch angle. Arbitrary small values for the target pitch angle were selected each time the neural pitch controller was activated. Figure 10 shows the results of the pitch controller tests. As demonstrated by these results, the OLNNET pitch controller performed well under both linear and nonlinear conditions.

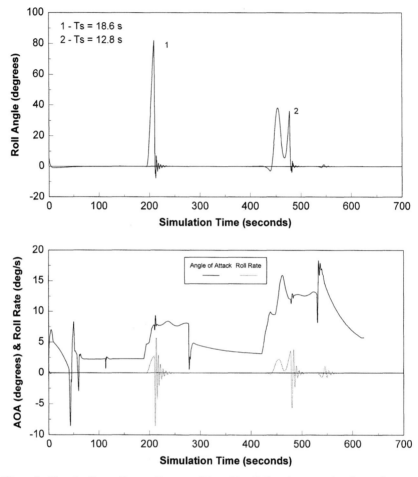

Figure 7   Neural roll autopilot, nonlinear conditions. Top: Roll angle versus time for nonlinear test runs. Bottom: Angle of attack and roll rate during runs.

C. CONCLUSIONS

This section has presented the results of an investigation demonstrating the feasibility of using on-line learning neural network controllers as a replacement for conventional gain scheduling-based control laws for the autopilot functions of a modern high-performance aircraft. The investigation was conducted in two phases.

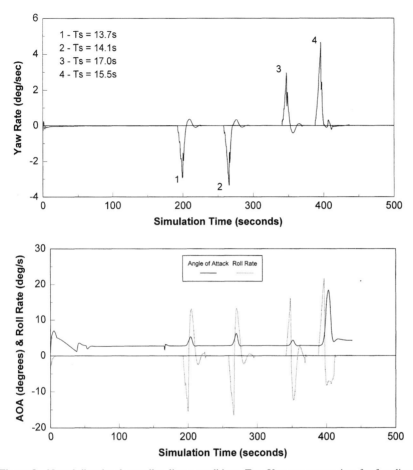

Figure 8 Neural directional autopilot, linear conditions. Top: Yaw rate versus time for four linear test runs (two at low roll angles and two at larger roll angles). Bottom: Angle of attack and roll rate during runs.

Phase I compared neural autopilots trained on-line with both the standard back-propagation and the extended back-propagation algorithms to the single-point designed conventional controllers included in the original AIAA controls design challenge software for the autopilot functions of altitude hold and airspeed hold at 18 flight conditions in the flight envelope. The results showed that both sets of neural autopilots were capable of performing well throughout the flight envelope. The results also demonstrated the generally superior performance of the EBP training algorithm over the SBP training algorithm.

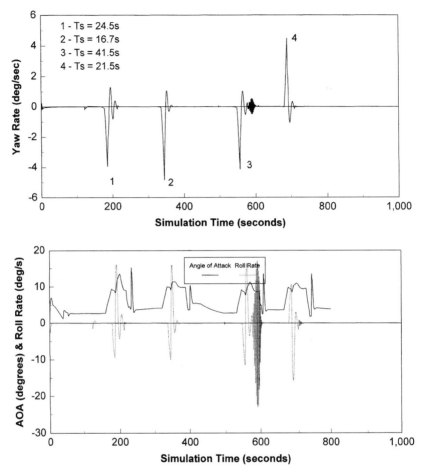

Figure 9 Neural directional autopilot, nonlinear conditions. Top: Yaw rate versus time for nonlinear test runs. Bottom: Angle of attack and roll rate during runs.

   In phase II, three neural autopilots trained with the EBP algorithm for the autopilot functions of roll angle hold, direction hold, and pitch angle hold were studied under both linear and nonlinear flight conditions. Although it may not be realistic to activate the autopilots under nonlinear conditions, it is important to investigate the capabilities of the neural control schemes under such conditions. The results showed acceptable performance for all of the tested neural controllers under both linear and nonlinear conditions.

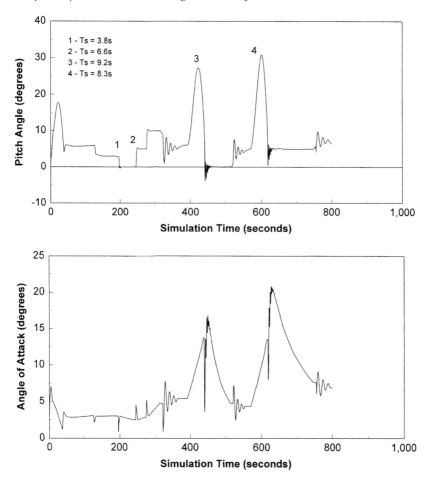

Figure 10 Neural pitch autopilot, linear and nonlinear conditions. Top: Pitch rate versus time for nonlinear test runs. Bottom: Angle of attack during runs.

V. NEURAL NETWORK COMMAND AUGMENTATION SYSTEMS

One of the most challenging and interesting areas of aircraft control theory is the design of flight control systems to augment the dynamic behavior of the aircraft. There has been and will continue to be increasing interest in flight control systems capable of changing the aircraft dynamics to follow multiple models for different mission situations. For example, an inherently dynamically unstable aircraft such as the B-2 or the F-117 may be required to respond in a highly stable manner during in-flight refueling or while performing bombing runs. Similarly, command augmentation systems can be used to artificially implement the dynamic behavior of a totally different aircraft to provide valuable pilot training lead time. An excellent example of this is the training of space shuttle flight crews. Shuttle pilots are trained in business jet class aircraft with a software-implemented flight control system designed to cause dynamic responses similar to those of the shuttle.

Traditionally, multiple dynamic modes have been introduced to the system and controlled through model reference adaptive control (MRAC). The MRAC control scheme can be defined by two methods: (1) a direct approach and (2) an indirect approach. In a direct MRAC scheme, the parameters of the controller are adjusted on-line. In an indirect MRAC scheme, a parameter identification algorithm is implemented with the controller parameters calculated based on the estimates of the actual system parameters. Both of these methods apply, in the classic modern control theory sense, to linear, time-invariant systems. Both controllers are single input–single output (SISO). Stability for the direct approach, for the above-mentioned systems, has been proved, but this cannot be said for the indirect approach. In addition, the indirect approach requires a high degree of input excitation for acceptable performance. Neither approach (direct or indirect) is designed to be applicable to either nonlinear or time-varying systems; however, through modifications to the recursive least-squares algorithm used in each, the MRAC approach can be applied to time-varying systems, but stability is not guaranteed [17].

A new approach to the design of CASs for high-performance aircraft is proposed and investigated here: on-line learning neural network (OLNNET) controllers. Neural controllers lend themselves to adaptive control applications through the network's inherent MIMO parallel processing structure. The OLNNET controllers were implemented in a direct approach, that is, actual sensor state feedback is used for the controller updates. In the classical direct approach the error signal to the controller is generated based on a controller output error. For this application, the controller output error is not available (as the model following the control sequence is not known *a priori*). The error signals for the network weight updates were provided as the weighted sum of the tracking error (defined as the difference between the desired and the actual response). This opens a potential problem area: if the correlation between the controller output and the tracking error is weak, then insufficient error information will be passed to the network. This would result in poor learning characteristics and unacceptable closed-loop performance. Several methods [18–20] for overcoming this have been proposed and successfully tested on nonlinear, SISO systems. However, because a strong correlation between the OLNNET controller output and the tracking error is assumed for the aircraft model and reference response models used here, and a major objective of this study was demonstration of the adaptation capabilities in a MIMO configuration, none of these methods were implemented.

In this investigation OLNNET CAS controllers were designed and implemented in the same aircraft model previously described for augmented control of the lateral-directional dynamics. The controllers were evaluated in the following areas:

- Learning capability
- Adaptation and robustness

Two MIMO OLNNET CAS controllers were implemented to track the desired model responses by minimizing the following performance indices (PIs) over a discrete time span:

$$J_{\text{LATERAL}} = \frac{1}{2} \sum_{k=1}^{N} \left[R_1 \big(p(k) - p_m(k) \big)^2 + Q_1 \beta(k)^2 \right] \qquad (31)$$

$$J_{\text{DIRECTIONAL}} = \frac{1}{2} \sum_{k=1}^{N} \left[R_2 p(k)^2 + Q_2 \big(\beta(k) - \beta_m(k) \big)^2 \right], \qquad (32)$$

where R_1, Q_1, R_2, and Q_2 are scalar, constant weighting terms, and N is the number of time steps during the maneuver.

As stated in previous sections of this paper, the performance of the OLNNET controllers in terms of classical design characteristics can be significantly influenced through selection of the performance index and associated weighting terms. Again, optimization of the PIs for each controller was not formally investigated. Brief parametric studies led to the PIs and weighting terms used in this study.

In the original configuration, the NASA 6-DOF model did not allow the user to input a pilot maneuver. The model software was modified to allow the user to program and save lateral and directional maneuvers that could be activated and run repeatedly during the same (or multiple) flight sessions. These preprogrammed maneuvers provided the basis for the evaluation of the OLNNET CAS's performance with regard to "learning" a particular maneuver repeated at a given flight condition and being able to track a particular model response throughout the aircraft flight envelope. Two pilot-commanded aileron maneuvers were used in this investigation. Both maneuvers contained a pair of aileron doublet commands. The first maneuver (Lateral 1) commanded an initial 0 degree aileron deflection for a short time prior to the doublet pair to allow the simulation to "settle" from the activation state. The second maneuver (Lateral 2) does not provide the initial 0 degree command. Each maneuver was flown with the roll rate CAS disengaged, and open-loop responses were generated. Figures 11 and 12 show the commanded aileron deflections and the open-loop responses for each lateral maneuver. Similarly, for the sideslip CAS, a single maneuver was programmed in. This maneuver

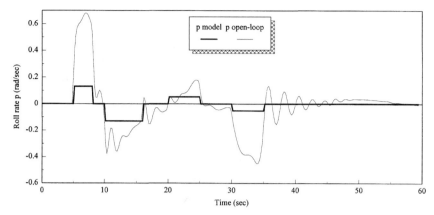

Figure 11 Commanded aileron deflection and open-loop roll-rate response for Lateral 1 maneuver.

consisted of a series of rudder steps. The sideslip maneuver commanded rudder deflection; the corresponding open-loop response is shown in Fig. 13. These maneuvers can be initiated by the user pilot starting from any flight condition.

Investigation of the OLNNET CAS performance was divided into three areas. The first was a study of the learning and adaptation performance of the lateral (roll rate) CAS under both linear and nonlinear conditions and the directional (sideslip) CAS under nonlinear conditions. According to MIL-F-8785 Level 1 flying qualities [21] specifications (Military Standard USAF 1987), a stability

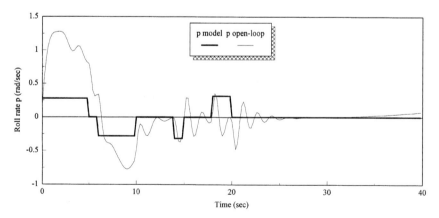

Figure 12 Commanded aileron deflection and open-loop roll-rate response for Lateral 2 maneuver.

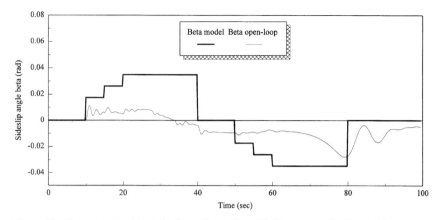

Figure 13 Commanded rudder deflection and open-loop sideslip response for directional maneuver.

augmentation system (which a CAS is a derivative of) should provide minimal cross-coupling during maneuvering through the flight envelope. Minimal cross-coupling is defined as

1. A pure lateral stick input results in a well-coordinated steady-state roll about the stability axes, that is, with negligible sideslip. The desired roll rate dynamic response should be first order and described by

$$\frac{p_m(s)}{\delta_{stick}(s)} = \frac{3}{s+3}. \tag{33}$$

2. A pure rudder pedal input results in a steady-state wing-level sideslip maneuver with negligible roll response. The desired sideslip dynamic response following a pilot rudder command should be second order and described by

$$\frac{\beta_m(s)}{\delta_{stick}(s)} = \frac{9}{s^2 + 4.8s + 9} \quad \text{which gives } \omega_n = 3 \text{ and } \xi = 0.8. \tag{34}$$

Equations 33 and 34 were used as response models for the OLNNET lateral and directional CASs.

A. PHASE I: STATISTICS OF LEARNING AND ADAPTATION

Two OLNNET architectures were designed and implemented for this phase of the study. The roll-rate CAS was a MIMO controller with sensor-determined aircraft states and pilot command-defined model responses as controller inputs and a three-output node configuration determining aileron deflection, rudder deflection, and differential elevator deflection. The linear roll-rate response model was a unity gain implementation of Eq. 33, with a model sideslip response of zero for any commanded aileron deflection. The lateral maneuver used in this study was Lateral 1 (Fig. 11). The performance index used for the controller free parameter updates is given by Eq. 35:

$$J_{\text{RollRate}} = \tfrac{1}{2} \sum_{k=1}^{N} \left[0.5 \left(p(k) - p_{\text{m}}(k) \right)^2 + 0.03 \beta(k)^2 \right]. \tag{35}$$

The sideslip CAS controller was also a MIMO configuration with an input set made up of aircraft states and desired responses. The controller has a two-node output layer controlling the aileron and rudder deflections. The performance index for the sideslip OLNNET controller is given by Eq. 36:

$$J_{\text{Sideslip}} = \tfrac{1}{2} \sum_{k=1}^{N} \left[0.03 p(k)^2 + 0.1 \left(\beta(k) - \beta_{\text{m}}(k) \right)^2 \right]. \tag{36}$$

As with the roll-rate CAS, the sideslip model used was a unity gain implementation of Eq. 34 with a desired roll-rate response of zero to the rudder command input. Figure 13 shows the desired sideslip response to the rudder command sequence. The complete OLNNET controller architectures are given in Table III.

The same OLNNET roll-rate controller was evaluated under nonlinear conditions. To generate responses in nonlinear aircraft dynamics, a desired roll-rate response with first-order characteristics given by Eq. 33 and magnitude equal to two times the commanded deflection was introduced.

The OLNNET CAS controllers were evaluated with regard to learning capability and adaptation capability. In general, the objective of an adaptive control

Table III

On-Line Learning Neural Network CAS Controller Architechtures

Parameter	Roll-rate CAS	Sideslip CAS
Input data set	p, p_{m}, β, β_{m}, δ_{A}, δ_{R}, δ_{DE}	p, p_{m}, β, β_{m}, δ_{A}, δ_{R}
Pattern	3	3
No. of hidden-layer nodes	12	12
No. of output-layer nodes	3	2
Learning rate	0.01/0.5	0.5
Performance index	$\text{PI} = 0.5(p - p_{\text{m}})$ $+ 0.025(\beta - \beta_{\text{m}})$	$\text{PI} = 0.05(p - p_{\text{m}})$ $+ 0.25(\beta - \beta_{\text{m}})$
Network outputs	δ_{A}, δ_{DE}, δ_{R}	δ_{A}, δ_{R}

scheme is to achieve some desired closed-loop response for a system with time-varying dynamics. Theoretically, the time dependence in the system may be due to system and/or measurement noise that is time dependent, or to the effects of un-modeled dynamics that are state dependent. Traditional adaptive controllers have no capability to distinguish between time and state dependencies in the system dynamics. Learning, on the other hand, implies the use of past information to determine the control sequence at the present time. A positive indication of learning would be the improvement in the closed-loop response to a given command sequence over time. To assess the learning performance of the OLNNET controllers, error statistics were computed for the mean and variance of the tracking error (Eqs. 37 and 38, respectively). The first study of this phase investigated the effect of the network learning rate on the controller performance. Recall that larger learning rates can lead to problems with local minima. The autopilot results (Section IV) showed that large learning rates allowed the controller to perform very well throughout the flight envelope, but a global learning trend was not clear. Sufficient and fast local learning gave the autopilot controllers excellent performance under all test conditions. The first phase of the CAS learning and adaptation study explores this global versus local issue.

The learning performance of the OLNNET MIMO roll-rate controller was evaluated in a fashion similar to that of the autopilots. The set of flight tests was performed twice with the roll-rate controller. On the first set of runs the controller learning rate was set to 0.01. On the second set of test runs the learning rate was set to 0.5. It is well documented that small learning rates aid the controller in mapping the global system and reduce the risk of local minima problems. The questions of interest are:

- Can a slow learning controller map the system quickly enough to give a reasonable transient response and gain control of the aircraft?
- Does the controller performance increase as the learning time increases?

Ten runs were made under the same flight conditions (20,000 feet and 800 feet/sec), where the command sequences for both the roll rate and sideslip OLNNET CASs were initiated at each run. Error statistics were computed for each run and for each CAS:

$$\text{MEAN}_{\text{TE}} = \mu_{\text{TE}} = \frac{\sum_{d=1}^{N}[y(k) - y_m(k)]}{N} \tag{37}$$

$$\text{VAR}_{\text{TE}} = \nu_{\text{TE}} = \frac{\sum_{k=1}^{N}[(y(k) - y_m(k)) - \mu_{\text{TE}}]^2}{N}, \tag{38}$$

where N is the number of samples during the maneuver and y and y_m are the actual and desired responses, respectively.

Figure 14 shows the results of runs 1, 3, and 10 of the 10 roll-rate runs under the nominal flight condition for the low learning rate case ($\eta = 0.01$) for the linear

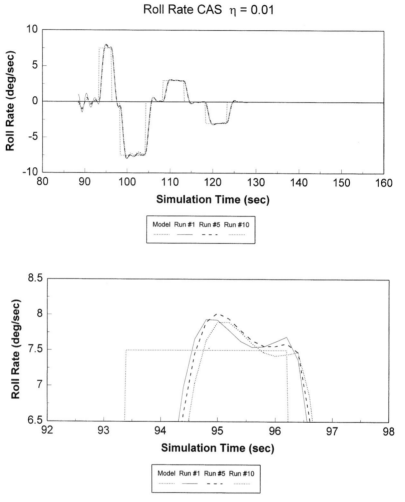

Figure 14 Top: Results for runs 1, 5, and 10 for the slow learning rate roll-rate CAS. Bottom: Zoomed view of first aileron maneuver peak.

roll-rate model. Included in each figure is the mean and variance of the tracking error for each run. As can be seen, the performance of the OLNNET roll-rate CAS is acceptable on each test run, but improvement over the set of runs (which would indicate learning) is not clear. The mean and variance of the tracking error were computed for each test run. Figure 15 shows graphically the trend of the mean and variance over the set of runs.

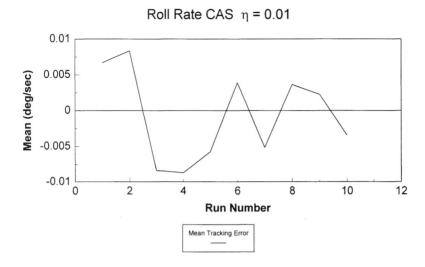

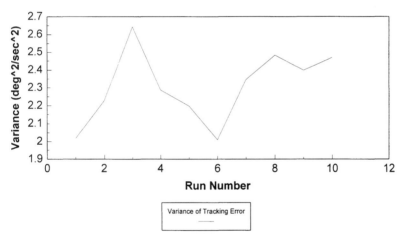

Figure 15 Top: Mean tracking error. Bottom: Variance of trancking error for the slow learning roll-rate CAS.

Next, the same linear response model was used to evaluate the learning statistics of the OLNNET roll-rate CAS with a larger learning rate ($\eta = 0.5$). Figure 16 shows the results of the faster learning controller. As can clearly be seen, in all plots the controller tracks the desired command, even on the initial run with no

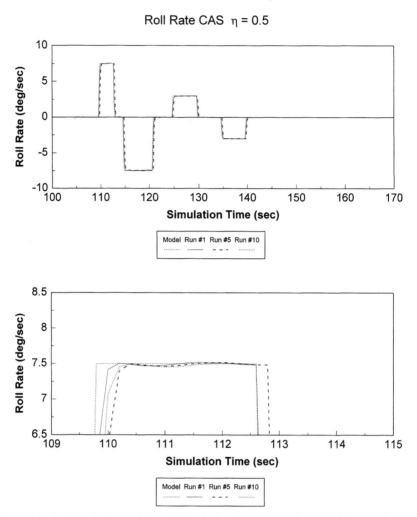

Figure 16 Top: Results for runs 1, 5, and 10 for the fast learning rate roll-rate CAS. Bottom: Zoomed view of first aileron maneuver peak.

a priori knowledge of the system. The performance is good because the control task is clear and the correlation between the network output error (the desired minus the actual aircraft response) and the control surface deflection is high. Figure 17 shows the trend of the mean and variance of the tracking error respectively for each test run. Again, no clear learning trend is observable. It should be noted that although neither controller demonstrated a clear global learning trend, the

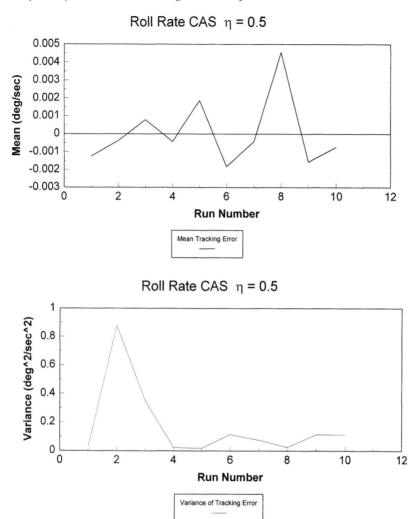

Figure 17 Top: Mean tracking error. Bottom: Variance of tracking error for the fast learning roll-rate CAS.

"faster" controller exhibited much better statistics. It should also be noted that the OLNNET CASs were active throughout the flight, regardless of whether the simulation was being "flown" manually via the keyboard or a preprogrammed maneuver. Between preprogrammed maneuvers, the simulation was flown manually to allow the OLNNET CASs to continue to "learn."

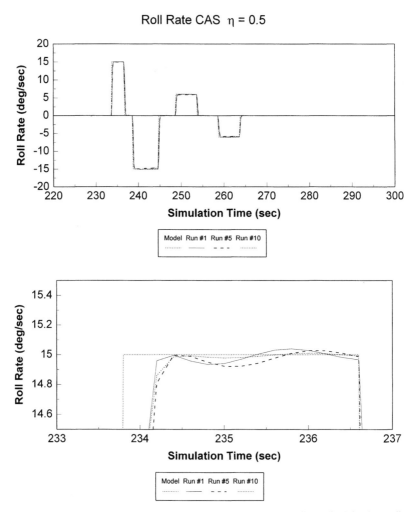

Figure 18 Top: Results for runs 1, 5, and 10 for the fast learning rate roll-rate CAS for the nonlinear response model runs. Bottom: Zoomed view of first aileron maneuver peak.

The OLNNET roll-rate CAS was next tested with a nonlinear response model. For this set of runs the learning was set to 0.5. Figure 18 shows the tracking results for test runs 1, 5, and 10. Again, the OLNNET roll-rate CAS was able to track the model response with little deviation. Figure 19 shows the mean and variance of the tracking error, respectively. As with the two previous set of runs, there is no clear indication of learning.

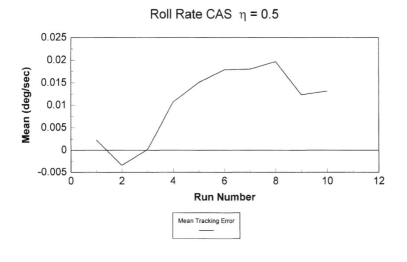

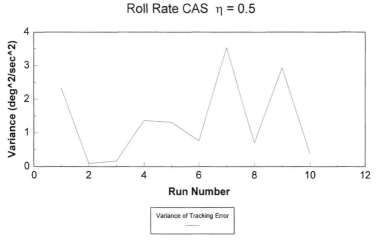

Figure 19 Top: Mean Tracking error. Bottom: Variance of tracking error for the fast learning roll rate and the nonlinear response model.

The final learning assessment was performed with the OLNNET sideslip CAS. The linear sideslip response model was used for the rudder command sequence. Figure 20 shows the results of test runs 1, 5, and 10, illustrating the OLNNET controller's ability to track the desired command. Figure 21 shows the error statistics for the OLNNET sideslip CAS. No indication of global learning is present.

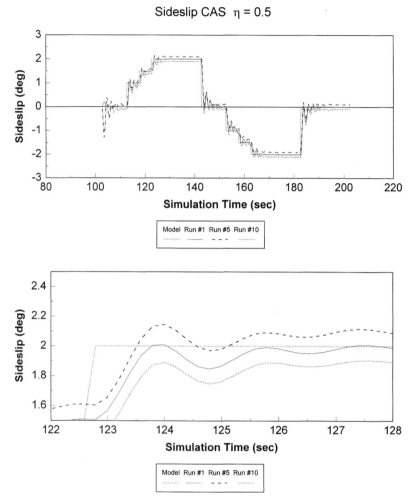

Figure 20 Top: Results for runs 1, 5, and 10 for the fast learning rate sideslip CAS for the linear response model runs. Bottom: Zoomed view of first aileron maneuver peak.

Analysis of the OLNNET CASs adaptation capability was performed by "flying" the aircraft model to 10 different points throughout the model's flight envelope. As in the autopilot investigation, it was a point of the CAS study to show the controllers' adaptability throughout the flight envelope, without the aid (or need) of gain scheduling. The same OLNNET controller architectures as used in the previous part of the CAS study were implemented, with learning rates of 0.5.

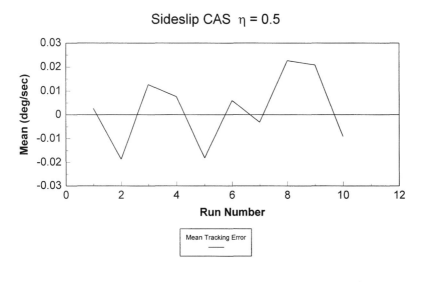

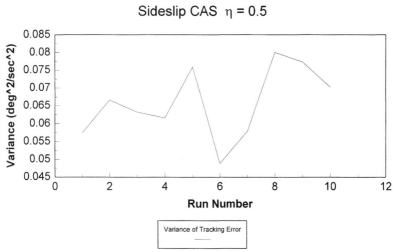

Figure 21   Top: Mean tracking error. Bottom: Variance of tracking error for the last learning sideslip CAS and the linear response model.

The OLNNET roll-rate CAS was evaluated for both linear and nonlinear models. The sideslip CAS was evaluated with the linear response model to the command sequence shown in Fig. 13. The preprogrammed maneuvers (both roll-rate and sideslip) were introduced at each flight condition. Error statistics were computed

Table IV

Tracking Error Statistics for Mean Tracking Error and Variance of Tracking Error for Both Linear and Nonlinear Roll-Rate Models for Each of the 10 Test Runs

Run	Linear ($\eta = 0.01$)		Linear ($\eta = 0.5$)		Nonlinear ($\eta = 0.5$)	
	Mean (deg)	Variance ($\deg^2/\sec^2$)	Mean (deg)	Variance ($\deg^2/\sec^2$)	Mean (deg)	Variance ($\deg^2/\sec^2$)
1	0.00669	2.0017	−0.00124	0.02585	0.00228	2.32940
2	0.00834	2.22754	−0.00372	0.87709	−0.00337	0.09490
3	−0.00842	2.64266	0.00794	0.35166	0.00166	0.17231
4	−0.00874	2.28720	0.00347	0.33145	0.00107	1.36255
5	−0.00575	2.19780	0.00187	0.01903	0.00151	1.31079
6	0.00392	2.01033	−0.00181	0.11471	0.00179	0.76788
7	−0.00517	2.34623	−0.00424	0.07549	0.00180	0.70088
8	0.00366	2.48317	0.00457	0.02526	0.00197	2.92942
9	0.00228	2.39841	−0.00156	0.11437	0.00123	0.67784
10	−0.00346	2.47072	−0.00735	0.11331	0.00131	0.38774

for each run (lateral and directional) at each flight condition. A listing of the 10 flight conditions is given in Table IV. Figures 22 and 24 show the results of the linear and nonlinear OLNNET roll-rate CAS runs, respectively, for flight conditions 1, 3, and 10. As can be seen, the OLNNET controller performance is excellent throughout the flight envelope, tracking both the linear and nonlinear responses with precision. Figures 23 and 25 show the mean and variance of the tracking error for both linear and nonlinear runs, respectively, at each flight condition. It should be noted here that all sideslip responses to these maneuvers were negligible and therefore are not presented.

The above flight tests were repeated for the linear sideslip model with error statistics computed. Again, the neural CAS demonstrated excellent adaptation capability by successfully tracking the desired response throughout the flight envelope. Figure 26 shows the OLNNET sideslip CAS performance at each flight condition. Figure 27 shows the mean and variance of the tracking error.

B. PHASE II: MULTIPLE MODEL FOLLOWING CAPABILITIES

The second phase of the investigation focused on the multiple model adaptive control capabilities of the OLNNET CASs. For this study a roll-rate CAS with exactly the same architecture and performance index as that of the previous

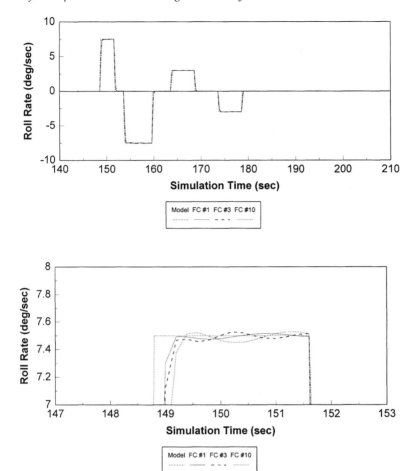

Figure 22 Adaptation analysis—linear model, roll-rate CAS $\eta = 0.5$. Top: Results for flight conditions 1, 3, and 10 for the fast learning rate roll-rate CAS for the linear response model runs. Bottom: Zoomed view of first aileron maneuver peak.

section was implemented. The NASA model was modified to allow the user to implement any one of three different response models manually at any time during the flight through keyboard commands. The response models chosen represented slow, moderate, and high roll-rate response to the preprogrammed command sequence used in phase I of the study. The aircraft was flown to the 10 regions of the flight envelope used in the previous study and listed in Table V. In each of these regions the preprogrammed maneuver was flown with each of the different

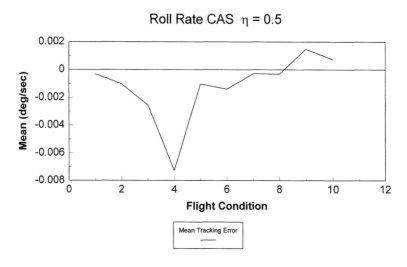

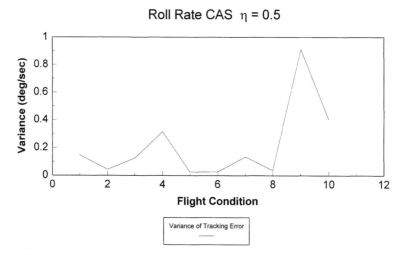

Figure 23   Top: Mean tracking error. Bottom: Variance of tracking error for the fast learning roll-rate CAS and the linear response model for the adaptation analysis.

model responses. Each model was activated in random order. Error statistics were computed for each run. An interesting result of this analysis is the learning performance (in terms of mean and variance of the tracking error) of the controllers when introduced to multiple models compared to the learning performance when introduced to a single model, as in the first phase.

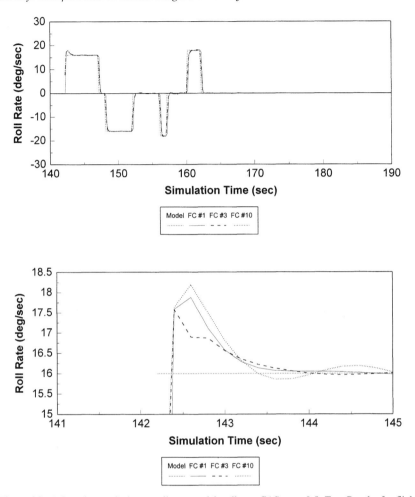

Figure 24 Adaptation analysis—non-linear model, roll-rate CAS $\eta = 0.5$. Top: Results for flight conditions 1, 3, and 10 for the fast learning rate roll-rate CAS for the nonlinear response model runs. Bottom: Zoomed view of first aileron maneuver peak.

Figure 28 shows the results for each roll rate response model at flight conditions 1, 3, and 10. Figure 29 shows zoomed plots of the first pulses of Fig. 28. The aircraft response, along with each desired model, is plotted against the maneuver time to allow for comparison of each model for each run. As is clearly illustrated,

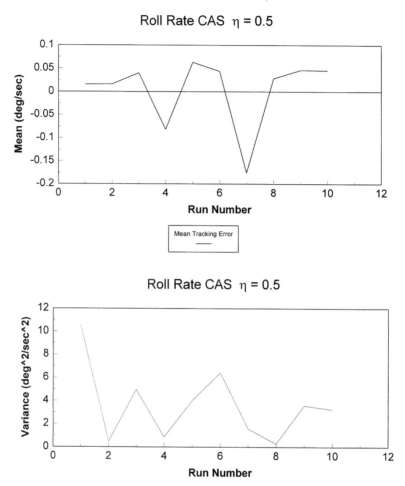

Figure 25 Top: Mean tracking error. Bottom: Variance of tracking error for the fast learning roll-rate CAS and the linear response model for the adaptation analysis.

the neural roll-rate CAS was able to track each model at each flight condition. Table VI presents the mean and tracking error for each model at each flight condition. A comparison of Tables IV and VI shows that there is no degradation in the network performance when multiple models are presented.

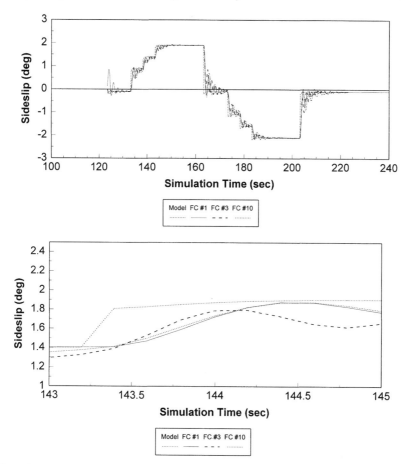

Figure 26 Adaptation analysis—linear model, sideslip CAS $\eta = 0.5$. Top: Results for flight conditions 1, 3, and 10 for the fast learning rate sideslip CAS for the linear response model runs. Bottom: Zoomed view of first aileron maneuver peak.

C. Conclusions

This section has presented the results of an investigation of the performance of a multiple-input multiple-output (MIMO) on-line learning neural network controller for command augmentation system (CAS) functions in a high-performance aircraft. OLNNET CASs were designed and studied for roll-rate and sideslip models following performance. The roll-rate controller was evaluated under both linear and nonlinear conditions. The sideslip controller was evaluated under linear

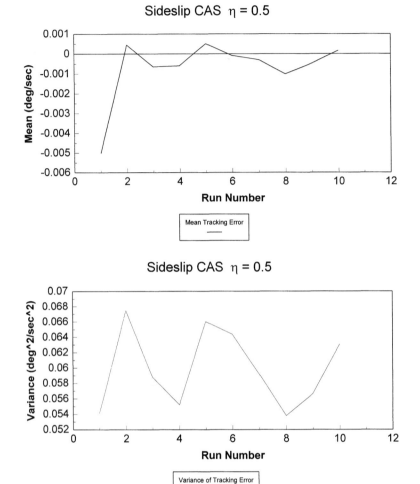

Figure 27 Top: Mean tracking error. Bottom: Variance of tracking error for the fast learning sideslip CAS and the linear response model for the adaptation analysis.

conditions. The investigation was conducted in three phases. The first phase evaluated the learning and adaptation characteristics of the neural controllers. In the
learning evaluation, the tracking performance of the neural controllers was tested
over 10 runs under the same flight conditions. Both OLNNET roll-rate controllers
were able to track the command sequence, with the faster learning (0.5) controller
performing better, as expected. Neither controller showed a clear learning trend

Table V

Ten Flight Conditions Used for the Neural CAS Investigation

No.	Alitude (ft)	Velocity (ft/sec)
1	15,000	700
2	18,000	800
3	20,000	1,200
4	22,000	1,400
5	24,000	800
6	22,000	1,300
7	26,000	1,600
8	30,000	1,400
9	28,000	1,800
10	31,000	1,600

over the set of runs. Similarly, the sideslip controller was able to track the sideslip command sequence with no clear global learning present. Although less than desirable, this absence of global learning is explainable as follows. The controller first "sees" the system when it is activated on-line. To get a stable closed-loop response, sufficient error information must be passed to the controller for updating. The large required updates cause a local learning effect. Because this type of aircraft control is inherently a local control problem, this effect does not cause a significant degradation in performance. The second part of the first phase investigated the OLNNET CAS's adaptation capability. The two OLNNET CAS controllers (lateral and directional) were tested with the same response models at 10 different flight conditions. The OLNNET CAS controllers showed excellent model tracking performance over the range of flight conditions. This was done without explicit gain scheduling or on adaptation strategy.

The second phase of the CAS investigation involved the inherent MMAC capability of the OLNNET CAS controllers. The different roll-rate response models were activated in a random order under each of three flight conditions. The OLNNET roll-rate CAS was able to accurately track each model under each flight condition, with no loss of performance in tracking any model. This phase of the study demonstrated that an OLNNET controller can adapt to track multiple models throughout the flight envelope with no *a priori* knowledge of the desired responses.

The final phase of the CAS investigation was a demonstration of the autoreconfiguration capability of the OLNNET roll-rate CAS. Linear and nonlinear roll-rate response models were flown under the same flight conditions for three runs. The first run provided a baseline controller response to the commanded maneuver.

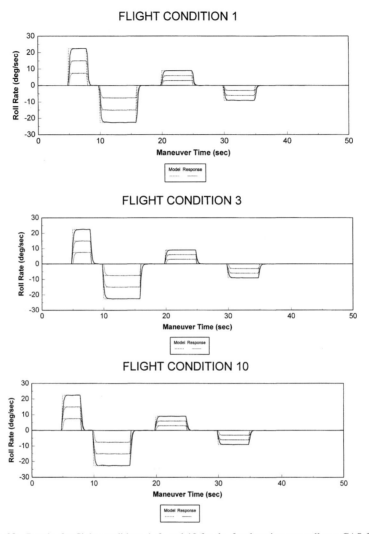

Figure 28 Results for flight conditions 1, 3, and 10 for the fast learning rate roll-rate CAS for the multiple model runs.

On the second run, the ailerons were "locked" at a 0 degree deflection at 18 sec into the maneuver. The OLNNET controller was able to adapt by changing the differential elevator deflection to accommodate the loss of aileron effectiveness and continue tracking the model with no loss of performance for both response

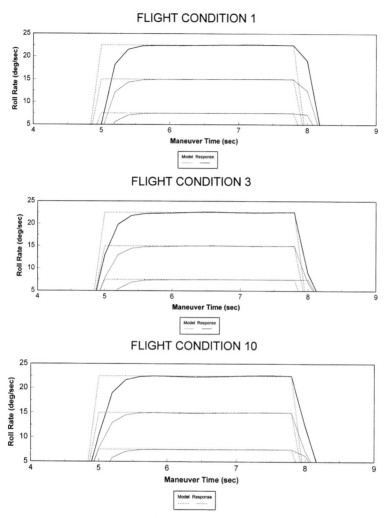

Figure 29 Results for flight conditions 1, 3, and 10 for the fast learning rate roll-rate CAS for the multiple model runs zoomed to show first peaks.

models. On the final run, the differential elevator was locked to a 0 degree deflection at 18 sec into the maneuver. Again, the OLNNET controller was able to accommodate the failure by changing the aileron deflection.

This investigation has demonstrated the potential for OLNNET controllers in the SAS and CAS functions of a high-performance aircraft. The controllers' in-

Table VI

Mean and Variance of Tracking Error for Three Models under Each of the 10 Flight Conditions

Flight condition	Model 1 (Low p)		Model 2 (Moderate p)		Model 3 (High p)	
	Mean (deg/sec)	Variance (deg²/sec²)	Mean (deg/sec)	Variance (deg²/sec²)	Mean (deg/sec)	Variance (deg²/sec²)
1	3.333e − 05	8.590e − 01	2.933e − 03	2.369e + 00	1.466e − 03	5.449e + 00
2	−1.260e − 17	3.807e − 01	5.700e − 03	2.155e + 00	3.666e − 04	1.603e + 00
3	−1.666e − 04	9.481e − 01	−2.566e − 03	8.375e − 01	−5.566e − 03	1.454e + 00
4	−3.000e − 04	3.287e − 01	−5.351e − 04	5.257e − 01	−3.366e − 03	1.953e + 00
5	2.666e − 04	1.031e − 01	−1.000e − 04	2.948e + 00	6.660e − 01	2.934e + 01
6	2.675e − 04	1.318e − 01	−1.066e − 03	6.798e − 01	−3.733e − 03	1.508e + 00
7	1.333e − 04	2.187e − 01	−1.133e − 03	8.659e − 01	−1.204e − 03	1.178e + 00
8	−1.333e − 03	1.025e − 01	1.400e − 03	2.042e + 00	−1.033e − 03	7.515e + 00
9	3.678e − 04	1.229e − 01	−1.033e − 03	8.803e − 01	2.333e − 04	2.819e + 00
10	6.666e − 05	6.460e − 01	−9.333e − 04	8.701e − 01	−6.666e − 05	2.505e + 00

herent capability to track multiple models under any flight conditions and autore-configure after a failure opens a world of possible control applications.

VI. CONCLUSIONS AND RECOMMENDATIONS FOR ADDITIONAL RESEARCH

The results presented in this chapter have demonstrated that neural controllers are viable alternative controllers. There are several areas, however, on which more information is needed. Section IV mentioned the effects of the performance index weights on performance specifications such as rise time, settling time, overshoot, etc. These effects need to be better understood. Furthermore, there is little formal methodology for the design of neural controllers (or neural networks in general). Another interesting topic would be the MIMO capabilities and limitations of OLNNET controllers. The CAS investigation used a three-output node controller for lateral–directional control. A question that arises is: Could a single, massive MIMO controller be used for both lateral–directional and longitudinal control? If so, what would the performance be compared to the controllers presented here? The autoreconfiguration capability illustrated in Section V warrants further investigation. A study looking at the speed at which the controller can adapt and the degree of failure that the controller can tolerate would provide useful information.

A. CONCLUSIONS

This section has presented the results of a feasibility study conducted with the goal of demonstrating that on-line learning neural network (OLNNET) controllers can replace conventional, gain scheduling-based controllers in the flight control systems of modern high-performance aircraft. Neural network controllers were designed, implemented, and tested for autopilot and command augmentation system functions. The closed-loop performance of the OLNNET controllers was simulated with the six-degree-of-freedom aircraft model distributed for the 1991 AIAA Controls Design Challenge. The model is representative of a modern jet fighter.

The results of the autopilot investigation showed that the OLNNET controllers were able to learn the local system dynamics with sufficient speed to provide acceptable performance throughout the aircraft flight envelope at both linear and nonlinear conditions. The command augmentation system (CAS) investigation demonstrated that MIMO OLNNET controllers were capable of tracking multiple-response models (linear and nonlinear) throughout the flight envelope. The CAS investigation also illustrated the autoreconfiguration capability that is inherent in an OLNNET controller.

Most current neural network investigations have focused on designing controllers that are similar in structure to more traditional controllers. Although these studies have generated insight into the behavior of neural networks in control applications and given credibility to the neural network field, they have largely overlooked the potential of on-line learning controllers. The inherent parallel processing structure of a neural network, coupled with the continuous adaptation capability of on-line learning, makes these controllers an attractive option for highly nonlinear, time-varying systems, or systems that contain a high degree of modeling uncertainty. As aerospace vehicle development shifts more to the private sector, cost will become more of a driving factor. A controller capable of learning and adapting on-line without complex reconfiguration and accommodation strategies could be of fiscal and operational value.

ACKNOWLEDGMENT

This work was supported by the NASA/West Virginia Space Consortium grant NGT-40047.

REFERENCES

[1] W. L. Baker and J. A. Farrell. Learning augmented flight control for high performance aircraft, AIAA paper 91-2836. In *Proceedings of the AIAA Guidance, Navigation, and Control Conference*, New Orleans, LA, August 1991.
[2] T. Troudet, S. Garg, and W. C. Merrill. Neural network application to aircraft control system design, AIAA paper 91-2715. In *Proceedings of the AIAA Guidance, Navigation, and Control Conference*, New Orleans, LA, August 1991.
[3] Department of Electrical and Computer Engineering. *Neural Networks: A Short Course*. West Virginia University, Morgantown, WV, February 1992.
[4] R. P. Lippmann. An introduction to computing with neural nets. *IEEE ASSP Magazine* April:124–131, 1987.
[5] R. Hecht-Nielsen. Neurocomputing—picking the human brain. In *Proceedings of the IEEE First International Conference on Neural Networks*, San Diego, CA, June 1987.
[6] P. K. Simpson. *Artificial Neural Systems*. Pergamon Press, Fairview Park, NY, 1990.
[7] Laurene Fausett. *Fundamentals of Neural Networks: Architectures, Algorithms, and Applications*. Prentice-Hall, Englewood Cliffs, NJ, 1994.
[8] C. M. Ha. Neural network approach to the AIAA control design challenge, AIAA paper 91-2672. In *Proceedings of the AIAA Guidance, Navigation, and Control Conference*, New Orleans, LA, August 1991.
[9] C. M. Ha, Y. P. Wei, and J. A. Bessolo. Reconfigurable aircraft flight control system via neural networks, AIAA paper 92-1075. In *Proceedings of the AIAA Aerspace Design Conference*, Irvine, CA, February 1992.
[10] Peter Milligan, Walter Baker, and Mark Koenig. *Control Augmentation Synthesis Via Adaptation and Learning*, AIAA-93-3728-CP. Autonomous Systems Group, The Charles Stark Draper Laboratory.

[11] M. R. Napolitano, C. Neppach, V. Casdorph, S. Naylor, M. Innocenti, and F. Bini. Sensor failure detection, identification, and accommodation using on-line learning neural architectures, AIAA Paper 94-3598. In *Proceedings of the AIAA Guidance, Navigation, and Control Conference*, Scottsdale, AZ, August 1994.

[12] M. R. Napolitano, C. I. Chen, and S. Naylor. Actuator failure detection and identification using neural networks. In *AIAA J. Guidance Control Dynamics* 16:999–1008, 1993.

[13] M. R. Napolitano, C. Neppach, V. Casdorph, and S. Naylor. On-line learning non-linear direct neuro controllers for restructurable flight control systems. *AIAA J. Guidance, Control Dynamics* 18:170–176, 1995.

[14] C. L. Chen and R. S. Nutter. An extended back-propagation algorithm by using heterogeneous processing units. In *Proceedings of the International Joint Conference on Neural Networks*, Baltimore, MD, June 1992.

[15] M. Hagiwara. Theoretical derivation of momentum term in back-propagation. In *Proceedings of the International Joint Conference on Neural Networks*, Baltimore, MD, June 1992.

[16] R. W. Brumbaugh. An aircraft model for the AIAA controls design challenge. NASA Contractor Report 186091, December 1991.

[17] K. J. Astrom and B. Wittenmark. *Adaptive Control*, Addison-Wesley, Reading, MA, 1989.

[18] D. Psaltis, A. Sideris, and A. Yamamura. A multilayered neural network controller. *IEEE Control Systems Magazine* 4:1988.

[19] V. C. Chen and Y. H. Pao. Leaning control with neural networks. In *Proceedings of the International Conference on Robotics and Automation* 3:1989.

[20] M. Kawato, F. Furukawa, and R. Suzuki. A hierarchical neural network model for control and learning of voluntary movement. *Biol. Cybernet.* 57.

[21] Military Standard: Flying Qualities of Piloted Vehicles, MIL-STD-1797 (USAF). ASD/ENES, Wright Patterson AFB, OH, March 31, 1987.

Nonlinear System Modeling

Shaohua Tan

Department of Electrical Engineering
National University of Singapore
Singapore 119260

Yi Yu

Institute of Systems Science
National University of Singapore
Singapore 119597

Johan Suykens

Department of Electrical Engineering
Katholieke Universiteit Leuven
Heverlee, Belgium

Joos Vandewalle

Department of Electrical Engineering
Katholieke Universiteit Leuven
Heverlee, Belgium

I. INTRODUCTION

Modeling nonlinear dynamical systems by using neural networks has increasingly been recognized as a distinct and important system identification paradigm. It initially appeared as a mere collection of constructive nonlinear modeling techniques that were loosely connected and *ad hoc* in nature [1–3]. Collective research efforts in the past years have made it a more coherent approach, with both clear theoretical content and practical relevance [4–6]. In a further extension of its relevance, recent research has shown that this paradigm fits into a larger picture of concerted (past and present) efforts, such as regression analysis, wavelets, fuzzy systems, etc., in tackling the nonlinear modeling problem in its general form [7, 8]. For ease of reference, we shall use "neural modeling" in this chapter as a generic term to refer to this paradigm.

Sharing a number of common features with other nonlinear modeling techniques, neural modeling can be viewed as an expansion technique for approximating an unknown nonlinear system. In doing so, however, it has its own features, focus, and methodology.

Control and Dynamic Systems

383

To highlight just one general feature, we note that neural modeling is aimed at developing constructive techniques that can be applied in practical applications. This is often directly reflected in the assumptions it makes in solving the modeling problem. Indeed, realistic assumptions such as finite number of data samples, nonavailability of relevant statistical distributions, etc., are often explicitly or implicitly made. These assumptions make the formulation of the modeling problem deterministic, thereby requiring new insight to develope techniques to solve it.

The current literature on neural modeling is vast. We refer the reader to [7] and [9] for an overview. The objective of the present chapter is not to provide yet another overview, but to supply the reader with a comprehensive description of current developments. In other words, we shall share with the reader our own experiences in formulating the neural modeling problem and developing relevant techniques to solve it. We shall especially emphasize the key issues and the pitfalls the reader will have to be aware of to understand and solve the problem.

The choice of model structure is a key issue in neural modeling. There are many such choices with various pros and cons. To be consistent with our objective and to narrow the scope of our treatment, we choose to consider only two general types of neural network model structure: one is the radial basis function (RBF) neural network with a single hidden layer; the other is the multilayer back-propagation neural network. This focused treatment allows us to present detailed formulation on the modeling problem and the modeling techniques, and to address precisely the critical issues of both problem formulation and algorithm development.

Accordingly, this chapter is literally divided into two parts. The first part, consisting of Sections II through IV, is concerned with the use of RBF neural networks for nonlinear modeling; the second part, made up of Sections V through VII, approaches the nonlinear modeling problem through a state–space formulation with the multilayer back-propagation neural network model structure. Section VIII contains two illustrative examples to show how the techniques developed can be applied to solve specific nonlinear modeling problems.

Because of the different natures of the two model structures, the underlying assumptions made in each part in formulating the neural modeling problem are also different. In the first part, we take a deterministic stand in assuming that, except for a finite number of data samples, no statistical knowledge of any sort is available about the unknown system. Consequently, the approach taken in this part becomes deterministic in nature. In the second part, we assume that noise statistics are known and an unlimited number of data samples are available. The approach is, therefore, more statistical. Such a difference in approach, however, should not create a conceptual division in understanding the notions and techniques that we have tried to develop as an integrated whole in this chapter.

II. RBF NEURAL NETWORK-BASED NONLINEAR MODELING

A. THE NEURAL MODELING PROBLEM

Given an input sequence $\{u_t\}$ and the corresponding output sequence $\{y_t\}$ ($t = 0, 1, \ldots$) of an unknown nonlinear discrete-time multivariable dynamical system, where u_t is the q-dimensional input vector, and y_t is the m-dimensional output vector of the system at discrete time t, suppose that the following model structure represents our prior knowledge about the unknown system:

$$y_{t+1} = f(y_t, u_t), \tag{1}$$

in which $f(\cdot)$ is some unknown nonlinear vector function. We are interested in constructing a neural network model that models the system. This is the input/output nonlinear modeling problem that we shall consider exclusively until Section IV; thereafter we shall look at the nonlinear modeling problem in the state space.

Following the standard linear system identification terminology, whatever is inside the function $f(\cdot)$ can be lumped together and called a regressor [10]. In the case of (1), the regressor is simply the vector $x_t = [y_t, u_t]^T$, which is a special case of a more general form of regressor

$$[y_t, y_{t-1}, \ldots, y_{t-k+1}, u_t, u_{t-1}, \ldots, u_{t-l+1}]^T,$$

where k and l are integers. This general regressor, along with the model structure (1), is known as the NARX model (see [3] for a proposed classification). One can also treat the NARX model as a special case of (1) by embedding the regressor into the dimensions of both u_t and y_t. But doing so will imbue these dimensions with the prior knowledge.

Apparently, our assumption of having prior knowledge about the system is crucial in fixing the precise form of the regressor. This assumption is also intended to make the modeling problem more focused on structure issues that are solvable in a neural network framework. It is worth noting that fixing the regressor for a practical nonlinear modeling problem can be a very difficult task, and may require such knowledge as physical insight into the unknown system.

Observe that with the above formulation, the model is nonrecurrent in the sense that the past model output does not enter as the known data for modeling. The recurrent modeling, which can also be viewed as the regressor containing the modeled output, is known to be a harder problem to solve. However, we shall comment toward the end of Section IV on how to adapt the developed modeling techniques to make them applicable to the recurrent case.

B. Two Basic Issues in Neural Modeling

As in any other modeling scheme, to use neural networks to solve the preceding nonlinear modeling problem, one must resolve two issues: the structural and parameter determinations. Translated into neural network terms, they become the two basic problems of finding an appropriate neural network structure and the weights within the structure. It is fair to say that various neural modeling techniques are developed as specific ways of addressing these two basic issues.

Of the two, the structral determination is of greater importance in deciding about the quality of the model. However, structural determination is known to be a very difficult task. Even now, the structures of quite a number of neural network models are fixed by either heuristics or trial-and-error means. By offering a relatively simple and analytically tractable functional structure, RBF neural networks serve to ease the difficulty of the structural determination.

Considered by many authors in the past [11–18], RBF neural networks can be regarded as a general function reconstruction scheme using radial-basis functions as the interpolating functions. As we shall see shortly, its structural feature allows rigorous determination of RBF neural network structures for the nonlinear modeling problem. The linear-in-the-weights characteristic of RBF neural networks also makes possible the derivation of efficient and convergent weight updating rules. In what follows, we shall first describe the RBF neural network structure before elaborating on the suitability of such a neural network as a generic nonlinear model structure.

C. RBF Neural Network Structure

A RBF neural network is simply a two-layer feedforward network structured as follows. The first layer of the network consists of RBF neurons and the second layer of linear neurons. The inputs to the neural network are directly connected to the RBF neurons in the first layer; the outputs of the RBF neurons are then connected to the linear neurons in the second layer through adaptable weights. For the sake of definiteness, let us assume that there are ℓ inputs to the RBF neural network, n RBF neurons in the first layer, and m linear neurons in the second layer (Fig. 1).

The transfer function of the ith RBF neuron in the first layer takes the form

$$z_i(x) = r\left(\left\| x - p_i \right\|^2 / \sigma_i^2\right),$$

where $z_i(x)$ is the output of the ith RBF neuron, and $r(\cdot)$ is a radial basis function and can be chosen as any suitable radial basis function. For reasons we shall elaborate when we discuss the effect of dimensionality, the Gaussian function

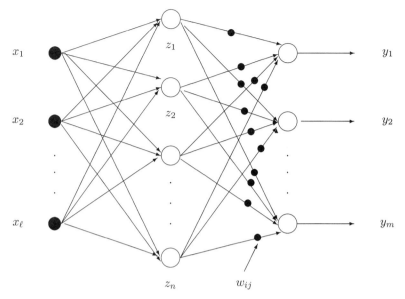

Figure 1   The configuration of RBF neural networks.

given below appears to be a sensible choice for $r(\cdot)$:

$$z_i(x) = e^{-\|x-p_i\|^2/\sigma_i^2}, \tag{2}$$

in which x is the input vector of dimension ℓ; p_i is the center vector of dimension ℓ of the ith RBF neuron; the positive scalar σ_i is the width of the ith RBF neuron. Both p_i and σ_i can be different for different RBF neurons, thus offering flexibility in locally shaping the RBF neural network structure.

The transfer function for the ith linear neuron is given by the following matrix notation:

$$y = Wz(x), \tag{3}$$

where $y = [y_1 \ y_2 \cdots y_m]^T$, which lumps the outputs of all of the linear neurons together to form a vector; $z(x) = [z_1(x) \ z_2(x) \cdots z_n(x)]^T$; and W is an $m \times n$ matrix with the ijth entry w_{ij} interpreted as the adaptable weight from the jth RBF neuron in the first layer to the ith linear neuron in the second layer.

Equation (3) apparently decomposes all of the parameters involved in a RBF neural network into two categories: the matrix W containing all of the adaptable weights, and the vector $z(\cdot)$, containing all such structural parameters as the number of neurons n, the center vectors p_i, and the widths σ_i.

D. Suitability of the RBF Neural Network for Nonlinear Modeling

There are a few reasons that make the above RBF neural network suitable for nonlinear modeling. These reasons and related issues will be enumerated in what follows.

1. The Universal Approximation Property

First is the so-called universal approximation property of RBF neural networks, as rigorously proved in [19]. This property states that for any ϵ and an arbitrary continuous nonlinear function $f(\cdot)$ defined on a compact (bounded and closed) set $\mathcal{C}$, there always exists a weight matrix $\overline{W}$ and a vector $\bar{z}(x)$, all of finite dimensions, such that

$$\max_{x \in \mathcal{C}} \left| f(x) - \overline{W}\bar{z}(x) \right| < \epsilon.$$

Using the nonlinear modeling language, this property tells us the following: As long as the unknown system (1) has a compact region of operation and $f(\cdot)$ is continuous over the region of operation, then there always exists a RBF neural network of the form (3) that will model (1) within an arbitrarily small error bound.

The property of universal approximation provides a rationale for RBF neural networks to be used as a generic model structure for nonlinear systems. However, this property should not be carried too far. In fact, there are four points that we must pay special attention to. The first is that the described universal approximation property is only an existence statement. It does not prescribe any specific way for actually finding all of the parameters of the model.

The second reason, which is more critical but tends to be neglected, is that the conditions for the universal approximation property and those of our modeling problem are very different. Indeed, scrutiny shows that the universal approximation property requires the full knowledge of a function $f(\cdot)$ to assert the existence of a RBF neural network model, whereas our modeling problem only assumes a knowledge of $f(\cdot)$ at discrete sample points (given by the input and output samples).

In a sense, to solve the modeling problem is like finding a constructive proof of the universal approximation property. Moreover, such a proof will have to be found based on the given discrete data samples. Nonlinear modeling is thus a more difficult problem to solve.

The third point is that the universal approximation property only holds for a compact set $\mathcal{C}$, which necessarily confines the modeling problem to a compact region of operation. In other words, we cannot choose an arbitrary and unbounded input sequence $\{u_t\}$ for modeling; furthermore, the system to be modeled will have to be BIBO (bounded input–bounded output) stable to ensure the existence of a compact region of operation. This is why the techniques to be developed shortly cannot be used to model unstable systems.

The final point is that of time invariance. One may think that it is straightforward to add time as an additional variable in $f(\cdot)$, so that the time-varying case can be accommodated. This idea, however, cannot work for the simple reason that time is an unbounded variable in dynamical systems. Careful thought reveals that unless we confine the modeling problem to a finite time interval (which is often not meaningful), there is no easy way to model time-varying systems within the current neural network modeling framework.

These discussions appear to have set validity boundaries on how far we can go with the current neural network modeling methodology.

2. The Link to Spatial Sampling and Reconstruction Framework

The second point mentioned above makes it necessary to consider the modeling problem along the lines of sampling and reconstruction. As the function $f(\cdot)$ is of high dimension, the sampling and reconstruction will have to be done in high dimension. We speak of spatial sampling and reconstruction in this case [20].

There are two aspects to a sampling and reconstruction theory. One has to do with how the samples should be collected at an adequate sampling rate for reconstruction. The other is has to do with the choice of reconstructing functions. It is shown in [15] that the RBF neural network with the Gausian function (2) is ideally suited as a reconstructing function. More importantly, when the Gaussian function is chosen as the RBF function, there is a relatively easy way of building the reconstructing function by deploying the Gaussian functions on a grid drawn in the system's region of operation. This result has provided a sound theoretical basis for developing systematic techniques for setting up the structure of a RBF neural model. As we shall show shortly, this result is the basis for the structure determination of a RBF neural network.

Actually, the above result can easily be extended to other RBFs that have a compact support, or such that they only have significant function value (above some threshold) for a compact set, like the Gaussian function. As there exist RBFs that are neither of these types, the above result cannot be extended to all types of RBFs. The Gaussian function has certainly provided a definite choice that meets the requirement. We shall show in the next section that the requirement for the compact support is the key to implementing a systematic structure adaptation algorithm.

3. The Linear-in-the-Weight Structural Feature

As seen in (3), the RBF neural network structure naturally separates the structural parameters of n, p_i, and σ_i from the weight matrix W. Therefore, if the structural parameters can be fixed by some means, the network simply becomes linear-in-the-weight. It is then relatively easily to apply existing linear identification methods, such as the least-squares method, or to develop other efficient algorithms to determine the weights.

E. EXCITATION-DEPENDENT MODELING

Let us consider yet another aspect of the nonlinear modeling problem to make our formulation more realistic.

As the neural modeling problem is essentially a sampling and reconstruction problem in a high-dimensional space, there is the obvious question of whether the given input and output data samples are adequate in representing the unknown system in the given region of operation. Or more precisely, we need to make sure that our choice of $\{u_t\}$ (if we are allowed to make such a choice) is such that $\{y_t, u_t\}$ will be dense enough in the modeling region $\mathcal{C}$ to allow the reconstruction of $f(\cdot)$. Here the term "dense enough" is defined in the sense of spatial sampling theorem (see [20] for more details on the theory).

This is an easy problem if the unknown system is a linear system. Indeed, the well-known persistent excitation condition ensures the adequacy for linear systems when $\{u_t\}$ is chosen to be white noise. Unfortunately, no such result is yet available for general nonlinear systems (1). The reason for this is apparent: although we can choose $\{u_t\}$, we cannot control $\{y_t\}$, so that the combined sequence $\{x_t\}$ will be dense everywhere in $\mathcal{C}$. The behavior of the combined sequence depends on the nature of the nonlinear function $f(\cdot)$. The issue is further complicated if $\{u_t\}$ has to come from an on-line situation.

There are quite a few options we can choose to proceed from here. We can limit the function $f(\cdot)$ to some special cases and develop excitation conditions for these cases for the modeling to be possible. Or we can ignore this problem, hoping that the reality will take care of the problem by itself, and go on to develop the modeling algorithm. Or we can modify the modeling problem formulation to take this fact into consideration. It is the last option we choose in our work.

The key idea we use in reformulating the modeling problem is very simple: rather than model the whole system at one time, we only model the part of the system (1) that is being excited by particular input sequences. This idea is called *excitation-dependent modeling*. To use an intuitive explanation: excitation-dependent modeling resembles the way a human being develops his or her knowledge of a subject. The knowledge acquired on the subject matter is based on experiences with it. No knowledge of a particular part of the subject matter will be acquired if that part has never been revealed to him or her before. Following this idea, only a partial model is being constructed from the available data. The complete model of the unknown system will only emerge when the data have become dense enough in the specified modeling region.

The excitation-dependent modeling, however, should not be misunderstood as being valid for only some particular input sequences and thus as not generalizable to other input sequences. In fact, if $\{x_t\}$ induced by a particular input sequence is dense enough in a closed subset C_s of C, then the model built on the input sequence will generalize well to any input sequences with x_t ($t = 1, 2, \ldots$) falling within C_s, but not to those with some x_t falling outside C_s.

F. RBF NEURAL NETWORK MODELING WITH A FIXED MODEL STRUCTURE

One straightforward idea of RBF neural modeling is to use an optimal search algorithm to find all of the parameters n, p_i, σ_i and the weight matrix W against a suitably defined cost function. The difficulty with such an idea is that the resulting optimization problem is nonlinear and high-dimensional. There is, therefore, no guarantee that a solution can be found in a reasonable amount of time. Even worse is the fact that there is no easy way to verify whether a solution has been found. It is not surprising that with this idea, one often ends up with a suboptimal solution at best after an extensive computational effort.

An alternative idea that will also lead to a suboptimal solution, but with much less effort, is to divide the problem into two separate problems of determining the structural parameters n, p_i, σ_i and determining the weight matrix W. It is this idea we shall follow in the present subsection to develop a specific RBF modeling technique. Modeling techniques based on other ideas will be presented in the ensuing sections.

1. Off-Line Determination of RBF Neural Model Structure

The RBF neural network model structure is traditionally fixed by heuristic methods [11, 13, 14]. A widely used heuristic idea suggests that n, p_i, and σ_i are closely related to the data clusters generated by x_t samples in S. Using this idea, n is chosen as the number of the data clusters in S, and p_i, σ_i as the center and radius, respectively, of the ith data cluster. The main concern is, therefore, to design efficient algorithms that can quickly locate the data clusters from a sequence of data.

A simple analysis reveals, however, that such an idea cannot be substantiated theoretically. One the one hand, in virtually all modeling situations, there are no clearly distinguishable data clusters that emerge from the data samples $\{x_t\}$, except for artificially superimposing such clusters. On the other hand, the concept of data clusters does not appear to have any connection with the location of RBF neurons. One can construct simple examples to show that even for artificially created data clusters, one RBF neuron is often not sufficient to represent the data at one cluster.

Another heuristic method uses the idea of covering for the determination of the structural parameters [21]. In this case, all of the centers are taken as the data points themselves, and are sufficiently spread in such a way that if they are considered the centers of hyperspheres with radius σ_i, any data sample x_t should at least be covered by one such hypersphere. This idea can be implemented as an efficient algorithm for progressively allocating a new RBF neuron if there is a data sample outside the region covered by the previous hyperspheres [18]. Again, this method lacks a theoretical foundation.

In what follows, let us present a simple method that can be theoretically justified.

To start, the region of operation C of the unknown system needs to be known. If it is not given, $\{x_t\}$ (for t going to a certain large value) can be used to estimate C. Then a hypercube S containing C is drawn up. Such a hypercube is often chosen to be slightly larger than C to cater to the possible boundary effect.

Next, a uniformly spaced grid is formed in S. The distance between two neighboring grid points (denoted by Δ and termed *grid size*) can be adjusted to change the fineness of the grid. To ensure the modeling accuracy, the grid size should be chosen such that it is smaller than the Nyquist sampling period of the unknown system. Because such a period is often unavailable, the common practice is to make an initial choice first and then to progressively refine it.

Once the grid is set up, the number of RBF neurons n is simply the number of grid points, and their centers p_i are the coordinates of the grid points. σ_i can finally be chosen to be around $\frac{2}{3}\Delta$ to Δ as the widths for all of the RBF neurons (see [16] for a more elaborate discussion).

The above procedure has a theoretical backing as detailed in [15]. Intuitively, the idea can be understood as placing RBF neurons at regular grid locations for function approximation. As long as the grid is fine enough and the RBF neurons are reasonably shaped at each locality, each local feature of the function $f(\cdot)$ can be approximated by one or more RBF neurons. Therefore, the RBF neurons will collectively approximate the whole function in the whole region of C.

For a high-dimensional modeling problem, the above procedure is likely to generate a neural network of a huge size. It is not uncommon to have a RBF neural network model with a few thousand RBF neurons just to approximate a two-dimensional model over a small region of operation. Even worse, the size will grow exponentially with the dimension. This is known as the "curse of dimensionality." The techniques presented in the ensuing sections are designed partially to address this issue.

2. Weight Update

When the structure is fixed, the weight matrix can be determined by using a linear identification technique (see, e.g., [22]). The best-known such technique is the recursive least-squares method (or its variations) [10]. There are, however, two potential problems with these methods. One is their high computational cost; the other is that because of the choice of the Gaussian functions, some matrices involved in the adaptation may have small values (in the interval of $(0, 1)$), and such matrices are likely to cause numerical problems during the computation [18].

These considerations justify further development of efficient weight updating algorithms. One algorithm developed in [16] and is briefly presented below. The key idea of this algorithm is to construct a one-step-ahead recursive predictor that will generate $\hat{y}_t$, the prediction of y_t. Then the prediction error is used to construct the updating rule for the weight matrix.

Let $W_t \in R^{m \times n}$ be the weight matrix, and $e_t = \hat{y}_t - y_t \in R^m$ be the prediction error at time t. Here the prediction is obtained by the following linear predictor:

$$\hat{y}_t = A\hat{y}_{t-1} - Ay_{t-1} + W_{t-1}z(x_{t-1}) - z^T(x_{t-1})z(x_{t-1})\tilde{e}_{t-1}, \qquad (4)$$

where $A \in R^{m \times m}$ is a diagonal matrix with all of its diagonal elements confined within $(-1, 1)$.

Defining the auxiliary error vector $\tilde{e}_t \in R^m$,

$$\tilde{e}_{t-1} = \frac{Ce_{t-1} + W_{t-1}z(x_{t-1}) - y_t}{1 + z^T(x_{t-1})z(x_{t-1})}, \qquad (5)$$

where $C \in R^{m \times m}$ is a constant diagonal matrix, we can introduce the following weight updating formula:

$$W_t = W_{t-1} - \tilde{e}_{t-1}z^T(x_{t-1}). \qquad (6)$$

It has been shown in [16] that this algorithm will converge.

3. Remarks

Efficiency appears to be the advantage of this weight updating algorithm when compared to other methods, such as the one used in [23]. In most cases, a training cycle of only a few thousands, compared to tens of thousands, is needed to stay within the mean-square modeling error (on the order of 1%), provided, of course, the RBF neural network structure is adequately chosen.

To end this section, we note that the structural determination and the weight adaptation are treated as two separate problems. It will be shown in the next section that RBF neural networks allow a natural way of combining these two to form structure-adaptive modeling algorithms.

III. ON-LINE RBF STRUCTURAL ADAPTIVE MODELING

A. THE NEED FOR ON-LINE STRUCTURAL ADAPTATION

The preceding off-line structural determination has several specific drawbacks. First, the knowledge of an unknown system is somehow required to set up a fine enough grid size.

Second, to lay a proper grid, this technique requires knowledge of the compact region of operation to start with, which is often not available for an unknown dynamical system [16].

Third, a fine enough uniform grid used in the technique corresponds to the worst-case scenario, and cannot accommodate the specific nonlinear profile of a system. As a result, there is the potential problem of overfitting, and the network size is often much larger than it is otherwise necessary.

All of these problems can be traced to the fact that the structural determination of the RBF neural model structure is an off-line process and is handled disjointedly from the weight determination. This implies that with limited prior knowledge about an otherwise unknown system, off-line structural determination can either overdetermine or underdetermine the structure, and there is almost no chance that it can produce a proper model structure.

A simple remedy appears to be a modification of the technique such that both the weights and the *structure* can be adapted at the same time. We shall pursue this idea in this section to develop a feasible technique.

The notion of structure adaptation in neural networks is not new. Lee [24] has provided a quite extensive discussion on various possible methods for carrying out the adaptation for multilayer neural networks. However, these methods are largely heuristic and often lead to unguided decisions on the choice of parameters. This section will present a RBF neural network-based structure-adaptation technique that can be rigorously substantiated. We shall show that there is guaranteed convergence with the technique.

The key idea of the technique is to view a RBF neural network as an assembly of standardized simple but localized nonlinear building blocks, the RBF neurons. Then the process of adapting the structure of an unknown system is carried out by selecting the relevant RBF neurons from a suitably defined pool of such building blocks to shape the current model structure. This procedure is easily combined with a weight updating formula to produce a partial model for the system that evolves with the coming inputs and outputs. In the following subsections, the details of the technique will be presented, along with discussions on the convergence and the rationale behind a number of choices.

B. ON-LINE STRUCTURE ADAPTATION TECHNIQUE

The algorithm consists of five basic steps: setting up the nominal grid, on-line structural generation, weight update, on-line grid adaptation, and aging. Let us provide an overview of how the algorithm works before describing each of the five steps in detail.

To generate the RBF neural network structure, initially a nominal grid of arbitrarily fixed size is drawn up on the whole linear space where $\mathcal{C}$ lies. The RBF neural model is initialized to contain no RBF.

When the ith data x_i comes, it is used to locate an activation region on the nominal grid. All of the RBF neurons and are selected within this region and are joined with those RBF neurons selected in the same way in the previous steps to form the structure of the partial model for the system at this stage. The weight update is then performed to modify the weights of the partial model.

There is also an aging process whereby selected RBF neurons will be eliminated from the partial model structure if their corresponding weights have not been updated for a given period of time.

The nominal grid adaptation is initiated when the current grid size does not allow the partial model to achieve the target modeling error within a prescribed period of time. The nominal grid size is simply adapted by reducing it to half of its current size, to ensure that the previous grid points correspond to the points on the new grid. After the grid adaptation, the structure adaptation will proceed again with the finer nominal grid.

At the end of the this procedure, the partial model will be such that the RBF neurons will be nonuniformly placed in the system's region of operation $\mathcal{C}$, depending on the input sequence, the system dynamics, and the choice of aging parameter.

Let us clarify one feature before going into the details of each step. We note that the nearly compact support of the Gaussian function is essential in developing the structure-adaptive technique. The local feature helps to isolate a model structural deficiency at a locality, and makes it easy to adapt that part of the structure by using local building blocks.

To understand the point precisely, let us note that if we let $\|x - p_i\| \geq \kappa \|\sigma_i\|$, where $\|\cdot\|$ is a vector norm and κ is a positive integer, then $z_i(x)$ decreases very rapidly with increasing κ. As $\|\sigma_i\| \leq \|\Delta\|$ for all i, we can be sure that $z_i(x)$ is negligible for any x at least $3\|\Delta\|$ away from p_i. κ is called the *structural radius*. Thus, for each x_t, those RBF neurons that are centered at least 3 grid points away from x_t will have an almost zero $z_i(\cdot)$, and thus will not affect y_{t+1} in (3).

Let p_t^* be the nearest grid point to x_t and $\mathcal{S}_t$ be the set of grid points that are no farther than κ grid points away from p_t^*. Then we can always choose a suitable κ so that y_{t+1} can be evaluated to the desired accuracy by using the RBF neurons in $\mathcal{S}_t$ only. Note that the number of RBF neurons in $\mathcal{S}$ is $(2\kappa + 1)^{m+l}$, which increases with increasing κ. But for all practical purposes, κ need not exceed 3, thus limiting the number of RBF neurons we must consider in evaluating y_{t+1} at the current locality as defined by x_t.

Note that this technique is not amendable to all types of RBF neural networks, because not all of them are locally (or nearly locally) supported. The technique also provides yet another reason for us to prefer the Gaussian function as an appropriate choice of RBF.

1. Nominal Griding

The nominal grid is simply set by choosing a grid size Δ to cover the whole $(m+l)$-dimensional space that contains $\mathcal{C}$. Any prior knowledge about the dimensional bandwidths of the underlying system can be incorporated in selecting Δ at this stage. If such knowledge is available, Δ can be arbitrarily chosen initially, and will be refined on-line by the grid adaptation process. Once Δ is chosen, the width σ can be chosen to be $\frac{2}{3}\Delta$, as suggested in the preceding section.

Note that the nominal griding is done for the whole linear space, and therefore extends to infinity. To fix its position, the origin of the space is required to be a grid point. Thus the nominal grid can be regarded as defining an integer-coordinate system on the linear space whereby every grid point can be referred to by an $(m + l)$-dimensional vector of integers. For convenience, the origin of the space is set to 0 in the grid-induced integer coordinate system.

Because it is laid on the entire space containing the region of operation $\mathcal{C}$, the nominal grid can also be treated as a grid laid only on $\mathcal{C}$. However, by gridding the whole space, we have managed to grid $\mathcal{C}$ without knowing either its shape or its size in the first place. This allows our modeling technique to operate without assuming any knowledge of the region of operation of the unknown system, other than the fact that it is compact.

Note that only the part of the grid on $\mathcal{C}$ will be used in the later structure adaptation. Therefore, it does not matter if the grid size extends to infinity. Also note that the grid is called "nominal" in the sense that all of the grid points on $\mathcal{C}$ are potential candidates to be selected to form the partial model structure.

As part of the initialization, a so-called aging period T must be selected, which can be from a small positive value all the way up to infinity. Its choice will drastically affect the final model structure. A small T tends to shape the structure more toward the region being excited by the current input sequence.

κ also must be chosen based on the given modeling error. As discussed, it can be 3 at most. Finally, a so-called adaptation period τ must be set to decide upon the time to adapt the nominal grid.

2. On-Line Structure Generation and Structural Aging

The structure for the partial model is shaped at this stage in two phases: generation and aging.

During the generation phase, for each x_t, the central grid point p_t^* is determined by dividing every component of x_t by Δ and rounding the result to the nearest integer. With p_t^* and κ, the set $\mathcal{S}_t$ of RBF neurons can be determined. We shall call $\mathcal{S}_t$ the *current structural updating set*. The weight updating at time t (to be discussed shortly) is only performed on those weights associated with the RBF neurons in $\mathcal{S}_t$. We also define another set $\mathcal{S}_d$ (called the *partial model set*) and let it be empty initially. After the weight updating, the neurons in $\mathcal{S}_t$ will join those in $\mathcal{S}_d$ to form an updated $\mathcal{S}_d$. The set $\mathcal{S}_d$ forms the partial model at time t.

During the aging phase, a RBF neuron will be eliminated from $\mathcal{S}_d$ if its corresponding weight has not been updated for a period equal to or greater than the aging period T. Obviously, T tunes the neural model structure more toward the region excited by the current input sequence. The value of T decides the final shape of the structure. Smaller T tends to squeeze the model structure, leading to slow convergence. Larger T, on the other hand, may result in a fat model structure.

3. Grid Adaptation

As the initial choice of Δ is arbitrary, the modeling error is likely to remain higher than the prescribed error modeling bound with the preceding operations carried out for the adaptation period τ. This is when the grid adaptation is activated to refine the nominal grid by reducing the grid size Δ by half. The new grid will be $\frac{1}{2}\Delta$. This is done in such a way that a new grid point is added exactly halfway between any two adjacent grid points on the previous grid. This method ensures that the existing partial model structure $\mathcal{S}_d$ is not affected by the grid adaptation operation. After the grid adaptation, another round of structural generation will be activated. The whole process repeats until the modeling error falls within the prescribed limit.

4. Weight Update

Using $\hat{y}_t$ to denote the neural model estimate of y_t, and using the partial model set $\mathcal{S}_d$ we have just built, above we can form the following partial neural model at time t for the unknown system:

$$\hat{y}_{t+1} = W_t z(x_t), \qquad (7)$$

where W_t is the weight matrix at t. Note that the dimension of the above system depends on the size of the partial model set.

Defining $e_t = \hat{y}_t - y_t$ as the modeling error at time t, the weight updating formula can be designed as

$$W_{t+1} = W_t - \frac{\alpha e_t z(x_t)}{\beta + z^T(x_t)z(x_t)}, \tag{8}$$

where α, β are two positive constants that can be chosen as $0 < \alpha < 2$ and $\beta > 0$. Note that in the preceding equations all of the RBF neurons $z_i(\cdot)$ $(i = 1, 2, \ldots, n)$ in the vector $z(\cdot)$ are centered at the grid points of the nominal grid.

The preceding technique can be analytically shown to be convergent, in the sense that $\lim_{t\to\infty} |e_t| = 0$. We refer the reader to [25] for the details on the analysis.

C. REMARKS

Let us summarize the three important features of this on-line structure adaptation scheme. First, at each time instant there is a fixed number of neurons to be dealt with, regardless of the size of the nominal grid. Second, because of the periodic deletion, the number of RBF neurons in the RBF neural model will be considerably lower compared to the off-line case for the same error requirement. Moreover, these RBF neurons will be located in the region of operation to capture input history and the dynamical nature of the unknown system. The third feature is that there is no need to have explicit knowledge of the compact region of operation C, apart from its existence. The existence of C guarantees that the structural set S_d will end up with a finite number of RBF neurons.

From the point of view of methodology, we note the following advantage of the on-line structure adaptation. In the off-line case, the decision of how fine to make the grid is always a difficult one. It represents a tradeoff between the model size and the modeling precision. In the on-line case, this tradeoff is absorbed into the structure adaptation process. Fixing the model size is not a single decision, but the result of an iterative process.

Let us reflect again, at this point, on the issue of excitation-dependent modeling. Apparently, the scheme builds a partial model in an on-line iterative process. When the given data sequence becomes sufficiently dense in C, the partial model becomes the model of the system on C. Otherwise, it simply models the part of the system revealed by the given data sequence. Note that this observation is valid when the data sequence is of finite length, and is totally independent of the underlying distributions of the data. We believe that the idea of excitation-dependent modeling is the key to deterministically formulating and solving the complex nonlinear modeling problem.

IV. MULTISCALE RBF MODELING TECHNIQUE

In this section we present yet another technique for solving the structure determination problem. This technique has to do with the notion of modeling on different scales.

A. BASIC MOTIVATION

A nonlinear system can be seen as having both global and local features. The global features are collective features over its whole region of operation C, and the local features are those pertaining to specific local areas. To draw an intuitive analogy, we can think of a gray-scale image and note that the brightness of the image is a global feature, whereas the gray level at a particular position is a local feature. The global and local features are apparently only relative, and have no clear boundary, as demonstrated in this example. They can be placed on an equal footing by using the idea of scale. Therefore, when the scale changes from coarse to fine, the feature changes from globle to local. This resembles the familiar process of zooming in.

The idea of global/local features can help to shape the formulation of the nonlinear modeling problem. Indeed, with a large modeling error, we can build the nonlinear model by using more global features and ignoring the local ones. We then add more local features to the model at increasingly finer scales to reduce the modeling error until the desired accuracy is reached. This is the key idea of what we call multiscale nonlinear system modeling. Obviously, the central issue in implementing such a modeling idea is to find suitable model structures that are both universally approximating and capable of multiscale representation.

Let us use the idea of a multiscale to examine a few recently developed nonlinear model structures, such as the back-propagation neural network [26], the RBF neural network as considered extensively in the preceding sections (see also [14–16]), the fuzzy system [25, 27] and so on. We note that although all of them are capable of universal approximation, none of them captures the features at different scales to represent a nonlinear system. A loose look at these model structures and the corresponding modeling schemes reveals that the nonlinear system representation is done either globally or locally without the decomposition in terms of scales.

To take the back-propagation neural network as an example: its layered feedforward structure with the globally supported sigmoid function appears to represent a nonlinear system profile from a global perspective. It requires a quite complex combination of sigmoid functions to represent delicate local features to a high precision. Therefore, a high modeling error requirement leads to complex network structure and extensive computation.

The RBF neural network and the fuzzy system, on the other hand, use the locally supported functions like the Gaussian function as its activation function. Such a structure tends to represent a nonlinear system from a local perspective. However, very often a large number of RBF neurons are needed to reasonably represent the system on a global scale.

It seems that a multiscale model structure can strike a balance and overcome the shortcomings of purely globally or locally based modeling approaches. Our key point in this section is to show that with appropriate choices of parameters, RBF neural networks offer a natural multiscale structure for nonlinear modeling. This provides further justification for the preference for RBF neural networks.

B. STRUCTURE OF A MULTISCALE RBF NEURAL NETWORK

It is easy to build the scale into the standard RBF neural network (3). The idea is to define appropriate width levels and the center locations of the RBF functions, so that they form a function set in a finite number of scales.

In matrix notation, a multiscale RBF neural network can be seen as a partitioned form of (3):

$$y = [W_0, W_1, \ldots, W_d] \begin{bmatrix} z_0(x) \\ z_1(x) \\ \vdots \\ z_d(x) \end{bmatrix} = \sum_{j=0}^{d} W_j z_j(x), \qquad (9)$$

where W_j is the submatrix of W at scale j with appropriate dimensions, $z_j(x)$ is the set of RBF neurons at scale j, and d represents the finest possible scale.

At scale j, the partitioned vector $z_j(x)$ is defined by

$$z_j(x) = \begin{bmatrix} z_{j1}(x) \\ z_{j2}(x) \\ \vdots \\ z_{jn_j}(x) \end{bmatrix}, \qquad (10)$$

in which

$$z_{ji}(x) = e^{-\|2^j x - p_{ji}\|^2/\sigma^2}, \qquad i = 1, 2, \ldots, n_j, \qquad (11)$$

where n_j is the number of RBF neurons at scale j. The centers p_{ji} can be chosen as the points of a grid at the current scale, and will be clarified in the algorithm to be developed shortly.

Observe that because of the scale change of 2^j, the actual width of the RBF neurons at scale j is $\sigma/2^j$, which is drastically decreasing with increasing j. Combined with the location of p_{ji}, it induces the scale for the RBF neurons.

The RBF network described above provides many of the features needed for implementing multiscale modeling. Let us show first that the multiscale RBF neural network is a universal approximator. It is a direct result from the following refinement property of the Gaussian function:

$$z_{ji}(x) = \sum_k a_k z_{(j+1)k}(x), \tag{12}$$

where a_k is a constant sequence. Denoting space V_j as span$\{z_{ji}, \ i = 1, 2, \ldots\}$, we have

$$V_j \subset V_{j+1}, \qquad \bigcap_j V_j = 0, \qquad \overline{\bigcup_j V_j} = L^2(\mathcal{C}), \tag{13}$$

which implies that the multiscale RBF neural network has the property of universal approximation (see [28] for more details).

Apart from the universal approximation, another important feature is that the multiscale RBF neural network (9) effectively decomposes the model structure into several partial models in terms of scales. This model structure can be used to capture the features of an unknown system over an $O(2^{-j})$ width interval centered at the point $2^{-j}i$. If the partial model can be built at each scale j, we can simply sum them up to obtain the whole system model.

C. COARSE-TO-FINE RESIDUE-BASED MODELING IDEA

With the preceding multiscale RBF neural network, we can divide the whole modeling task into several submodeling tasks at different scales induced by the model structure. More globally supported RBF neurons are used at coarser scales to capture global characteristics of the unknown system, whereas more locally supported RBF neurons are used at finer scales to fill in the local features.

Our modeling technique will start from the coarsest scale and proceed to finer ones to gradually build a nonlinear model [29]. At each scale, a partial model at that scale is built to model the residue portion left from the previous scale. The recursive least-squares algorithm is used to find weights of the partial model at this scale. This process continues until the residual drops below a prescribed modeling error bound. The whole nonlinear system model is then obtained by bringing the partial models together. To further reduce the model structure, a reduction technique is also developed to delete some of the RBF neurons at each scale.

The idea of multiscale modeling for nonlinear systems has already been looked upon in the context of wavelet networks [30, 31]. However, the basic idea in these works is simply to make use of the orthogonality property of wavelet functions to enhance existing neural modeling techniques. The structure issue is often overlooked, and the multiresolution nature of wavelet networks is not explored for efficient and effective modeling. Indeed, the existing techniques can often be interpreted as a fine-to-coarse algorithm. As a result, the wavelet model structure determination becomes a trial-and-error process in finding the finest possible scale to build the entire wavelet network.

To determine the optimal structure for a wavelet network, Coifman and Wickerhauser [32] have written entropy-based algorithms to select the best bases from a wavelet packet to reduce the size of the wavelet network. Developed in a general context, their idea is to list heuristically all of the possible wavelets at all possible scales and put them into a packet. Then the entropy of each wavelet in the packet is examined. A wavelet in the packet will be selected if it does not lead to the increase in the entropy of the whole network. This is a typical bottom-up approach to structure optimization. The obvious difficulties of this approach lie in finding all of the possible scales and in the choice of a proper entropy function. Moreover, the extensive base expansions and comparisons are too time-consuming to be of practical interest. By adopting a top-down approach, we have managed to overcome the aforementioned difficulties by uncovering the network structure one scale at a time from coarse to fine, with stops at a fine enough scale as guided by the modeling precision requirement. This resembles peeling an onion: we start from the outmost layer and precede one layer at a time until we reach the layer that is fine enough for our purpose; whereas a bottom-up approach will have to cut the onion in two and start searching at the finest possible layer first.

D. CONSTRUCTION OF MULTISCALE RBF MODEL

Let M data samples $x_t = [y_t^T, u_t^T]^T$ ($t = 1, 2, \ldots, M$) of (1) be given. The modeling procedure starts at the coarsest scale $j = j_0$ by laying down a grid of an arbitrary coarse size on $\mathcal{C}$ and choosing σ to be two-thirds this size. Here j_0 can be any finite positive integer to start with. The common choice is 0. Using the number of grid points (say, n_{j_0}), their locations and the chosen σ, the partial model at scale j_0 can be defined as

$$\hat{y}_{t+1}^{j_0} = \sum_{i=1}^{n_{j_0}} w_{j_0 i} z_{j_0 i}(x_t), \tag{14}$$

where $\hat{y}_{t+1}^{j_0}$ denotes the modeled output at instant $t + 1$.

The least-squares algorithm is used next to determine the weight matrix W_{j_0} by using the given M data samples. After this partial model is constructed, we use it to form the following residue data:

$$r_{t+1}^{j_0} = \hat{y}_{t+1}^{j_0} - y_{t+1}.$$

If the maximum norm $\max_t |r_{t+1}^{j_0}|$ (the max operation is over all t) is less than a prescribed error bound, the procedure is terminated. Otherwise, we go to scale $j_1 = j_0 + 1$, where the grid size is halved, resulting in a finer grid. We build the partial model on this new grid by using the number of grid points and their locations, but unlike the structure adaptive algorithm, σ will not be changed. As we mentioned above, the width change is built into the multiscale RBF structure. This partial model takes the following form:

$$\hat{y}_{t+1}^{j_1} = \sum_{i=1}^{n_{j_1}} w_{j_1 i} z_{j_1 i}(x), \tag{15}$$

which has the same form as (14), except that the number, locations, and width of the RBF neurons are different. The least-squares algorithm is used again to determine the weight matrix W_{j_1}. But this time, instead of $\{y_{t+1}\}$, the residue $\{r_{t+1}^{j_0}\}$ is used. The model constructed at this scale is of the residue rather than the actual system.

After this partial model is built, we form the residue set at the next scale as follows:

$$r_{t+1}^{j_1} = \hat{y}_{t+1}^{j_1} - r_{t+1}^{j_0}.$$

If the maximum norm $\max_t |r_{t+1}^{j_1}|$ is less than the error bound, the procedure stops. Otherwise, we go on to the next scale and continue with the same operation at each scale until the maximum residue norm drops below the error bound. The whole system model is then obtained as

$$\hat{y}_{t+1} = \sum_{j=j_0}^{d} \hat{y}_{t+1}^{j}, \tag{16}$$

where d is the finest possible scale reached to achieve the modeling accuracy.

Redundancy in the partial model may exist at each scale, which can be explored for further model structure reduction at each scale. The reduction procedure along with other operational details is described below.

1. Weight Updating

The least-squares operation can be done off-line. However, the problem is that there is a matrix inversion computation that is likely to be inefficient and may suffer from numerical condition problems. Therefore, the recursive least-squares algorithm is actually used instead in the computation [33].

The stop criterion for the least-squares algorithm is that there will be no further improvement between two recursions, or

$$\left\| W_j^{k+1} - W_j^k \right\| < \frac{\epsilon}{2}. \tag{17}$$

To prevent numerical problems, the above equation is sometimes evaluated over a range of k values, and their average is taken.

Because of the local nature of RBFs, even for the partial model, the recursive mode of weight updating will only be able to update a limited number of weights each time a new data point is presented. Therefore, we can arrange the computational procedure so that only a small number of weights are actually updated. This idea has served as a basis for developing the RBF structure adaptive modeling scheme as detailed in the above section.

2. Structure Reduction

The number of RBF neurons at each scale is roughly $O(2^j)$, which can be exceedingly large at very fine scales. Many of these neurons, however, are redundant and can be removed without affecting the precision of the model. We present one reduction technique that can be performed after the least squares updating.

After the weight update at each scale j, those weights in the partial model $\hat{y}_{t+1}^j$ will be deleted if they do not contribute significantly to the modeling accuracy, or if

$$\| w_{ji} \| < \epsilon. \tag{18}$$

The corresponding RBF neuron can be removed from the partial model, where w_{ji} is some entry in the W_j matrix. This thresholding process will eliminate a substantial number of redundant RBF neurons, especially at fine scales.

To make the modeling scheme more efficient, the modeling accuracy at local data samples will be monitored as well. Any input/output data sample where modeling accuracy has already been reached will no longer be involved in future updating at finer scales. In other words, if a residue entry at scale j is such that $\| r^j \| < \epsilon$, it will be removed from the residue data set to avoid using it for future operations. In this way, the data set will be drastically decreasing at finer scales, resulting in far less computation.

E. MODEL VALIDATION

It can be rigorously shown that the preceding coarse-to-fine multiscale modeling scheme is convergent [33]. The theoretical analysis states that for any continuous, time-invariant nonlinear system (1), the residual $\|r^j\|$ defined above is such that

$$\lim_{j \to +\infty} \|r^j\| = 0. \tag{19}$$

In other words, given an arbitrary modeling error $\epsilon > 0$, there always exists a fine enough scale d such that $\|r^d\| \leq \epsilon$.

The above property only ensures the modeling accuracy of the given observed data. However, the analysis has also been performed to understand the generalizability of this modeling scheme. Using the theory in [34], there is a so-called VC dimension h_j for the model structure at each scale j with the following property: $h_0 < h_1 < h_2 < \ldots$. The structural risk minimization (SRM) principle states that the larger the VC dimension of a model structure, the poorer its generalizability. As our multiscale RBF modeling scheme is from coarse to fine, the model we build has the largest generalizability [34].

Generalization is based on the fact that the unknown systems to be modeled are smooth: small changes in some inputs result in a corresponding small change in the output. If nothing else is known about the unknown system except for a limited number of observed data, the only reasonable assumption about the system is its high degree of smoothness. Therefore, the objective of the modeling has two parts: one is to ensure that the modeling accuracy of the observed data is within the prescribed error bound, which is often guaranteed by the convergence analysis; the other is to have as smooth a model as possible. The problem can be further examined by finding a model that minimizes a cost function consisting of two terms. The first term measures the mean-squares error between the model and the actual outputs for the observed data; the second term measures the cost associated with the smoothness of the model.

However, it is a typical nonlinear minimization problem and it is very difficult to find an optimal model because of the potential problem of local minima. The multiscale RBF networks modeling, in contrast, is a linear approach. It minimizes the mean-squares error at each scale and carries the residue to the finer scale. As a result, it builds a model with maximum smoothness (see [35] for more extensive discussion).

F. EXTENSION TO RECURRENT NEURAL MODELING

Let us briefly discuss the possible extension of all of the modeling techniques developed in the preceding sections to the recurrent case. One idea is based on the following well-known conventional result in differential equations. If two Lipschitz continuous differential equations differ only by ϵ in function norm, then starting from the same initial condition, their respective trajectories will differ by $\epsilon \rho(t)$, where $\rho(t)$ is exponentially related to time t (see [36]). This result also has a discrete-time equivalent.

Interpreting the result, we note that the trajectory deviation will only be unbounded when t goes to infinity. Therefore, if we restrict the modeling to trajectories of finite duration, say T, the recurrent modeling can be done by fixing an error bound for the trajectory and adjusting the function approximation accuracy to achieve the error bound indirectly.

Thus, as long as the modeling is confined to finite trajectory duration, the preceding techniques can always be applied. The only difference this makes is that the trajectory error bound may translate to a very high accuracy requirement for the model function, which is, sometimes, only achievable with a RBF neural network of very large size.

V. NEURAL STATE–SPACE–BASED MODELING TECHNIQUES

A. BASIC MOTIVATION

Starting from the present section, we shall develop techniques for neural modeling by using a neural state–space approach. The type of recurrent neural networks to be used in state–space form makes use of multilayer perceptrons for the state and output equation.

The noise is also considered in the problem formulation. To take process noise into account, a Kalman gain is used in the nonlinear predictor. Neural state–space models have a format similar to that of the Kalman filter, with steady-state Kalman gain for linear systems, but with nonlinear dependency on the state and input of the model.

Narendra's dynamic back-propagation is applied for system modeling, and expressions for the sensitivity model are given. Neural state–space models can be regarded as a brute-force black-box approach to modeling general nonlinear systems. It is possible, however, to gain more insight into the representations, thanks to the uncertain linear system interpretation. A convex hull and LFT representation is discussed. This insight makes it possible, for example, to obtain more specific parameterizations when more is known about the system (a priori) and to use linear models as the starting point in the dynamic back-propagation procedure.

An advantage of the state–space approach is that NL_q stability theory [6] can be applied to identify models that are guaranteed to be globally asymptotically stable. Another advantage is with respect to model-based control design. Linear or neural controllers can be designed based on identified neural state–space models, within the framework of NL_q theory [6].

B. NEURAL STATE–SPACE MODELS

Let us consider nonlinear systems of the form

$$\begin{cases} x_{k+1} = f_0(x_k, u_k) + \varphi_k \\ y_k = g_0(x_k, u_k) + \psi_k \end{cases}, \tag{20}$$

where $f_0(\cdot)$ and $g_0(\cdot)$ are continuous nonlinear mappings, $u_k \in R^m$ is the input vector, $y_k \in R^l$ is the output vector, $x_k \in R^n$ is the state vector, and $\varphi_k \in R^n$, $\psi_k \in R^l$ are the process noise and measurement noise, respectively. The last two are assumed to be zero mean white Gaussian with covariance matrices

$$E\left\{ \begin{bmatrix} \varphi_k \\ \psi_k \end{bmatrix} [\varphi_s^T \; \psi_s^T] \right\} = \begin{bmatrix} Q & S \\ S^T & R \end{bmatrix} \delta_{ks}.$$

In [37] the following predictor with parameter vector θ has been proposed:

$$\begin{cases} \hat{x}_{k+1} = f(\hat{x}_k, u_k; \theta) + K(\theta)\epsilon_k; & \hat{x}_0 = x_0 \text{ given} \\ \hat{y}_k = g(\hat{x}_k, u_k; \theta) \end{cases}, \tag{21}$$

where $f(\cdot)$ and $g(\cdot)$ are continuous nonlinear mappings and $\epsilon_k = y_k - \hat{y}_k$ denotes the prediction error. The steady-state Kalman gain K is parameterized directly, instead of indirectly through a Riccati equation as the result of extended Kalman filtering. For a linear system, a predictor (2) with linear mappings for $f(\cdot)$ and $g(\cdot)$ corresponds to the Kalman filter, with white noise innovations input ϵ_k [38].

In [6, 39] it was proposed to parameterize $f(\cdot)$ and $g(\cdot)$ in (2) by multilayer perceptrons, leading to the neural state space model:

$$\begin{cases} \hat{x}_{k+1} = W_{AB} \tanh(V_A \hat{x}_k + V_B u_k + \beta_{AB}) + K\epsilon_k; & \hat{x}_0 = x_0 \\ \hat{y}_k = W_{CD} \tanh(V_C \hat{x}_k + V_D u_k + \beta_{CD}) \end{cases}, \tag{22}$$

with $V_A \in R^{n_{hx} \times n}$, $V_B \in R^{n_{hx} \times m}$, $\beta_{AB} \in R^{n_{hx}}$, $W_{CD} \in R^{l \times n_{hy}}$, $V_C \in R^{n_{hy} \times n}$, $V_D \in R^{n_{hy} \times m}$, $\beta_{CD} \in R^{n_{hy}}$, $K \in R^{n \times l}$, where n_{hx}, n_{hy} are the number of hidden neurons of the multilayer perceptrons. This recurrent neural network architecture is shown in Fig. 2. For deterministic identification, one has $K = 0$.

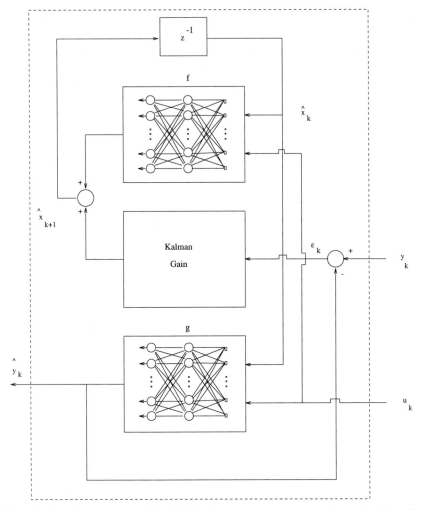

Figure 2 Neural state–space model, which is a discrete-time recurrent neural network with multi-layer perceptrons for the state and output equation and a Kalman gain for taking process noise into account.

Remark. It is well known that linear state–space models are only unique up to a similarity transformation. For the neural state–space model, this is at least up to a similarity transformation and sign reversal of the hidden neurons [6]. It can indeed be shown that the following two neural state–space models produce the same I/O behavior:

$$\begin{cases} \hat{z}_{k+1} = W_{AB} \tanh(V_A \hat{z}_k + V_B u_k + \beta_{AB}) + K\epsilon_k \\ \hat{y}_k = W_{CD} \tanh(V_C \hat{z}_k + V_D u_k + \beta_{CD}) \end{cases}$$

and

$$\begin{cases} \hat{x}_{k+1} = \widehat{W}_{AB} \tanh\left(\widehat{V}_A \hat{x}_k + \widehat{V}_B u_k + \hat{\beta}_{AB}\right) + \widehat{K}\epsilon_k \\ \hat{y}_k = \widehat{W}_{CD} \tanh\left(\widehat{V}_C \hat{x}_k + \widehat{V}_D u_k + \hat{\beta}_{CD}\right) \end{cases},$$

where $\widehat{W}_{AB} = S^{-1}W_{AB}T_1$, $\widehat{V}_A = T_1^{-1}V_A S$, $\widehat{V}_B = T_1^{-1}V_B$, $\hat{\beta}_{AB} = T_1^{-1}\beta_{AB}$, $\widehat{K} = S^{-1}K$, $\widehat{W}_{CD} = W_{CD}T_2$, $\widehat{V}_C = T_2^{-1}V_C S$, $\widehat{V}_D = T_2^{-1}V_D$, $\hat{\beta}_{CD} = T_2^{-1}\beta_{CD}$, where $T_i = P_i J_i$ ($i = 1, 2$), P_i is a permutation matrix, and $J_i = \text{diag}\{\pm 1\}$ and $\hat{z}_k = S\hat{x}_k$, where S is a similarity transformation.

VI. DYNAMIC BACK-PROPAGATION

Let us consider the neural state–space model as a special case of a predictor of the form

$$\begin{cases} \hat{x}_{k+1} = \Phi(\hat{x}_k, u_k, \epsilon_k; \alpha); & \hat{x}_0 = x_0 \text{ given} \\ \hat{y}_k = \Psi(\hat{x}_k, u_k; \beta) \end{cases}, \quad (23)$$

where $\Phi(\cdot)$, $\Psi(\cdot)$ are twice continuously differentiable functions and α, β are elements of the parameter vector θ, to be identified from a number of N input/output training data $Z^N = \{u_k, y_k\}_{k=1}^{k=N}$ as

$$\theta^* = \arg\min_\theta V_N(\theta, Z^N) = \frac{1}{N} \sum_{k=1}^{N} l(\epsilon_k(\theta)). \quad (24)$$

A typical choice for $l(\epsilon_k)$ in this prediction error algorithm is $\frac{1}{2}\epsilon_k^T \epsilon_k$. Let us denote $\theta = [\theta_d; \theta_s]$, where θ_d is the deterministic part and θ_s the stochastic part of the parameter vector:

$$\theta_d = \left[W_{AB}(:); V_A(:); V_B(:); \beta_{AB}; W_{CD}(:); V_C(:); V_D(:); \beta_{CD}\right],$$
$$\theta_s = \left[K(:)\right], \quad (25)$$

and : denotes a columnwise scanning of a matrix. Instead of (5), another approach is to first find the deterministic part and second to identify the Kalman gain while keeping the deterministic part constant:

$$\theta_d^* = \arg\min_{\theta_d} V_N\left(\theta_d, Z^N\right)$$

$$\theta_s^* = \arg\min_{\theta_s} V_N\left(\theta_d, \theta_s, Z^N\right)\big|_{\theta_d=\theta_d^*} \tag{26}$$

$$\theta^* = \left[\theta_d^*; \theta_s^*\right].$$

To solve (5) or (7) by means of a gradient-based optimization algorithm (such as a steepest descent, conjugate gradient, or quasi-Newton method [40, 41]), one computes the gradient

$$\frac{\partial V_N}{\partial \theta} = \frac{1}{N} \sum_{k=1}^{N} \epsilon_k^T \left(-\frac{\partial \hat{y}_k}{\partial \theta}\right). \tag{27}$$

Dynamic back-propagation according to Narendra and Parthasarathy [1, 42] makes use then of the sensitivity model (Fig. 3):

$$\begin{cases} \dfrac{\partial \hat{x}_{k+1}}{\partial \alpha} = \dfrac{\partial \Phi}{\partial \hat{x}_k} \cdot \dfrac{\partial \hat{x}_k}{\partial \alpha} + \dfrac{\partial \Phi}{\partial \alpha} \\[2mm] \dfrac{\partial \hat{y}_k}{\partial \alpha} = \dfrac{\partial \Psi}{\partial \hat{x}_k} \cdot \dfrac{\partial \hat{x}_k}{\partial \alpha} \\[2mm] \dfrac{\partial \hat{y}_k}{\partial \beta} = \dfrac{\partial \Psi}{\partial \beta} \end{cases} \tag{28}$$

to generate the gradient of the cost function. The sensitivity model is a dynamical system with state vector $\partial \hat{x}_k/\partial \alpha \in R^n$, driven by the input vector, consisting of $\partial \Phi/\partial \alpha \in R^n$, $\partial \Psi/\partial \beta \in R^l$, and at the output $\partial \hat{y}_k/\partial \alpha \in R^l$, $\partial \hat{y}_k/\partial \beta \in R^l$ are generated. The Jacobians $\partial \Phi/\partial \hat{x}_k \in R^{n\times n}$ and $\partial \Psi/\partial \hat{x}_k \in R^{l\times n}$ are evaluated around the nominal trajectory.

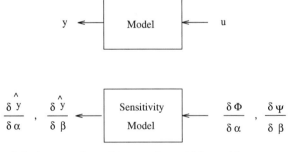

Figure 3 Dynamic back-propagation making use of a sensitivity model, to generate the gradient of the cost function.

To write the derivatives, let us take an elementwise notation:

$$\begin{cases} \hat{x}^i := \sum_j w_{AB}{}^i_j \tanh\left(\sum_r v_{A_r}^j \hat{x}^r + \sum_s v_{B_s}^j u^s + \beta_{AB}{}^j\right) + \sum_j \kappa^i_j \epsilon^j \\ \hat{y}^i = \sum_j w_{CD}{}^i_j \tanh\left(\sum_r v_{C_r}^j \hat{x}^r + \sum_s v_{D_s}^j u^s + \beta_{CD}{}^j\right), \end{cases} \tag{29}$$

where $\{\cdot\}^i$ and $\{\cdot\}^i_j$ denote, respectively, the ith element of a vector and the ijth entry of a matrix. The time index k is omitted after introducing the assignment operator $:=$. Defining

$$\begin{aligned} \phi^l &= \sum_r v_{A_r}^l \hat{x}^r + \sum_s v_{B_s}^l u^s + \beta_{AB}^l \\ \rho^l &= \sum_r v_{C_r}^l \hat{x}^r + \sum_s v_{D_s}^l u^s + \beta_{CD}^l, \end{aligned} \tag{30}$$

one obtains (see [6, 39])

$$\frac{\partial \Phi}{\partial \alpha} : \begin{cases} \dfrac{\partial \Phi^i}{\partial w_{AB}{}^j_l} = \delta^i_j \tanh(\phi^l) \\[2mm] \dfrac{\partial \Phi^i}{\partial v_{A_l}^j} = w_{AB}{}^i_j \left(1 - \tanh^2(\phi^j)\right)\hat{x}^l \\[2mm] \dfrac{\partial \Phi^i}{\partial v_{B_l}^j} = w_{AB}{}^i_j \left(1 - \tanh^2(\phi^j)\right) u^l \\[2mm] \dfrac{\partial \Phi^i}{\partial \beta_{AB}{}^j} = w_{AB}{}^i_j \left(1 - \tanh^2(\phi^j)\right) \\[2mm] \dfrac{\partial \Phi^i}{\partial \kappa^j_l} = \delta^i_j \epsilon_l \end{cases} \tag{31}$$

$$\frac{\partial \Psi}{\partial \beta} : \begin{cases} \dfrac{\partial \Psi^i}{\partial w_{CD}{}^j_l} = \delta^i_j \tanh(\rho^l) \\[2mm] \dfrac{\partial \Psi^i}{\partial v_{C_l}^j} = w_{CD}{}^i_j \left(1 - \tanh^2(\rho^j)\right) \hat{x}^l \\[2mm] \dfrac{\partial \Psi^i}{\partial v_{D_l}^j} = w_{CD}{}^i_j \left(1 - \tanh^2(\rho^j)\right) u^l \\[2mm] \dfrac{\partial \Psi^i}{\partial \beta_{CD}{}^j} = w_{CD}{}^i_j \left(1 - \tanh^2(\rho^j)\right) \end{cases}$$

$$\frac{\partial \Phi}{\partial \hat{x}_k} : \frac{\partial \Phi^i}{\partial \hat{x}^r} = \sum_j w_{AB}{}^i_j (1 - \tanh^2(\phi^j)) v_{A_r}^j$$

$$\frac{\partial \Psi}{\partial \hat{x}_k} : \frac{\partial \Psi^i}{\partial \hat{x}^r} = \sum_j w_{CD}{}^i_j (1 - \tanh^2(\rho^j)) v_{C_r}^j.$$

Hence calculating the gradient of the neural state–space model requires as many simulations of the sensitivity model as the number of elements in the parameter vector θ. Parallelization of the algorithm at this level is then straightforward.

To identify neural state–space models, one follows the usual approach in neural networks: one divides the available data in a training and test set and optimizes until the minimum error on the test set is obtained. In this way overfitting is avoided in case too many hidden neurons have been chosen [5]. The choice of the number of hidden neurons n_{hx}, n_{hy} is a matter of trial and error. Aspects of model validation, including higher order correlation tests are described, for example, in [43].

In [6, 39] one may find examples of identification of nonlinear systems with hysteresis, corrupted by process and observation noise, and identification of a glass furnace. An example of identification of chaotic systems, which is known to be a difficult task for recurrent neural networks, is given in [6, 44]. A special learning strategy, making use of an increasing time horizon and training in packets, is advantageous for that purpose.

VII. PROPERTIES AND RELEVANT ISSUES IN STATE–SPACE NEURAL MODELING

A. UNCERTAIN LINEAR SYSTEM REPRESENTATIONS

The neural state–space model can be written as (see [6, 39])

$$
\begin{cases}
\hat{x}^i := \sum_j w_{AB}{}^i_j \gamma_{AB}{}^j_j \phi^j + \sum_j \kappa^i_j \epsilon^j \\
\hat{y}^i = \sum_j w_{CD}{}^i_j \gamma_{CD}{}^j_j \rho^j,
\end{cases}
\tag{32}
$$

in the elementwise notation of (10) and (11), with

$$
\begin{aligned}
\gamma_{AB}{}^j_j &= \begin{cases} \tanh(\phi^j)/\phi^j, & (\phi^j \neq 0) \\ 1, & (\phi^j = 0); \end{cases} \quad j = 1, \dots, n_{hx} \\
\gamma_{CD}{}^j_j &= \begin{cases} \tanh(\rho^j)/\rho^j, & (\rho^j \neq 0) \\ 1, & (\rho^j = 0); \end{cases} \quad j = 1, \dots, n_{hy}.
\end{aligned}
\tag{33}
$$

The fact that the γ elements are equal to 1 if the argument of the activation function becomes zero is easily seen by applying l'Hôpital's rule or by using the Taylor expansion for $\tanh(\cdot)$. It is essential that the activation function be a static nonlinearity that belongs to the sector [0, 1] [45]. The γ elements have the property that they belong to bounded intervals: $\gamma_{AB}{}^j_j \in [0, 1]$, $\gamma_{CD}{}^j_j \in [0, 1]$ for all j. In matrix-vector notation, this corresponds to

$$\begin{cases} \hat{x}_{k+1} = A(\hat{x}_k, u_k)\hat{x}_k + B(\hat{x}_k, u_k)u_k + v(\hat{x}_k, u_k) + K\epsilon_k \\ \hat{y}_k = C(\hat{x}_k, u_k)\hat{x}_k + D(\hat{x}_k, u_k)u_k + w(\hat{x}_k, u_k) \end{cases}, \qquad (34)$$

with

$$\begin{aligned} A(\hat{x}_k, u_k) &= W_{AB}\Gamma_{AB}(\hat{x}_k, u_k)V_A & B(\hat{x}_k, u_k) &= W_{AB}\Gamma_{AB}(\hat{x}_k, u_k)V_B \\ C(\hat{x}_k, u_k) &= W_{CD}\Gamma_{CD}(\hat{x}_k, u_k)V_C & D(\hat{x}_k, u_k) &= W_{CD}\Gamma_{CD}(\hat{x}_k, u_k)V_D \\ v(\hat{x}_k, u_k) &= W_{AB}\Gamma_{AB}(\hat{x}_k, u_k)\beta_{AB} & w(\hat{x}_k, u_k) &= W_{CD}\Gamma_{CD}(\hat{x}_k, u_k)\beta_{CD} \end{aligned}$$
(35)

and diagonal matrices

$$\begin{aligned} \Gamma_{AB} &= \mathrm{diag}\{\gamma_{AB_1^1}, \dots, \gamma_{AB_{n_{hx}}^{n_{hx}}}\} \\ \Gamma_{CD} &= \mathrm{diag}\{\gamma_{CD_1^1}, \dots, \gamma_{CD_{n_{hy}}^{n_{hy}}}\}. \end{aligned}$$
(36)

From these expressions it is clear that the γ elements of the Γ matrices are a measure of the "hardness" of the nonlinearity of the system. When all γ elements in the neural state–space model are close to 1, the system behaves linearly. Representation (15) also gives more insight into the "brute force," black-box approach of neural state–space models. By using more specific parameterizations than in (3), for example, a model whose structure is known but the precise form of the nonlinearity of which is unknown, this insight might be used to estimate parametric uncertainties for the elements.

Representation (15) can then be further interpreted as a convex hull or an LFT representation (see [6, 39]).

1. Convex Hull Interpretation

For (16) one can write

$$\begin{aligned} A &= \sum_{i=1}^{n_{hx}} \gamma_{AB_i} A_i & A_i &= W_{AB}(:,i)V_A(i,:) \\ B &= \sum_{i=1}^{n_{hx}} \gamma_{AB_i} B_i & B_i &= W_{AB}(:,i)V_B(i,:) \\ v &= \sum_{i=1}^{n_{hx}} \gamma_{AB_i} v_i & v_i &= W_{AB}(:,i)\beta_{AB}(i) \\ C &= \sum_{i=1}^{n_{hy}} \gamma_{CD_i} C_i & C_i &= W_{CD}(:,i)V_C(i,:) \\ D &= \sum_{i=1}^{n_{hy}} \gamma_{CD_i} D_i & D_i &= W_{CD}(:,i)V_D(i,:) \\ w &= \sum_{i=1}^{n_{hy}} \gamma_{CD_i} w_i & w_i &= W_{CD}(:,i)\beta_{CD}(i), \end{aligned}$$
(37)

where $(:,i)$ and $(i,:)$ denote the ith column and ith row, respectively. As a result, the matrices belong then to the convex hull,

$$[A\ B\ v] \in \mathrm{Co}\{[A_1\ B_1\ v_1], \dots, [A_r\ B_r\ v_r]\}, \qquad (38)$$

which is a convex polytope with 2^r vertices (corners) and $r = n_{hx}$, and

$$[C \ D \ w] \in \text{Co}\{[C_1 \ D_1 \ w_1], \dots, [C_s \ D_s \ w_s]\}, \tag{39}$$

a convex polytope with 2^s vertices and $s = n_{hy}$.

2. LFT Representation

In modern control theory uncertainty is often represented by means of an LFT (linear fractional transformation) [46, 47], corresponding to a nominal linear system with feedback perturbation. For representation (15) this can be done as follows. Depending on the input and state vector sequences, the elements $\gamma_{AB}{}_j^j$ and $\gamma_{CD}{}_j^j$ belong to intervals

$$\begin{aligned}
\gamma_{AB}{}_j^j &\in \left[\gamma_{AB}{}^-_j{}^j, \gamma_{AB}{}^+_j{}^j\right] \subset [0, 1] \\
\gamma_{CD}{}_j^j &\in \left[\gamma_{CD}{}^-_j{}^j, \gamma_{CD}{}^+_j{}^j\right] \subset [0, 1].
\end{aligned} \tag{40}$$

The nominal γ values are then defined as the midpoint of these intervals. According to [48], one defines

$$\begin{aligned}
\gamma_{AB\,j}^{(0)\,j} &= \left(\gamma_{AB}{}^-_j{}^j + \gamma_{AB}{}^+_j{}^j\right)/2 & s_{AB}{}_j^j &= \left(\gamma_{AB}{}^+_j{}^j - \gamma_{AB}{}^-_j{}^j\right)/2 \\
\gamma_{CD\,j}^{(0)\,j} &= \left(\gamma_{CD}{}^-_j{}^j + \gamma_{CD}{}^+_j{}^j\right)/2 & s_{CD}{}_j^j &= \left(\gamma_{CD}{}^+_j{}^j - \gamma_{CD}{}^-_j{}^j\right)/2
\end{aligned} \tag{41}$$

and

$$\begin{aligned}
\Gamma_{AB}^{(0)} &= \text{diag}\{\gamma_{AB\,1}^{(0)\,1}, \dots, \gamma_{AB\,n_{hx}}^{(0)\,n_{hx}}\} & S_{AB} &= \text{diag}\{s_{AB}{}_1^1, \dots, s_{AB}{}_{n_{hx}}^{n_{hx}}\} \\
\Gamma_{CD}^{(0)} &= \text{diag}\{\gamma_{CD\,1}^{(0)\,1}, \dots, \gamma_{CD\,n_{hy}}^{(0)\,n_{hy}}\} & S_{CD} &= \text{diag}\{s_{CD}{}_1^1, \dots, s_{CD}{}_{n_{hy}}^{n_{hy}}\}.
\end{aligned} \tag{42}$$

One obtains

$$\begin{aligned}
\Gamma_{AB} &= \Gamma_{AB}^{(0)} + S_{AB}\Delta_{AB} & \Delta_{AB} &= \text{diag}\{\delta_{AB}{}_1^1, \dots, \delta_{AB}{}_{n_{hx}}^{n_{hx}}\} \\
\Gamma_{CD} &= \Gamma_{CD}^{(0)} + S_{CD}\Delta_{CD} & \Delta_{CD} &= \text{diag}\{\delta_{CD}{}_1^1, \dots, \delta_{CD}{}_{n_{hy}}^{n_{hy}}\},
\end{aligned} \tag{43}$$

with $\delta_{AB}{}_j^j \in [-1, 1]$ and $\delta_{CD}{}_j^j \in [-1, 1]$, such that $\|\Delta_{AB}\| \le 1$ and $\|\Delta_{CD}\| \le 1$. One can then write

$$\begin{aligned}
A(\hat{x}_k, u_k) &= A^{(0)} + A_\delta(\delta_{AB}) & B(\hat{x}_k, u_k) &= B^{(0)} + B_\delta(\delta_{AB}) \\
C(\hat{x}_k, u_k) &= C^{(0)} + C_\delta(\delta_{CD}) & D(\hat{x}_k, u_k) &= D^{(0)} + D_\delta(\delta_{CD}) \\
v(\hat{x}_k, u_k) &= v^{(0)} + v_\delta(\delta_{AB}) & w(\hat{x}_k, u_k) &= w^{(0)} + w_\delta(\delta_{CD}),
\end{aligned} \tag{44}$$

with

$$A^{(0)} = W_{AB}\Gamma_{AB}^{(0)}V_A \qquad A_\delta(\delta_{AB}) = W_{AB}S_{AB}\Delta_{AB}V_A$$

$$B^{(0)} = W_{AB}\Gamma_{AB}^{(0)}V_B \qquad B_\delta(\delta_{AB}) = W_{AB}S_{AB}\Delta_{AB}V_B$$

$$C^{(0)} = W_{CD}\Gamma_{CD}^{(0)}V_C \qquad C_\delta(\delta_{CD}) = W_{CD}S_{CD}\Delta_{CD}V_C$$

$$D^{(0)} = W_{CD}\Gamma_{CD}^{(0)}V_D \qquad D_\delta(\delta_{CD}) = W_{CD}S_{CD}\Delta_{CD}V_D \tag{45}$$

$$v^{(0)} = W_{AB}\Gamma_{AB}^{(0)}\beta_{AB} \qquad v_\delta(\delta_{AB}) = W_{AB}S_{AB}\Delta_{AB}\beta_{AB}$$

$$w^{(0)} = W_{CD}\Gamma_{CD}^{(0)}\beta_{CD} \qquad w_\delta(\delta_{CD}) = W_{CD}S_{CD}\Delta_{CD}\beta_{CD}.$$

One obtains then the following LFT (Fig. 4) for the neural state–space model:

$$\hat{y} = \mathcal{F}_u(G, \Delta) \begin{bmatrix} u \\ \epsilon \\ 1 \end{bmatrix}, \tag{46}$$

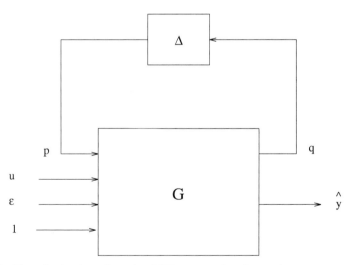

Figure 4 Linear fractional transformation (LFT) representation of a neural state–space model, with nominal model G and Δ uncertainty block, due to the nonlinearity of the activation function $\tanh(\cdot)$.

with state–space representation

$$
\begin{cases}
G: & \begin{bmatrix} \hat{x}_{k+1} \\ \hline \hat{y}_k \\ q_k \end{bmatrix} = \begin{bmatrix} A^{(0)} & B^{(0)} & K & v^{(0)} & [W_{AB}S_{AB}0] \\ \hline C^{(0)} & D^{(0)} & 0 & w^{(0)} & [0W_{CD}S_{CD}] \\ \begin{bmatrix} V_A \\ V_C \end{bmatrix} & \begin{bmatrix} V_B \\ V_D \end{bmatrix} & 0 & \begin{bmatrix} \beta_{AB} \\ \beta_{CD} \end{bmatrix} & 0 \end{bmatrix} \cdot \begin{bmatrix} \hat{x}_k \\ u_k \\ \epsilon_k \\ 1 \\ p_k \end{bmatrix}, \\
\Delta: & p_k = \Delta \cdot q_k, \ \Delta = \mathrm{diag}\{\Delta_{AB}, \Delta_{CD}\}, \ \|\Delta\| \le 1.
\end{cases}
$$

$$(47)$$

The uncertainty block is real diagonal, and its dimension is determined by the number of hidden neurons.

B. LINEAR MODELS AS THE STARTING POINT

Often the system to be identified is only weakly nonlinear, and a linear model is already available. In that case one might use these linear models as starting points for dynamic back-propagation of neural state–space models, instead of random starting points, to improve the accuracy of the model.

Indeed, from our insight into the uncertain linear system interpretation of neural state–space models, we know that for $\Gamma_{AB} = I$, $\Gamma_{CD} = I$, one has

$$
\begin{bmatrix} A & B \\ C & D \end{bmatrix} = \begin{bmatrix} W_{AB}V_A & W_{AB}V_B \\ W_{CD}V_C & W_{CD}V_D \end{bmatrix}.
$$

$$(48)$$

Suppose that, for example, the following Kalman filter [38] model is available:

$$
\begin{cases}
\hat{x}_{k+1} = A\hat{x}_k + Bu_k + K\epsilon_k \\
\hat{y}_k = C\hat{x}_k + Du_k
\end{cases}.
$$

$$(49)$$

Then by taking

$$
W_{AB} := \frac{1}{\alpha_1}[I_n R_1], \qquad [V_A V_B] := \alpha_1 \begin{bmatrix} A & B \\ 0 & 0 \end{bmatrix},
$$

$$
\beta_{AB} := 0; \qquad (n_{hx} \ge n)
$$

$$
W_{CD} := \frac{1}{\alpha_2}[I_l R_2], \qquad [V_C V_D] := \alpha_2 \begin{bmatrix} C & D \\ 0 & 0 \end{bmatrix},
$$

$$
\beta_{CD} := 0; \qquad (n_{hy} \ge l),
$$

$$(50)$$

and letting $\alpha_1, \alpha_2 \to 0$, one forces the neural state–space model to behave linearly. R_1, R_2 are arbitrary matrices of appropriate dimension. Subspace algorithms are also useful for generating these starting points, because of their one-shot, noniterative nature [49].

C. Imposing Stability

In [6, 50] sufficient conditions for global asymptotic stability and I/O stability with finite L_2-gain of discrete-time multilayer recurrent neural networks (NL$_q$ theory) have been derived. In [51] the conditions for global asymptotic stability have been used to modify the classical dynamic back-propagation procedure with an NL$_q$ stability constraint. In this way one can obtain identified models that are guaranteed to be globally asymptotically stable. In certain applications one indeed has the *a priori* knowledge that the true system is globally asymptotically stable or one is interested in such an approximator. In [51] it has been illustrated, for example, how a little process noise can cause limit cycle behavior in the identified model, instead of global asymptotic stability, as in the true system.

Let us consider the autonomous case of (3), without noise and with zero bias terms:

$$\hat{x}_{k+1} = W_{AB} \tanh(V_A \hat{x}_k). \tag{51}$$

After introducing a new state variable $\xi = \tanh(V_A \hat{x}_k)$, one obtains the NL$_1$ representation:

$$\begin{cases} \hat{x}_{k+1} = W_{AB} \xi_k \\ \xi_{k+1} = \tanh(V_A W_{AB} \xi_k) \end{cases}. \tag{52}$$

Defining the matrix

$$V_{tot} = \begin{bmatrix} 0 & W_{AB} \\ 0 & V_A W_{AB} \end{bmatrix}, \tag{53}$$

it is shown in [6] that a sufficient condition for global asymptotic stability of (33), with the origin as a unique equilibrium point, is to find a diagonal matrix D such that

$$\left\| D V_{tot} D^{-1} \right\|_2 < 1. \tag{54}$$

Hence dynamic back-propagation can be modified as follows:

$$\min_\theta V_N(\theta, Z^N) = \frac{1}{N} \sum_{k=1}^{N} l(\epsilon_k(\theta)) \quad \text{such that } \min_D \left\| D V_{tot}(\theta) D^{-1} \right\|_2 < 1.$$

$$\tag{55}$$

This constraint can be expressed as the LMI (linear matrix inequality):

$$V_{tot}^T D^2 V_{tot} < D^2. \tag{56}$$

Finding such a matrix D for a given matrix V_{tot} corresponds to solving a convex optimization problem [6, 52–55] and is known as "diagonal scaling" in the area of robust control theory [46, 47, 56]. Taking (37) as a constraint to dynamic back-propagation yields

$$\min_{\theta,D} V_N(\theta, Z^N) = \frac{1}{N} \sum_{k=1}^{N} l(\epsilon_k(\theta)) \qquad \text{such that } V_{tot}(\theta)^T D^2 V_{tot}(\theta) < D^2. \tag{57}$$

The cost function is differentiable, but the constraint becomes nondifferentiable when the two largest eigenvalues of the matrix $V_{tot}(\theta)^T D^2 V_{tot}(\theta) - D^2$ coincide [57]. Convergent algorithms for such nonconvex, nondifferentiable optimization problems have been described, for example, by Polak and Wardi [57]. The gradient-based optimization method makes use of the concept of a generalized gradient for the constraint.

Because the diagonal scaling condition can be conservative, the following condition might be used for the NL_1 (33). A sufficient condition for global asymptotic stability of (33) is to find a full rank matrix P such that

$$\kappa(P) \| P V_{tot} P^{-1} \|_2 < 1, \tag{58}$$

where $\kappa(P)$ denotes the condition number of P. In practice it is often sufficient to impose local stability at the origin and enlarge its basin of attraction by minimizing the condition number:

$$\min_P \kappa(P) \qquad \text{such that} \quad \| P V_{tot} P^{-1} \|_2 < 1. \tag{59}$$

Even when condition (39) is not satisfied, procedure (40) normally leads to globally asymptotically stable systems. This principle has been demonstrated for stabilizing and controlling systems with one or multiple equilibria, periodic and quasi-periodic behavior, and chaos [6]. The modified dynamic back-propagation algorithm then becomes

$$\min_{\theta,Q} V_N(\theta, Z^N) = \frac{1}{N} \sum_{k=1}^{N} l(\epsilon_k(\theta)) \qquad \text{such that} \begin{cases} V_{tot}(\theta)^T Q V_{tot}(\theta) < Q \\ I < Q < \alpha I \end{cases}. \tag{60}$$

The latter LMI corresponds to $\kappa(P_i) < \alpha_i$ [52] and $Q = P^T P$.

VIII. ILLUSTRATIVE EXAMPLES

A. RBF NEURAL NETWORK MODELS

We use the following single-input and single-output (SISO) dynamical system as an example:

$$y_{t+1} = \frac{u_t y_t}{1 + y_t^2} + \begin{cases} 0.6 \sin(32\pi u_t), & 0.3 \le |u_t| \le 0.4 \\ 0.6 \cos(2\pi u_t), & 0 \le |u_t| < 0.3 \text{ or } 0.4 < |u_t| \le 1 \end{cases}. \quad (61)$$

We shall confine the input u_t of the system to be bounded between -1 and 1. The corresponding output of the system can be found to cover the same range $[-1, 1]$. The system (61) is clearly continuous, time-invariant, and BIBO stable over the above region of operation. Geometrically, this system, as shown in Fig. 5, can be seen as being smooth over the whole region of $[-1, 1; -1, 1]$, except at two small local regions, where there are highly fluctuating local features.

To generate the data samples for modeling, the following input sequence is introduced:

$$u_{1t} = \cos(2\pi t/200), \quad (62)$$

where u_{1t} obviously falls within $[-1, 1]$ for all t. Let us first use the proposed RBF neural networks with a fixed uniform grid to model the part of the system

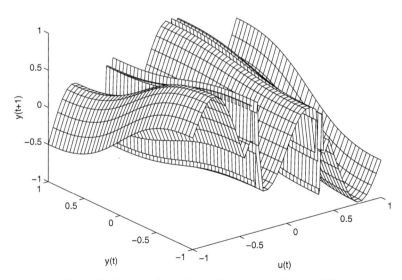

Figure 5 Output surface of the nonlinear dynamical system (61).

that is being excited by this particular excitation sequence. The modeling error is set at $\epsilon = 0.01$.

We first use a coarse grid with the size $\Delta_1 = \Delta_2 = 0.4$. Figure 6 shows the convergence trend of the modeling scheme, and the performance of the model is shown in Fig. 7a. Observe that the modeling result is fair in most of the region, except for those two local fluctuating features. Moreover, the number of RBF neurons is only 52.

To validate the model, the following new input sequence is chosen for this purpose:

$$u_{2t} = 0.6\cos(2\pi t/200) + 0.4\sin(8\pi t/200), \tag{63}$$

which contains two frequency components that are different from those in u_{1t}. Figure 7b shows the modeling performance of the above RBF model under this new input excitation. Although there is still some modeling error, the result is acceptable. It implies that the RBF model with a coarse grid has good general-

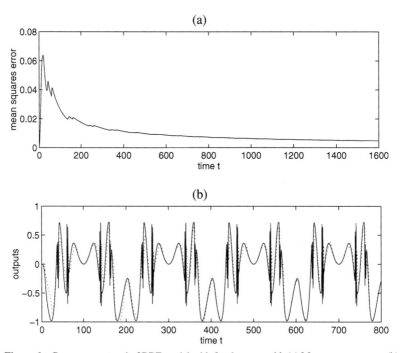

Figure 6 Convergence trend of RBF model with fixed coarse grid. (a) Mean-squares error. (b) Solid line: system output; dotted line: model output.

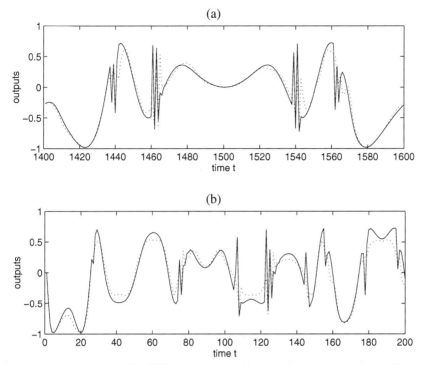

Figure 7 (a) Performance of the RBF model corresponding to excitation u_{1t}. (b) Model validation with new excitation u_{2t}. Solid line: system output; dotted line: model output.

izability. Moreover, the modeling error can be reduced dramatically after further training using the new excitation u_{2t}.

As we mentioned in Section II, there is a tradeoff between the modeling accuracy of the observed data and its generalizability. We illustrate this point by using a RBF neural network with a very fine grid. In this case, we choose $\Delta_1 = \Delta_2 = 0.01$ and use the same input sequence u_{1t}. The model is very accurate for the given input/output data, as shown in Fig. 8a. However, when we use u_{2t} to validate the model, it is extremely poor in generalization, which is shown in Fig. 8b. Moreover, the number of neurons is 1385, which is very large.

To achieve a model with both the accuracy of the observed data and good generalizability, we build the RBF model by the structure-adaptation technique.

The technique starts with an arbitrary coarse grid with size $\Delta_1 = \Delta_2 = 0.4$. Figure 9 shows its convergence trend. Observe that it takes longer to converge than in the fixed grid case. However, the convergence is quite smooth, and no jump occurs during the structure adaptation. The reason for this is quite simple.

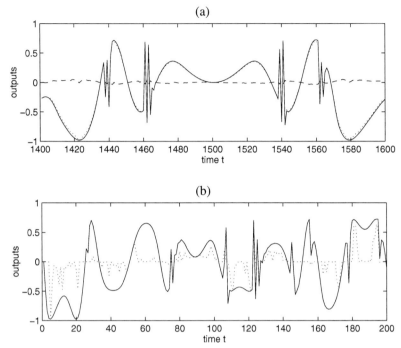

Figure 8 (a) Performance of the RBF model with fixed fine grid corresponding to excitation u_{1t}. (b) Model validation with new excitation u_{2t}. Solid line: system output; dotted line: model output; dashed line: modeling error.

After each grid halving operation, the new weight update is based on the weights of previous structure. Therefore, it retains and fine-tunes the part that has already been trained. Figure 10 shows the performance of the RBF model with structure adaptation for input sequences u_{1t} and u_{2t}, respectively. It achieves both high modeling accuracy and good generalization with a small number of neurons—218 in this case.

The above observations can be further explained as follows. With certain specific excitations, we can extract only part of the unknown system based on the observed data. Figure 11a shows the input/output samples of (61) with the excitation u_{1t}. Based on these input/output data, Fig. 11b–d shows model outputs for the whole range $[-11; -11]$ of the above three different types of RBF neural networks. The RBF model with fixed coarse grid is quite smooth. But it is unable to capture the local features. In contrast, the RBF model with fixed fine grid is not smooth and results in overfitting. It is the RBF model with structure adaptation that is able to capture both the global and local features of the system.

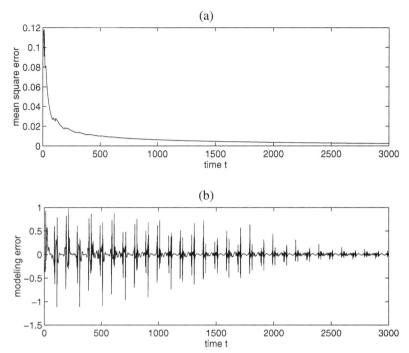

Figure 9 Convergence trend of structure-adaptation RBF model. (a) Mean-squares error. (b) Modeling error.

Compared to the structure-adaptation technique, which is slow and has difficulty choosing some of parameters, the multiscale RBF modeling technique appears to be more efficient. Here we only illustrate its scale decomposition property. After using the multiscale modeling scheme with dense enough samples of (61), the multiscale RBF model constructed has three scales, $j = 0$, 1 and 4, with a total of 280 RBF neurons.

Starting from scale $j = 0$, the performance of the partial model built at scales up to 1 by using the multiscale modeling algorithm is shown in Fig. 12a and b. Observe that the partial model at this scale can capture the global feature of the system but not those two local features. This is due to the fact that more globally supported activation functions at coarse scales are more suited to capturing the global features of the system.

At scales 2 and 3, there is no significant improvement in the modeling accuracy when the RBF neurons at those scales are considered. This fact is revealed by applying the structure reduction rule, which eliminates all of the RBF neurons at these scales.

424 Shaohua Tan et al.

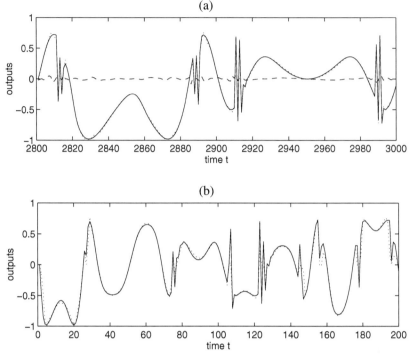

Figure 10 (a) Performance of the RBF model with fixed fine grid corresponding to excitation u_{1t}. (b) Model validation with new excitation u_{2t}. Solid line: system output; dotted line: model output; dashed line: modeling error.

Only at scale $j = 4$ has the modeling accuracy been reached; Fig. 12c and d shows the performance. Note that the multiscale RBF model obtained can approximate the unknown system very well both globally and locally.

B. NEURAL STATE–SPACE MODELS

In this section, we present an example of the use of neural state–space models for modeling a highly nonlinear system that is corrupted by process noise and measurement noise [6, 39]. For additional examples we refer the reader to [6].

The system to be identified in this example is an interconnected system, consisting of two dynamical subsystems and two static nonlinearities: a hysteresis curve $f_1(\cdot)$ and a hyperbolic tangent function $f_2(\cdot) = \tanh(\cdot)$. The linear systems L and M are both SISO of order 2 and 1 with state vectors x_k and z_k, respectively.

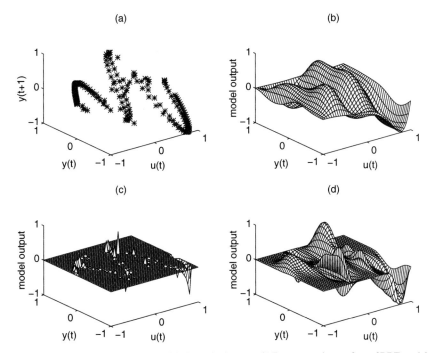

Figure 11 (a) Input/output samples with the excitation u_{1t}. (b) Reconstruction surface of RBF model with fixed coarse grid. (c) Reconstruction surface of RBF model with fixed fine grid. (d) Reconstruction surface of structure-adaptation RBF model.

The interconnected system with input u_k, output y_k, and state vector $[x_k; z_k]$ has the following form:

$$\begin{cases} x_{k+1} = A_L x_k + b_L u_k + \begin{bmatrix} 1 \\ 0 \end{bmatrix} v_k \\ z_{k+1} = a_M z_k + b_M f_1(c_L^T x_k) \\ y_k = f_2(c_M z_k + d_M f_1(c_L^T x_k)) + w_k \end{cases}, \tag{64}$$

with v_k, w_k as zero mean white Gaussian noise processes. The I/O data were generated by a random input signal, uniformly distributed in the interval $[-1, 1]$. Process noise v_k and measurement noise w_k both have a standard deviation of 0.01. The system matrices for L and M are $a_M = 0.7$, $b_M = c_M = d_M = 1$,

$$A_L = \begin{bmatrix} 0.1 & -0.2 \\ 1 & 0.3 \end{bmatrix}, \qquad b_L = \begin{bmatrix} 0 \\ 1 \end{bmatrix}, \qquad c_L = \begin{bmatrix} 1 \\ 0 \end{bmatrix}, \qquad d_L = 0.$$

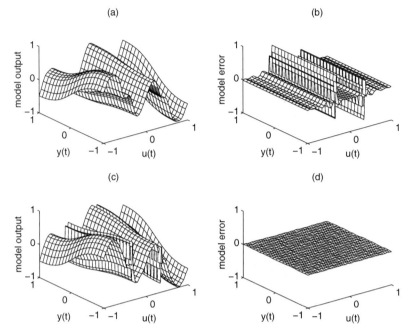

Figure 12 Performance of multiscale RBF model at scale 1. (a) Reconstruction surface. (b) Modeling error. At scale 4: (c) reconstruction surface; (d) modeling error.

The nonlinearity $f_1(\cdot)$ is shown in Fig. 13 and is defined by Table I. The right or left part of the curve is selected depending on the sign of x_2. In total, 2000 data points were generated with $c = 1$, $d = 0.2$ in f_1 and an zero initial state. This data set is split into two parts: a training set containing the first 1000 data points,

Table I

Definition of $f_1(\cdot)$

	$x_2 > 0$	$x_2 \leq 0$
$-c - d \leq x_1 \leq -c + d$	$f_1 = -c$	—
$-c + d \leq x_1 \leq c + d$	$f_1 = x_1 - d$	—
$c - d \leq x_1 \leq c + d$	—	$f_1 = c$
$-c - d \leq x_1 \leq c - d$	—	$f_1 = x_1 + d$
$-c - d \leq x_1$	$f_1 = -c$	$f_1 = -c$
$x_1 \leq c + d$	$f_1 = c$	$f_1 = c$

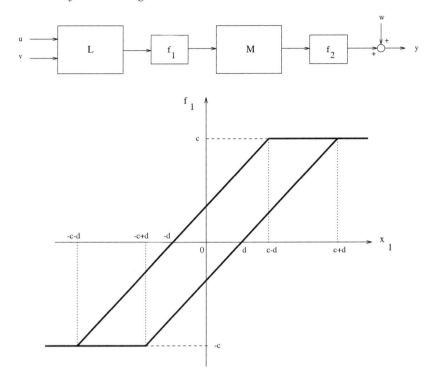

Figure 13 (Top) Nonlinear interconnected system consisting of two linear dynamic systems L and M, respectively, of order 2 and 1, and two static nonlinearities: a hysteresis curve $f_1(\cdot)$ and $f_2(\cdot) = \tanh(\cdot)$. The system is corrupted with process noise v and measurement noise w. (Bottom) Hysteresis curve $f_1(x_1)$. Depending on the sign of x_2, the right or left part of the curve is selected.

and a test set consisting of the following 1000 data points, which are fresh data for testing the obtained models. The corresponding fitting error $V_{N_{\text{fit}}}$ and the generalization error $V_{N_{\text{gen}}}$ are defined on these sets. As a predictor a neural state–space model was taken with $n = 3$, $n_{hx} = n_{hy} = 7$. The number of hidden neurons is chosen on a trial-and-error basis, with the understanding that too many hidden neurons may lead to overfitting, with bad generalization of the model [5]. To minimize the cost function, a quasi-Newton method with BFGS (Broyden–Fletcher–Goldfarb–Shanno) updating of the Hessian and a mixed quadratic and cubic line search was used (function fminu of Matlab's optimization toolbox). Simulation of the neural state–space model and its corresponding sensitivity model, needed to generate the gradient of the cost function, were both written in C code, making

428

Shaohua Tan et al.

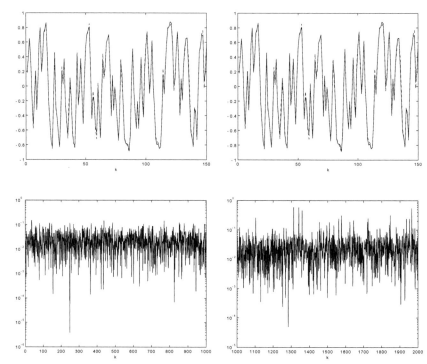

Figure 14 Nonlinear system identification using neural state–space models. (Top) Output of the plant (full line) and output of the neural state space model (dashed line) through time, shown for part of the training data (left) and part of the test data (right). (Bottom) Absolute value of the prediction error through time, for the training data (left) and the test data (right).

use of Matlab's *cmex* facility. The best local minimum after taking 100 different starting points (according to a random Gaussian distribution with standard deviation 0.5) was $V_{N_{\text{fit}}} = 6.3848e-04$. This model also has a minimum generalization error equal to $V_{N_{\text{gen}}} = 1.5803e - 03$. Simulation results for this model are shown in Fig. 14. Model validation tests were done according to [43] and are presented in Fig. 15 for the training data.

For the interpretation of the system as an uncertain linear system, the intervals $[\gamma_{AB_j}^{-\,j}, \gamma_{AB_j}^{+\,j}]$, $[\gamma_{CD_j}^{-\,j}, \gamma_{CD_j}^{+\,j}]$, were calculated based on the training data and the optimal model, with $\gamma_{AB_j}^{+\,j} = 1$, $\gamma_{CD_j}^{+\,j} = 1$, and

$$\gamma_{AB}^- = [0.0562\ 0.8942\ 0.7313\ 0.7659\ 0.8904\ 0.8306\ 0.3841]^T$$

$$\gamma_{CD}^- = [0.1785\ 0.6271\ 0.5012\ 0.6022\ 0.0244\ 0.1877\ 0.2284]^T$$

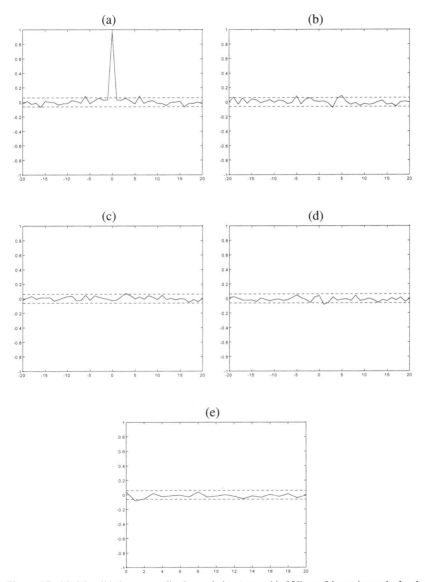

Figure 15 Model validation: normalized correlation tests with 95% confidence intervals for the model with minimum fitting and generalization error, evaluated on the training set: (a) $\hat{\phi}_{\epsilon\epsilon}(\tau)$, (b) $\hat{\phi}_{u\epsilon}(\tau)$, (c) $\hat{\phi}_{u^{2'}\epsilon}(\tau)$, (d) $\hat{\phi}_{u^{2'}\epsilon^2}(\tau)$, (e) $\hat{\phi}_{\epsilon(\epsilon u)}(\tau)$. The identified neural state–space model is valid in this sense.

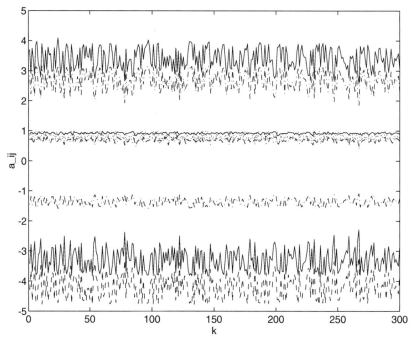

Figure 16 Parametric uncertainties of the elements of the 3×3 matrix $A(\hat{x}_k, u_k)$. The elements are shown through time, evaluated for part of the training set, according to the expression $A(\hat{x}_k, u_k) = W_{AB} \, \Gamma_{AB}(\hat{x}_k, u_k) \, V_A$. This figure is just intended to give an impression of the variation of these elements. Similar plots can be made for $B(\hat{x}_k, u_k)$, $C(\hat{x}_k, u_k)$, and $D(\hat{x}_k, u_k)$.

as the results, which indicates that the nonlinearity of the underlying system is rather "hard." The variation on the elements of the matrix $A(\hat{x}_k, u_k)$ according to the uncertain linear system interpretation is shown in Fig. 16. Similar plots can be made for $B(\hat{x}_k, u_k)$, $C(\hat{x}_k, u_k)$, $D(\hat{x}_k, u_k)$.

ACKNOWLEDGMENTS

This research was partially supported by EU grant KIT 124 SYSIDENT; a fund from the Belgian Programme on Interuniversity Poles of Attraction, initiated by the Belgian State, Prime Minister's Office for Science, Technology and Culture (IUAP-17); and a fund from Concerned Action Project MIPS (Model-based Information Processing Systems) of the Flemish Community.

REFERENCES

[1] K. S. Narendra and K. Parthasarathy. Identification and control of dynamical systems using neural networks. *IEEE Trans. Neural Networks* 1:4–27, 1990.

[2] S. Chen, S. A. Billings, and P. M. Grant. Nonlinear system identification using neural networks. *Internat. J. Control* 51:1191–1213, 1990.

[3] S. Chen and S. Billings. Neural networks for nonlinear dynamic system modelling and identification. *Internat. J. Control* 56:319–346, 1992.

[4] M. Brown and C. Harris. *Neurofuzzy Adaptive Modeling and Control*. Prentice Hall, New York, 1994.

[5] J. Sjöberg, H. Hjalmarsson, and L. Ljung. Neural networks in system identification. In *Preprints from the 10th IFAC Symposium on System Identification*, Vol. 2, pp. 49–72. Copenhagen, 1994.

[6] J. Suykens, J. Vandewalle, and B. De Moor. *Artificial Neural Networks for Modeling and Control of Non-linear Systems*. Kluwer Academic Publishers, Boston, MA, 1995.

[7] J. Sjöberg, Q. Zhang, L. Ljung, A. Benveniste, B. Delyon, P. Glorennec, H. Hjalmarsson, and A. Juditsky. Nonlinear black-box modeling in system identification: a unified overview. *Automatica* 31:1691–1724, 1995.

[8] A. Juditsky, H. Hjalmarsson, A. Benveniste, B. Delyon, L. Ljung, J. Sjöberg, and Q. Zhang. Nonlinear black-box models in system identification: mathematical foundations. *Automatica* 31:1725–1750, 1995.

[9] K. J. Hunt, D. Sbarbaro, R. Zbikowski, and P. J. Gawthrop. Neural networks for control systems—a survey. *Automatica* 28:1083–1112, 1992.

[10] L. Ljung and T. Söderström. *Theory and Practice of Recursive Identification*. MIT Press, Cambridge, MA, 1983.

[11] D. S. Broomhead and D. Lowe. Multivariable functional interpolation and adaptive networks. *Complex Systems* 2:321–355, 1988.

[12] T. Poggio and F. Girosi. Networks for approximation and learning. *Proc. IEEE* 78:1481–1497, 1990.

[13] J. Moody and C. Darken. Fast learning in networks of locally-tuned processing units. *Neural Comput.* 1:281–294, 1989.

[14] J. Platt. A resource-allocating network for function interpolation. *Neural Comput.* 3:213–225, 1991.

[15] R. M. Sanner and J. E. Slotine. Gaussian networks for direct adaptive control. *IEEE Trans. Neural Networks* 3:837–863, 1992.

[16] S. Tan, J. Hao, and J. Vandewalle. Stable and efficient neural network modeling of discrete multichannel signals. *IEEE Trans. Circuits Systems I* 41:829–840, 1994.

[17] S. Qian and D. Chen. Signal representation using adaptive normalized Gaussian functions. *Signal Process.* 36:1–11, 1994.

[18] S. Tan, J. Hao, and J. Vandewalle. Efficient identification of RBF neural net models for nonlinear discrete-time multivariable dynamical systems. *Neurocomputing* 9:11–26, 1995.

[19] J. Park and I. W. Sandberg. Universal approximation using radial-basis-function networks. *Neural Comput.* 3:246–257, 1990.

[20] D. P. Peterson and D. Middleton. Sampling and reconstruction of wave number limited functions in *n*-dimensional Euclidean spaces. *Inform. and Control (Shenyang)* 5:279–323, 1962.

[21] S. Lee and R.M. Kil. A Gaussian potential function network with hierarchical self-organizing learning. *Neural Networks* 4:207–224, 1991.

[22] M. T. Musavi, W. Ahmed, K. H. Chan, K. B. Faris, and D. M. Hummels. On the training of radial basis function classifiers. *Neural Networks* 5:595–603, 1992.

[23] S. Chen, C. Cowan, and P. Grant. Orthogonal least squares learning algorithm for radial basis function networks. *IEEE Trans. Neural Networks* 2:302–309, 1991.

[24] T.-C. Lee. *Structure Level Adaptation for Artificial Neural Networks*. Kluwer Academic, Boston, 1991.
[25] S. Tan and Y. Yu. Adaptive fuzzy modeling of nonlinear dynamical systems. *Automatica* 32:637–643, 1996.
[26] S. Haykin. *Neural Networks: A Comprehensive Foundation*. Macmillan, New York, 1994.
[27] L. X. Wang and J. M. Mendel. Fuzzy basis functions, universal approximation, and orthogonal least-squares learning. *IEEE Trans. Neural Networks* 3:807–814, 1992.
[28] I. Daubechies. *Ten Lectures on Wavelets*. SIAM, Philadelphia, 1992.
[29] Y. Yu, S. Tan, J. Vandewalle, and E. Deprettere. Near-optimal construction of wavelet networks for nonlinear system modeling. In *Proceedings of IEEE International Symposium on Circuits and Systems (ISCAS'96)*, Vol. 3, pp. 48–51, 1996.
[30] Q. Zhang and A. Benveniste. Wavelet networks. *IEEE Trans. Neural Networks* 3:889–898, 1992.
[31] Y. C. Pati and P. S. Krishnaprasad. Analysis and synthesis of feedforward neural networks using affine wavelet transformations. *IEEE Trans. Neural Networks* 4:73–85, 1993.
[32] R. R. Coifman and M. V. Wickerhauser. Entropy-based algorithms for best-bases selection. *IEEE Trans. Inform. Theory* 38:713–718, 1992.
[33] Y. Yu, S. Tan, J. Vandewalle, and E. Deprettere. Multi-scale modeling of nonlinear systems using wavelet networks. Technical Report, ESAT Lab, Katholieke Universiteit Leuven, 1996.
[34] V. Vapnik. *Estimation of Dependences Based on Empirical Data*. Springer-Verlag, New York, 1982.
[35] W. Lawton, S. L. Lee, Y. Yu, and S. Tan. Fast algorithm for multi-scale wavelet modeling. Technical Report, ISS, NUS, 1996.
[36] M. W. Hirsch and S. Smale. *Differential Equations, Dynamical Systems, and Linear Algebra*. Academic Press, New York, 1974.
[37] G. Goodwin and K. Sin. *Adaptive Filtering, Prediction and Control*. Prentice Hall, Englewood Cliffs, NJ, 1984.
[38] L. Ljung. Asymptotic behavior of the extended Kalman filter as a parameter estimator for linear systems. *IEEE Trans. Automat. Control* AC-24:36–50, 1979.
[39] J. Suykens, B. De Moor, and J. Vandewalle. Nonlinear system identification using neural state space models, applicable to robust control design. *Internat. J. Control* 62:129–152, 1995.
[40] R. Fletcher. *Practical Methods of Optimization*, 2nd Ed., Wiley, Chichester and New York, 1987.
[41] P. E. Gill, W. Murray, and M. H. Wright. *Practical Optimization*. Academic Press, London, 1981.
[42] K. S. Narendra and K. Parthasarathy. Gradient methods for the optimization of dynamical systems containing neural networks. *IEEE Trans. Neural Networks* 2:252–262, 1991.
[43] S. Billings, H. Jamaluddin, and S. Chen. Properties of neural networks with applications to modeling non-linear dynamical systems. *Internat. J. Control* 55:193–224, 1992.
[44] J. Suykens and J. Vandewalle. Teaching a simple recurrent neural state space model to behave like Chua's double scroll. *IEEE Trans. Circuits Systems I Fund. Theory Appl.* 42:499–502, 1995.
[45] M. Vidyasagar. *Nonlinear Systems Analysis*. Prentice Hall, Englewood Cliffs, NJ, 1993.
[46] J. M. Maciejowski. *Multivariable Feedback Design*. Addison-Wesley, Reading, MA, 1989.
[47] A. Packard and J. Doyle. The complex structured singular value. *Automatica* 29:71–109, 1993.
[48] M. Steinbuch, J. Terlouw, O. Bosgra, and S. Smit. Uncertainty modeling and structured singular-value computation applied to an electromechanical system. *IEE Proc. D* 139:301–307, 1992.
[49] P. Van Overschee and B. De Moor. N4SID: subspace algorithms for the identification of combined deterministic-stochastic systems. *Automatica* 30:75–93, 1994.
[50] J. Suykens, B. De Moor, and J. Vandewalle. NL_q theory: a neural control framework with global asymptotic stability criteria. *Neural Networks*, to appear.
[51] J. Suykens, J. Vandewalle, and B. De Moor. NL_q theory: checking and imposing stability of recurrent neural networks for nonlinear modeling. Technical Report, ESAT Lab, Katholieke Universiteit Leuven, 1996.

[52] S. Boyd, L. El Ghaoui, E. Feron, and V. Balakrishnan. Linear matrix inequalities in system and control theory. *Stud. Appl. Math.* 15, 1994.

[53] P. Gahinet and A. Nemirovskii. General-purpose LMI solvers with benchmarks. In *Proceedings of the Conference on Decision and Control*, pp. 3162–3165, 1993.

[54] Y. Nesterov and A. Nemirovskii. Interior point polynomial algorithms in convex programming. *Stud. Appl. Math.* 13, 1994.

[55] M. L. Overton. On minimizing the maximum eigenvalue of a symmetric matrix. *SIAM J. Matrix Anal. Appl.* 9:256–268, 1988.

[56] E. Kaszkurewicz and A. Bhaya. Robust stability and diagonal Liapunov functions. *SIAM J. Matrix Anal. Appl.* 14:508–520, 1993.

[57] E. Polak and Y. Wardi. Nondifferentiable optimization algorithm for designing control systems having singular value inequalities. *Automatica* 18:267–283, 1982.

Index

A

Absence of local minima in the cost function surface for orthogonal activation function-based neural networks, 18

Adaptive fuzzy systems in nonlinear system identification, 196

Aircraft autopilots utilizing neural networks, 341–353

Algorithms of neural network systems, 336–341

Architecture
for optimal tracking neurocontrollers, 162
of four-layer recurrent neural network for feedback controller synthesis, 92
of four-layer recurrent neural network for state estimator synthesis, 97
of two-layer recurrent neural network for feedback controller synthesis, 88
of two-layer recurrent neural network for state estimator synthesis, 95

B

Backward simulators for generalized backward through time algorithms, 175

Block diagram
for an optimal tracking neurocontroller, 162
for training feedforward controllers, 169
for training neural network system identifiers for control systems, 165
of error filtering identification model utilizing radial basis function networks, 215

of regressor filtering identification models developed using radial basis function networks, 221

Boltzmann neural networks for recurrent high order neural networks for the identification of unknown nonlinear dynamical systems, 294–298

Boolean function identification using Fourier series neural networks, 21

C

Case studies of neurocontrollers, 176–187

Computational parallelism of orthogonal activation function-based neural networks, 15

Continuous-time identification of nonlinear systems, 210

Conventional nonlinear autoregressive moving average modeling techniques for nonlinear dynamic systems with exogenous inputs, 235–236

Convergent learning laws for the identification of general nonlinear dynamic systems, 289–294

D

Describing function identification by means of Fourier series neural networks, 34

Design procedures for neurocontrols for systems with unknown dynamics, 313–318

Desired network properties for system identification and control, 5

Direct adaptive controllers utilizing orthonormal activation function-based neural networks, 58
Direct control techniques with both modeling errors and dynamic uncertainties by utilizing neural networks, 139
Direct techniques to control unknown nonlinear dynamical systems using dynamic neural networks, 127
Dynamic back-propagation techniques in state–space modeling for dynamic systems, 409–412
Dynamic neural network models, 130
Dynamic uncertainty determination in systems by utilization of neural networks, 138

F

Feedback neurocontrollers, 170
Feedforward neurocontrollers, 168
Forward simulator for generalized backpropogation through time, 174
Fourier series neural network architecture, 9
Fourier series neural network-based adaptive control systems, 35
Frequency domain applications using Fourier series neural networks, 25
Frequency spectrum identification, 26
  scheme using the Fourier series neural network, 27
Function approximation capability of orthogonal functions for approximating nonlinear functions, 14
Function based approximation using orthogonal activation function-based neural networks, 23

G

Generalized backpropogation through time algorithms
  neural network systems for receding horizon optimal control of nonlinear dynamic systems, 157
Geometric significance of performance surface of a cost function for a neural network design, 13
Gradient descent learning algorithm, 12

H

High-order neural networks in system identification, 279
High-order neural network systems techniques for the identification of dynamic systems, 279

I

Identifiability and local model fitting for dynamic systems, 271–273
Identification neural network in receding horizon controllers, 164
Identification of dynamical systems, applications of neural networks, 298–304
Identification techniques utilizing neural network systems, 132
Indirect control system techniques utilizing neural networks, 132
Indirect techniques to control unknown nonlinear dynamical systems using dynamic neural networks, 127

L

Learning algorithms for nonlinear system identification, 200
Learning stability for the gradient descent rule for orthogonal activation function-based neural networks, 15
Linearly parametrized approximators for nonlinear system identification, 206
Linear parameter determination for dynamic systems, 259–271

M

Model generation and interpretation for orthogonal activation function-based neural networks, 20
Modeling for nonlinear dynamic systems by neural network methods including multilayer perceptrons, radial basis function networks, fuzzy basis function networks, and recurrent neural networks, 246–254
Model validation methods for nonlinear dynamic systems, 245–246
Multi-input–multi-output architecture of an orthogonal activation function-based neural network, 7
Multi-input–single-output Fourier series neural network architecture, 11

Multilayer feedforward neural network controllers, structure, 163

Multilayer neural networks in model approximation methods, 195, 223

Mutlilayer recurrent neural networks in linear control system design, 75

Multiscale radial basis function neural networks systems dynamic systems modeling techniques, 399–406

Multivariable nonlinear models for dynamic systems using neural networks, 231

N

Neural model reference adaptive system
  hardware layout, 45
  using Fourier series neural network estimators, 42
  techniques, 41

Neural network
  command augmentation systems for aircraft, 353–373
  nonlinear identifier systems, 47
  spectrum analyzer, 25

Neural networks
  for controller synthesis, 85
  for observer synthesis, 93
  with orthogonal activation functions, 2

Neural self-tuning regulators
  by means of Fourier series neural networks, 37
  using Fourier series neural network estimators, 37

Neural system identification and control, 4

Neurocontrol methods for systems with unknown dynamics, test cases, 309–312, 318–330

Neurocontrol system design, 163

Neurocontrol techniques for systems with unknown dynamics, 307–309

Nonlinear autoregressive moving average models for dynamic systems with exogenous inputs, 232

Nonlinear dynamic system modeling by means of neural networks, 383–384

Nonlinear parameter methods for nonlinear dynamic systems, 254–259

Nonlinear system modeling by various classes of neural network systems, illustrative examples, 419–430

Nonlinear system representation for dynamic systems, 233–235

O

Observer synthesis by neural networks, 93

On-line approximation for nonlinear system identification by neural networks, 191

On-line learning neural algorithms for aircraft control systems, 333

On-line radial basis function neural network system structural adaptive modeling, 394–398

Optimal tracking neurocontrollers for discrete time nonlinear dynamic systems, 157

Orthogonal activation function-based neural network, 8
  properties, 14
  rates of convergence, 19

Orthogonal functions for system identification and control, 1

Orthogonal neural network architecture, 7

Orthonormal activation functions, 9–10

Overfitted model problems, 232

P

Parameter convergence for the parameter weights utilized in a network using orthogonal activation functions, 18

Parameter determination in control systems utilizing neural networks, 134, 236

Parameter plus dynamic uncertainty determination in control systems utilizing neural networks, 138

Parsimonious principles for models of least complexity, 231

Performance evaluation of orthogonal activation function-based neural networks as nonlinear identifiers, 52

Performance surface of cost function
  for nonorthonormal activation function, 13
  for orthonormal activation functions, 13

Preliminary performance evaluation of orthogonal activation function-based neural networks, 21

R

Radial basis function
  network models for nonlinear model iden-
    tification, 213
  neural networks in nonlinear model ap-
    proximation, 195
  neural network systems for nonlinear dy-
    namic systems modeling, 385–393
Receding horizon optimal tracking
  control problem formulation, 159
  control problem of a nonlinear system, 159
  neurocontrollers for nonlinear dynamic
    systems, 157
Recurrent high-order neural networks in the
  identification of dynamical systems, 279,
  281–285
Recurrent neural networks in control system
  design, 73

S

Series-parallel method for nonlinear system
  identification, 211
State–space–based neural modeling system
  techniques for dynamic systems,
  406–409

State–space modeling of dynamic systems by
  neural network systems, properties,
  412–418
  relevant issues, 412–418
Structure determination of nonlinear dynamic systems, 236–244
Systems identification and control, 1

T

Time domain applications for system identification and control, 47
Transfer function identification by means of
  the neural network systems analyzer, 27
True versus instantaneous gradient descent
  for cost function of neural network design performance index, 14

U

Universal approximation of nonlinear dynamical systems, 197
Universal approximators for nonlinear systems identification, 194